Redox Metabolism in Ecophysiology and Evolution

Redox Metabolism in Ecophysiology and Evolution

Editors

Marcelo Hermes-Lima
Daniel Carneiro Moreira
Tania Zenteno-Savín

MDPI

Basel • Beijing • Wuhan • Barcelona • Belgrade • Novi Sad • Cluj • Manchester

Editors
Marcelo Hermes-Lima
Department of Cell Biology
University of Brasilia
Brasilia, Brazil

Daniel Carneiro Moreira
Faculty of Medicine
University of Brasilia
Brasilia, Brazil

Tania Zenteno-Savín
Programa de Planeación
Ambiental y Conservación
Centro de Investigaciones
Biológicas del Noroeste
La Paz, Mexico

Editorial Office
MDPI
St. Alban-Anlage 66
4052 Basel, Switzerland

This is a reprint of articles from the Special Issue published online in the open access journal *Antioxidants* (ISSN 2076-3921) (available at: https://www.mdpi.com/journal/antioxidants/special_issues/redox_ecophysiology_evolution).

For citation purposes, cite each article independently as indicated on the article page online and as indicated below:

Lastname, A.A.; Lastname, B.B. Article Title. *Journal Name* **Year**, *Volume Number*, Page Range.

ISBN 978-3-0365-9012-7 (Hbk)
ISBN 978-3-0365-9013-4 (PDF)
doi.org/10.3390/books978-3-0365-9013-4

Cover image courtesy of Ricardo Giesecke

Contents

antioxidants

Editorial

Redox Metabolism in Ecophysiology and Evolution

Daniel C. Moreira [1,*], Tania Zenteno-Savín [2] and Marcelo Hermes-Lima [3]

1 Research Center in Morphology and Applied Immunology, Faculty of Medicine, University of Brasilia, Brasilia 70910-900, Brazil
2 Centro de Investigaciones Biológicas del Noroeste, La Paz 23096, Mexico; tzenteno04@cibnor.mx
3 Department of Cell Biology, University of Brasilia, Brasilia 70910-900, Brazil; hermes@unb.br
* Correspondence: moreiradc@unb.br

Aerobic organisms have developed a complex system of endogenous antioxidants to manage the reactivity of oxygen and its byproducts. This interplay between reactive oxygen/nitrogen species (RONS) and defensive mechanisms has profound implications, extending well beyond mere interactions of attack and defense. Redox metabolism, encompassing non-radical redox metabolites (e.g., hydrogen peroxide and glutathione), redox-sensitive transcription factors (e.g., Nrf2 and FOXOs), and redox-sensitive proteins (e.g., thioredoxins and peroxiredoxins), forms a network of signaling pathways with extensive effects on various processes in aerobic organisms, spanning from circadian rhythms to the regulation of lifespan.

Oxidative stress, stemming from an imbalance between RONS production and cellular antioxidant capacity, is now recognized as a pivotal determinant of the life history of organisms. Environmental stressors have the potential to disrupt the equilibrium of redox reactions, eliciting compensatory adaptive responses. Modulations in redox metabolism have been documented in a wide range of species, spanning various phylogenetic groups and exposed to diverse environmental stressors such as fluctuations in temperature, changes in water availability, shifts in oxygen levels, exposure to UV radiation, pollutants, and more.

This Special Issue is dedicated to investigating the responses of redox metabolism in organisms experiencing alterations in individual or combined environmental factors, whether of biotic or abiotic origin. Comprehending these responses to stress inducers is critical for unraveling the role of oxidative stress in the survival and fitness of organisms. By examining how organisms confront the challenges posed by these changing factors, valuable insights can be gained into the interplay between redox metabolism and the overall health of ecosystems. This research not only enhances our comprehension of the underlying mechanisms but also illuminates potential strategies for mitigating the repercussions of environmental changes on biodiversity and ecological stability.

Arango et al. [1] investigated possible physiological costs associated with olive ridley sea turtles' nesting behavior, either solitary or in "arribadas" (mass aggregations). The study examined metabolic differences between the nesting modes and found that arribada nesters were larger and had higher thyroid hormone levels. Moreover, turtles that presented arribada behavior had metabolic pathways related to phospholipid and amino acid metabolism, as well as catabolic processes, upregulated compared with those reproducing solitarily. However, arribada nesters exhibited higher levels of oxidative damage in terms of lipid peroxidation and protein oxidation, suggesting a trade-off between fitness benefits and oxidative stress associated with arribada nesting.

Cassier-Chauvat et al. [2] examined the significance of the glutathione (GSH) system in defending cells against various stressors across phylogenetically diverse organisms. They underscored the crucial roles of glutathione-dependent systems in maintaining redox balance, detoxifying harmful substances, and regulating iron metabolism. The study delved into multiple pathways, including GSH synthesis, degradation, recycling, conjugation,

Citation: Moreira, D.C.; Zenteno-Savín, T.; Hermes-Lima, M. Redox Metabolism in Ecophysiology and Evolution. *Antioxidants* **2023**, *12*, 1769. https://doi.org/10.3390/antiox12091769

Received: 10 September 2023
Accepted: 13 September 2023
Published: 16 September 2023

glutathionylation/deglutathionylation, and iron–sulfur cluster [Fe-S] synthesis and repair. The authors ultimately suggested that cyanobacteria, regarded as the originators of the GSH system, are the ideal model organisms for further research into its components.

Cubillos et al. [3] investigated the impact of experimental radiation exposure on the burrowing behavior and oxidative damage in *Anthopleura hermaphroditica*, an intertidal anemone. They exposed adults and juveniles to different radiation treatments, including photosynthetically active radiation (PAR), ultraviolet A radiation (UVA), and ultraviolet B radiation (UVB), both with and without sediment. Animals were exposed either to PAR alone, PAR + UVA, or PAR + UVA + UVB. The results showed that exposure to PAR + UVA + UVB radiation led to faster burrowing responses in both adults and juveniles compared to PAR alone. Juveniles had higher basal levels of oxidative stress markers (protein carbonyl and malondialdehyde) than those in adults. Exposure to PAR + UVA + UVB further increased oxidative damage to protein and lipids in juveniles, regardless of the presence or not of sediment. However, the presence of sediment protected adults against various radiation treatments. The research found that *A. hermaphroditica* burrows into sediment as a behavioral response to UVR radiation, indicating their evasion of UVB radiation and UVR-induced oxidative stress.

Ferreira-Cravo et al. [4] explored the role of glutathione (GSH) in managing redox homeostasis during anoxia and reoxygenation in *Helix aspersa*, a model for anoxia tolerance. The researchers depleted the total GSH pool inhibiting GSH synthesis with L-buthionine-(S, R)-sulfoximine (BSO) and subjected snails to anoxia. BSO alone led to GSH depletion but no other significant changes in antioxidants and oxidative stress markers, while anoxia triggered an increase in glutathione peroxidase (GPX) activity. However, GSH depletion before anoxia raised the GSSG/tGSH ratio during anoxia, indicating a disrupted redox balance. Therefore, GSH depletion does not directly induce oxidative stress in snails but impairs their ability to manage oxidative challenges during anoxia. The study indicates the importance of GSH in countering oxidative challenges during hypoxia and reoxygenation in land snails.

Jacobs et al. [5] investigated the impact of aridification on wild common mole rats and their oxidative status along an aridity gradient. Liver oxidative status showed no significant change with aridity levels. However, aridity did affect the total antioxidant capacity (TAC) and oxidative stress index (OSI) of the kidney, with the most arid habitats showing the highest TAC. These findings were interpreted in the context of the kidney's role in water balance and retention. The authors conclude that social common mole rats in arid environments prioritize and protect their kidneys with higher antioxidants, mitigating the impact of aridification on oxidative stress. Nevertheless, more research is needed to further explore the effects of aridity on mammalian oxidative ecology, particularly in the context of climate change.

Malagoli et al. [6] reviewed the critical role of reactive oxygen species (ROS) in mollusks' physiological functions and their relevance in both environmental adaptation and immune responses. The authors focused on highly invasive mollusks (e.g., *Mytilus galloprovincialis*, *Dreissena polymorpha*, *Achatina fulica*, and *Pomacea canaliculata*) and their ability to manage ROS production during challenging conditions. The review emphasizes that the capacity to maintain redox homeostasis under oxidative pressure can be advantageous when facing environmental and immunological challenges, potentially making it a trait associated with invasiveness.

Ramirez et al. [7] investigated the effect of lipoic acid, a mitochondrial coenzyme that can act as an antioxidant or pro-oxidant, on energy metabolism and redox balance. Using *Artemia* sp. nauplii as a model organism, the authors found that lipoic acid treatment positively influenced the activity of the electron transport system, particularly at higher concentrations and longer exposure times. Additionally, lipoic acid promoted glucose catabolism, decreased ROS production, and influenced protein levels and total ammoniacal nitrogen production under specific conditions.

Yang et al. [8] investigated the response of *Cherax destructor*, a freshwater crustacean, to different temperatures. After 8 weeks of exposure to 30 °C, 25 °C, 20 °C, 15 °C, or 10 °C, levels of components of the glutathione redox system (GSH, GSSG, GST, and GR) and transcript levels of proteins that respond to temperature stress (heat-shock proteins and cold shock-domain proteins) were quantified in the hepatopancreas. With a focus on the effect of low temperatures, the study showed that growth indicators such as weight gain and length gain decreased at lower temperatures. Exposure to 10 °C also decreased GST activity and GSSG/GSH ratio, while increasing GSH levels in hepatopancreas when compared with animals maintained at 25 °C. Different temperatures modulated the expression of temperature stress-responsive proteins, among which HSP60, HSP70, and CSP were upregulated at 10 °C. Transcriptome sequencing identified differentially expressed genes related to endocrine disorders, glucose metabolism, antioxidant defense, and immune responses. The findings suggest that low temperature inhibited basal metabolism and immune ability but increased redox buffering capacity and temperature–stress response in the crayfish.

Zhang et al. [9] focused on the effects of mercury (as $HgCl_2$) toxicity on *Procambarus clarkii*, a type of crayfish commonly found in aquatic environments worldwide. The study examined the acute impact of Hg on various aspects, including biochemical responses, histopathology, hepatopancreatic transcriptome, and intestinal microbiome. Hg exposure led to significant changes in redox metabolism markers, including increases in malondialdehyde (MDA) content and a general trend of decreasing antioxidant levels. Structural damage was observed in the hepatopancreas and intestines. RNA-seq identified differentially expressed genes (DEGs) mainly associated with redox metabolism, ion transport, drug metabolism, immune response, and apoptosis. The study also showed that Hg exposure altered the composition of the intestinal microbiome. The findings provide insights into the mechanisms of Hg-induced toxicity in aquatic crustaceans at various levels, including tissue, cellular, molecular, and microbial levels.

Jacobs et al. [10] explored several factors that could be associated with redox balance and oxidative stress, such as phylogeny, social behavior, captivity, environmental aridity, resting metabolic rate (RMR), and maximum lifespan potential (MLSP).

The authors found that higher levels of oxidative markers in plasma are associated with higher MLSP by comparing several species of African naked mole rats. Such association was more evident in social species than in solitary species. Additionally, oxidative stress marker levels decrease with a higher aridity index, and wild-caught mole rats show higher antioxidant levels. The concept of hormesis, which involves a biphasic response to stimuli, is proposed as a potential mechanism contributing to longevity in certain members of the Bathyergidae family.

These studies provide further evidence of the role of oxidative stress as a key factor in the life history of living organisms. Changes in RONS production and antioxidant defenses in response to different biotic and abiotic factors (individually and/or combined) contribute to regulating signaling pathways that dictate whether an aerobic species will survive. Studies such as that of Jacobs et al. [5] highlight the role of combined responses at biochemical, physiologic, behavioral, and/or population levels in responding to those factors. Further, the studies included in this Special Issue suggest compensatory adaptive responses may modulate redox metabolism in phylogenetically diverse species. Closer analyses of these results will provide clues to understanding the evolution of antioxidant defense systems, the underlying mechanisms, the subtle connections of the redox metabolism with other vital systems, such as the immune system, and its participation in shaping existent biodiversity. Results from studies of redox metabolism with a comparative focus can provide clues towards dealing with environmental and climate change, provide viable solutions for the conservation of species, and maintain the balance between environmental, animal, and human health in a sustainable manner.

Author Contributions: The authors included in this study have each made significant and integral intellectual contributions to the article. Furthermore, all authors have provided their approval for the publication of this work. All authors have read and agreed to the published version of the manuscript.

Conflicts of Interest: The authors declare no conflict of interest.

References

1. Arango, B.G.; Ensminger, D.C.; Moreno-Santillán, D.D.; Harfush-Meléndez, M.; López-Reyes, E.M.; Marmolejo-Valencia, J.A.; Merchant-Larios, H.; Crocker, D.E.; Vázquez-Medina, J.P. Oxidative Stress Is a Potential Cost of Synchronous Nesting in Olive Ridley Sea Turtles. *Antioxidants* **2022**, *11*, 1772. [CrossRef] [PubMed]
2. Cassier-Chauvat, C.; Marceau, F.; Farci, S.; Ouchane, S.; Chauvat, F. The Glutathione System: A Journey from Cyanobacteria to Higher Eukaryotes. *Antioxidants* **2023**, *12*, 1199. [CrossRef] [PubMed]
3. Cubillos, V.M.; Álvarez, J.A.; Ramírez, E.; Cruces, E.; Chaparro, O.R.; Montory, J.; Spano, C.A. Effects of Ultraviolet Radiation on Sediment Burial Parameters and Photo-Oxidative Response of the Intertidal Anemone Anthopleura Hermaphroditica. *Antioxidants* **2022**, *11*, 1725. [CrossRef] [PubMed]
4. Ferreira-Cravo, M.; Moreira, D.C.; Hermes-Lima, M. Glutathione Depletion Disrupts Redox Homeostasis in an Anoxia-Tolerant Invertebrate. *Antioxidants* **2023**, *12*, 1197. [CrossRef] [PubMed]
5. Jacobs, P.J.; Hart, D.W.; Merchant, H.N.; Janse Van Vuuren, A.K.; Faulkes, C.G.; Portugal, S.J.; Van Jaarsveld, B.; Bennett, N.C. Tissue Oxidative Ecology along an Aridity Gradient in a Mammalian Subterranean Species. *Antioxidants* **2022**, *11*, 2290. [CrossRef] [PubMed]
6. Malagoli, D.; Franchi, N.; Sacchi, S. The Eco-Immunological Relevance of the Anti-Oxidant Response in Invasive Molluscs. *Antioxidants* **2023**, *12*, 1266. [CrossRef] [PubMed]
7. Buitrago Ramírez, J.R.; Marreiro Gomes, R.M.; De Sousa Araujo, A.C.; Muñoz Buitrago, S.A.; Piraine Souza, J.; Monserrat, J.M. The Effects of Lipoic Acid on Yolk Nutrient Utilization, Energy Metabolism, and Redox Balance over Time in *Artemia* sp. *Antioxidants* **2023**, *12*, 1439. [CrossRef] [PubMed]
8. Yang, Y.; Xu, W.; Jiang, Q.; Ye, Y.; Tian, J.; Huang, Y.; Du, X.; Li, Y.; Zhao, Y.; Liu, Z. Effects of Low Temperature on Antioxidant and Heat Shock Protein Expression Profiles and Transcriptomic Responses in Crayfish (Cherax Destructor). *Antioxidants* **2022**, *11*, 1779. [CrossRef] [PubMed]
9. Zhang, L.; Zhou, Y.; Song, Z.; Liang, H.; Zhong, S.; Yu, Y.; Liu, T.; Sha, H.; He, L.; Gan, J. Mercury Induced Tissue Damage, Redox Metabolism, Ion Transport, Apoptosis, and Intestinal Microbiota Change in Red Swamp Crayfish (Procambarus Clarkii): Application of Multi-Omics Analysis in Risk Assessment of Hg. *Antioxidants* **2022**, *11*, 1944. [CrossRef] [PubMed]
10. Jacobs, P.J.; Hart, D.W.; Merchant, H.N.; Voigt, C.; Bennett, N.C. The Evolution and Ecology of Oxidative and Antioxidant Status: A Comparative Approach in African Mole-Rats. *Antioxidants* **2023**, *12*, 1486. [CrossRef] [PubMed]

antioxidants

MDPI

Article

Oxidative Stress Is a Potential Cost of Synchronous Nesting in Olive Ridley Sea Turtles

B. Gabriela Arango [1,*], David C. Ensminger [1,2], Diana Daniela Moreno-Santillán [1], Martha Harfush-Meléndez [3], Elpidio Marcelino López-Reyes [3,4,5], José Alejandro Marmolejo-Valencia [6], Horacio Merchant-Larios [6], Daniel E. Crocker [7] and José Pablo Vázquez-Medina [1,*]

1 Department of Integrative Biology, University of California, Berkeley, Berkeley, CA 94720, USA
2 Department of Biological Sciences, San José State University, San Jose, CA 95192, USA
3 Centro Mexicano de la Tortuga, Oaxaca 70947, Mexico
4 Campamento Tortuguero Palmarito, Oaxaca 70984, Mexico
5 Facultad de Medicina Veterinaria y Zootecnia, Universidad Autónoma Benito Júarez de Oaxaca, Oaxaca 68120, Mexico
6 Instituto de Investigaciones Biomédicas, Universidad Nacional Autónoma de México, Ciudad de México 04510, Mexico
7 Department of Biology, Sonoma State University, Rohnert Park, CA 94928, USA
* Correspondence: bg.arango@berkeley.edu (B.G.A.); jpv-m@berkeley.edu (J.P.V.-M.)

Abstract: Olive ridley sea turtles, *Lepidochelys olivacea*, exhibit a polymorphic reproductive behavior, nesting solitarily or in mass aggregations termed "arribadas", where thousands of individuals nest synchronously. Arribada nesting provides fitness benefits including mate finding during nearshore aggregations and predator satiation at the time of hatching, but it is unknown if such benefits come with a physiological cost. We used plasma metabolite profiling, stable isotope analysis, biochemical and endocrine assays to test whether metabolic parameters differ between nesting modes, and if arribada nesting is associated with increased levels of oxidative damage compared to solitary nesting. Arribada nesters were bigger and had higher circulating thyroid hormone levels than solitary nesters. Similarly, pathways related to phospholipid and amino acid metabolism, catabolic processes, and antioxidant defense were enriched in individuals nesting in arribada. Stable isotope signatures in skin samples showed differences in feeding zones with arribada nesters likely feeding on benthic and potentially more productive grounds. Arribada nesters had increased levels of plasma lipid peroxidation and protein oxidation products compared to solitary nesters. These results suggest that metabolic profiles differ between nesting modes and that oxidative stress is a trade-off for the fitness benefits associated with arribada nesting.

Keywords: reproduction; life history theory; metabolic cost; arribada; oxidative damage

Citation: Arango, B.G.; Ensminger, D.C.; Moreno-Santillán, D.D.; Harfush-Meléndez, M.; López-Reyes, E.M.; Marmolejo-Valencia, J.A.; Merchant-Larios, H.; Crocker, D.E.; Vázquez-Medina, J.P. Oxidative Stress Is a Potential Cost of Synchronous Nesting in Olive Ridley Sea Turtles. *Antioxidants* **2022**, *11*, 1772. https://doi.org/10.3390/antiox11091772

Academic Editors: Marcelo Hermes-Lima, Daniel Carneiro Moreira and Tania Zenteno-Savín

Received: 12 August 2022
Accepted: 3 September 2022
Published: 8 September 2022

1. Introduction

Life history theory posits that reproductive effort negatively affects survival, but there is conflicting evidence about the proximal costs of reproduction [1]. One of the leading hypotheses suggests that oxidative stress is a trade-off for reproductive investment [2–4], but other studies do not support this idea [5,6]. Work in wild vertebrates shows that diversification of reproductive strategies is associated with sexual differences in oxidative stress [7–9]. In male free-ranging macaques [10], male northern elephant seals [11], and female North American red squirrels [12], reproductive activities increase oxidative damage. Therefore, the role of oxidative stress as a proximal cost of reproduction is likely species- and sex-specific and also varies among different reproductive strategies.

Sea turtles of the genus *Lepidochelys* spp. exhibit a polymorphic reproductive behavior, nesting both solitarily and in massive aggregations termed "arribadas" [13]. During arribada nesting, thousands of individuals nest synchronously over a two to seven day period [13–15]. Notably, the same individual can nest interchangeably in arribadas or solitarily; however, the factors that determine whether an individual joins the arribada or nests in solitary remain unknown [14,16].

Arribada nesting provides a social context that promotes multiple mating and paternity, increasing genetic exchange by three-fold compared to solitary nesting [14,17–19]. Arribada nesting also increases early hatchling dispersal by overwhelming or satiating predators [20]. Arribada nesting, however, does not promote hatchling success as increased organic matter from egg saturation and high nesting density in arribada beaches decrease hatchling success compared to solitary nesting [15,21,22]. Similarly, arribada nesters produce smaller hatchlings with lower male-to-female ratio than solitary nesters [13,23]. Thus, while both nesting modes are likely important for the population, it is unknown if the fitness benefits associated with arribada nesting come with a physiological cost.

Here, we compared circulating concentrations of reproductive and metabolic hormones, primary metabolites, stable isotope signatures, and markers of oxidative stress between olive ridley sea turtles (*Lepidochelys olivacea*) nesting solitarily and in arribada. We found that arribada nesters are bigger and likely have higher metabolic activity than solitary nesters. We also found higher circulating levels of lipid peroxidation and protein oxidation products in arribada than in solitary nesters. Our results suggest that oxidative stress is a potential cost for the fitness benefits associated with arribada nesting in olive ridley sea turtles.

2. Materials and Methods

2.1. Animals and Sample Collection

Animal handling protocols were approved by Sonoma State University. Samples were collected under permit SGPA/DGVS/12915/16 and imported to the US under permits CITES MX88143 and CITES 19US85728C/9. Nesting olive ridley sea turtles were sampled at the marine protected area of *La Escobilla*, Oaxaca, Mexico (15°47′ N; 96°44′ W) throughout the arribada (n = 13). Solitarily nesting individuals were sampled at *Campamento Tortuguero Palmarito*, Puerto Escondido, Oaxaca, México (15°53′26.3″ N; 97°07′52.2″ W), and *La Escobilla* (n = 10). Samples were collected after the animals had dug their nest, during the 'trance-nesting period' [24]. None of the animals were disturbed from nesting, nor returned to the ocean without laying their eggs. Mass was estimated using a regression from published olive ridley morphometric data (n = 59, mass = −47.44 + 1.13 × straight carapace length (SCL), r2 = 0.70, $p < 0.001$; [25]). Animals sampled during solitary nesting were also weighed using a hand-held scale (±0.1 kg) to assess the validity of the mass-estimation method. The equation used to estimate mass predicted the weight of solitary nesters with a mean error of 4%. ~25 mL of blood were collected from the cervical vein into chilled Vacutainer tubes [24,26]. Plasma and serum were prepared by centrifugation onsite, frozen, and subsequently stored at −20 °C until laboratory analysis. Epidermal tissue samples were collected using a 6.0 mm diameter biopsy punch (Miltex, York, PA, USA) and stored at −20 °C.

2.2. Biochemical Assays

Hormones: progesterone (P4), estradiol (E2), thyroxine (T4), triiodothyronine (T3) and testosterone were measured in serum using commercially available kits (P4: MP Biomedical ELISA catalog number 07BC1113, E2: MP Biomedical ELISA catalog number 07BC1111, T3: MP Biomedical RIA catalog number 06B-254215, T4: MP Biomedical RIA catalog number 06B-254011 (MP Biomedical, Irvine, CA, USA), testosterone: Enzo ELISA catalog number ADI-900-065 (Enzo, Farmingdale, NY, USA)). The assays were validated for use in olive ridley sea turtles. Serially diluted pooled samples (1:2 to 1:16) exhibited parallelism to the standard curve after log-logit transformation. Mean recovery of hormone added to serum pools was 101.5 ± 3.5%, 98.3 ± 6.5%, 99.4 ± 4.7%, and 101.3 ± 5.1% (r2 > 0.98) for P4, E2, T3 and T4, respectively. The testosterone kit was previously validated for use in sea turtle blood [27] and was further validated in our samples via parallelism and spike recovery.

Assays were conducted following the manufacturers' instructions. The glucose, lactate, and corticosterone concentrations included in our correlation analyses were measured in samples obtained from the same animals in our previous study [28].

Oxidative damage: Two markers of lipid peroxidation (4-hydroxynonenal: 4-HNE, malondialdehyde: MDA), a marker of protein oxidation (protein carbonyls), and a marker of protein nitration (3-nitrotyrosine) were measured in plasma using ELISA kits (Cell Biolabs catalog numbers: STA-838, STA-832, STA-310, STA-305, Cell Biolabs, San Diego, CA, USA). Assays were conducted following the manufacturer's instructions with minor modifications as described in our previous work [29].

Plasma lipids and total protein: Plasma triglycerides (TG) and non-esterified fatty acids (NEFA) were measured using colorimetric kits (TG: Cayman Chemical catalog number 10010303 (Cayman Chemical, Ann Arbor, MI, USA), NEFA: Wako Chemicals, HR Series NEFA-HR (2) catalog number: 999-34691 (Wako Chemicals, Richmond, VA, USA)). Total protein content was measured using a Rapid Gold BCA protein assay kit (Thermo Fisher Scientific catalog number: A53227 (Thermo Fisher Scientific, Waltham, MA, USA)). Oxidative damage values were normalized to total plasma protein levels.

All samples were analyzed in duplicate in a single assay. The average intra-assay coefficient of variation was <6%.

2.3. Metabolite Profiling

Analysis of primary metabolism by ALEX-CIS gas chromatography-mass spectrometry (GC-TOF MS) was conducted in plasma at the UC Davis West Coast Metabolomics Center following the methods of Fiehn et al. [30]. Raw data were pre-processed and stored as apex masses, exported to a data server with absolute spectra intensities, and further filtered with an algorithm implemented in the BinBased database [30]. Spectra were cut to a 5% base peak abundance and matched to a database entry. Quantification was stored as peak height using the unique ion as default for all database entries that are positively detected in more than 10% of the unidentified metabolites.

2.4. Stable Isotopes

Epidermal tissue samples were rinsed with DI water and dried at 60 °C for 48 h. Samples were grounded, weighted to ~1mg, and packed into tin capsules (3.5 × 5 mm, #041060, Costech Analytical Technologies, Valencia, CA, USA). Samples were analyzed for %C, $\delta^{13}C$, %N, and $\delta^{15}N$ by continuous flow dual isotope analysis using a CHNOS Elemental Analyzer interfaced to an IsoPrime100 mass spectrometer at the UC Berkeley Center for Stable Isotope Biogeochemistry. Stable isotope ratios are expressed in δ notation as parts per thousand (‰). Long-term external precision for C and N isotope determinations was ±0.10‰ and ± 0.20‰, respectively. C:N ratios in both groups were <3.5, validating $\delta^{13}C$ values by indicating a low lipid content.

2.5. Statistical Analyses

Biochemical assays: statistical analyses were conducted using JMP Pro 15 (SAS Institute, Cary, NC, USA). Equality of variances was assessed using Levene's test. P4, TG and NEFA data were log10 transformed to meet model assumptions. Arribada and solitary nesting groups were compared using two-sample *t*-tests. A nominal logistic regression was used to analyze 4-HNE data since the response variable was binary. Correlation analyses between hormones, mass, TG, NEFA, glucose and lactate were conducted by calculating Spearman rank correlation coefficients. The glucose, lactate, and corticosterone concentrations included in our correlation analyses were measured in samples obtained from the same animals in our previous study [28]. Statistical significance was considered at $p \leq 0.05$.

Metabolite profiling: statistical analysis was conducted using MetaboAnalyst 5.0 (Xia et al.; Main server: https://www.metaboanalyst.ca, 6 September 2022). Metabolite peaks were normalized using the median of all peak heights for all identified metabolites. To minimize noise, low quality peaks (0–5% of the mean) were filtered out [31]. Normalized peaks were log10 transformed to meet model assumptions. Data were compared between arribada and solitary using multiple *t*-tests with a 5% FDR adjustment. A heatmap was constructed using Pearson distance correlation [32] and the average clustering method for the top 50 most abundant metabolites. Quantitative enrichment analysis was conducted in MetaboAnalyst using the small molecule pathway database (SMPDB) [33].

Stable isotopes: Variables were compared between nesting modes using two-sample *t*-tests. Ratios of $\delta^{13}C$ and $\delta^{15}N$ were plotted against each other with standard ellipses with 40% of the data using ggplot2 in R.

3. Results

3.1. Biochemical Profiles

We compared circulating TG, NEFA, P4, E2, testosterone, T3 and T4 concentrations between solitary and arribada nesters to determine whether hormonal and biochemical profiles differ between nesting modes. While neither TG nor NEFA were different between arribada and solitary nesters (TG: solitary 14.46 ± 9.52 vs. arribada 20.47 ± 10.03 mg/mL, t = 1.94, df = 21, n = 23, p = 0.07; NEFA: solitary 0.93 ± 0.22 vs. arribada 1.08 ± 0.55 mM, t = 0.445, df = 22, n = 24, p = 0.66, Appendix A Figure A1a,b), P4 was higher in arribada than in solitary nesters (3.58 ± 1.71 vs. 2.10 ± 1.56 ng/mL, t = 2.612, df = 21, n = 23, p = 0.02; Figure 1a). E2 did not differ between nesting conditions (solitary 4.47 ± 2.79 vs. arribada 5.7 ± 2.24 pg/mL, t = −1.17, df = 21, n = 23, p = 0.25; Figure 1b). Similarly, testosterone was not different between solitary and arribada nesters (70.43 ± 42.88 vs. 75.08 ± 35.99 pg/mL, t = −0.29, df = 22, n = 24, p = 0.78; Figure 1c). Both T3 and T4 were higher in individuals nesting in arribada than in solitary nesters (T3: 48 ± 19.54 vs. 29.3 ± 11.86 ng/dL, t = 2.66, df = 21, n = 23, p = 0.015; T4: 1.45 ± 0.27 vs. 1.12 ± 0.13 ug/dL, t = 3.62, df = 21, n = 23, p = 0.002; Figure 1d,e). Mass was also higher in arribada than solitary nesters (29.20 ± 5.20 vs. 24.76 ± 4.37 kg, t = 2.17, df = 21, n = 23, p = 0.042, Appendix A Figure A1c). These results show that arribada nesters are heavier and have higher T3, T4 and P4 but not E2 or testosterone levels than solitary nesters.

We then conducted correlation analyses using mass, hormones, TG, NEFA, glucose, and lactate [28]. Glucose and lactate showed a positive association in both nesting modes; however, this association was stronger in arribada (r_s = 0.82, p = 0.0006) and not significant in solitary nesters (r_s = 0.60, p = 0.067) (Figure 2a,b). The strongest positive associations were observed between corticosterone and glucose (r_s = 0.87, p < 0.001, corticosterone and lactate (rs = 0.73, p = 0.004), and TG and NEFA (r_s = 0.75, p = 0.003) in arribada nesters (Figure 2b). These associations were not observed in solitary nesters, which showed positive associations between progesterone and estradiol (r_s = 0.71, p = 0.020), TG and mass (r_s = 0.77, p = 0.009), TG and T4 (r_s = 0.64, p = 0.048), and a negative correlation between TG and lactate (r_s = −0.66, p = 0.037) (Figure 2a). These results show that arribada and solitary nesters have different blood biochemical profiles. Furthermore, these results suggest that metabolic activity might be higher in arribada than in solitary nesters.

3.2. Metabolite Profiling

We conducted an analysis of primary metabolism by GC-TOF MS to further explore potential differences in circulating metabolites between nesting modes. Our analysis detected 481 metabolites including 155 known metabolites which clustered based on nesting mode (Figure 3a), further suggesting that metabolism differs between solitary and arribada nesters. Of the identified known metabolites, 10 were significantly expressed (FDR 5%) between nesting modes, with five downregulated and five upregulated in arribada compared to solitary nesters (Figure 3b). The top downregulated metabolite in arribada nesters was phosphoethanolamine, while the top upregulated metabolite was glutamic acid. Accordingly, enrichment analysis showed over-representation of pathways related to phospholipid (phosphatidylcholine and phosphatidylethanolamine biosyntehesis, sphingolipid metabolism), and amino acid metabolism (beta-alanine, tryptophan, tyrosine, and glutamate metabolism), catabolic processes including malate-aspartate shuttle and the glucose-alanine cycle, and glutathione (GSH) metabolism. Of note, two of the most down-regulated metabolites in arribada compared to solitary nesters were the antioxidants uric acid and α-tocopherol. These results further suggest that arribada nesters have increased metabolic activity compared to solitary nesters. Moreover, these results suggest that there are differences in redox metabolism between nesting modes.

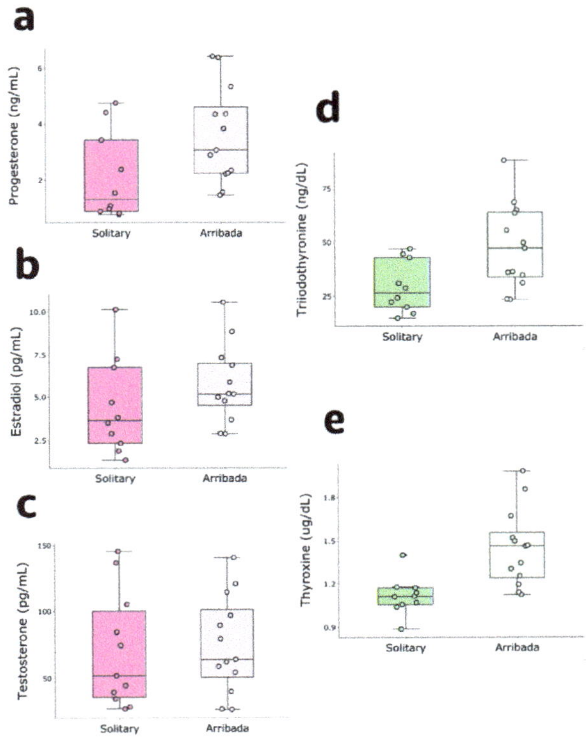

Figure 1. Reproductive and metabolic hormone levels in solitary and arribada nesters. (**a**) Progesterone ($p = 0.02$), (**b**) estradiol ($p = 0.25$), (**c**) testosterone ($p = 0.78$), (**d**) triiodothyronine ($p = 0.015$), and (**e**) thyroxine ($p = 0.002$).

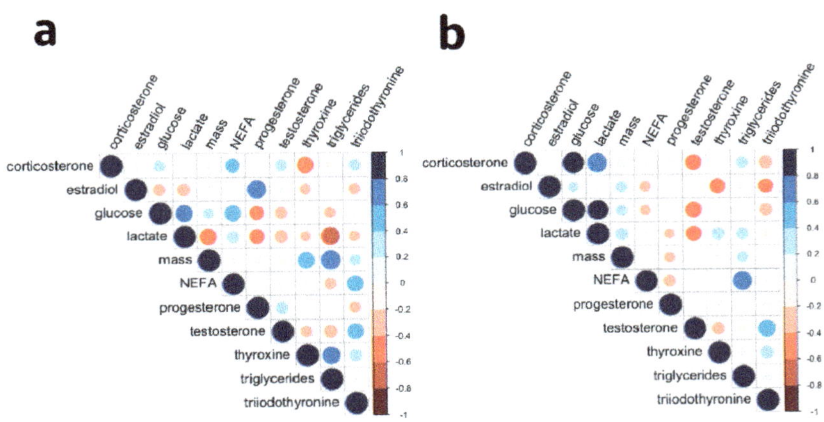

Figure 2. Correlation analysis between hormones and metabolites in (**a**) solitary and (**b**) arribada nesters.

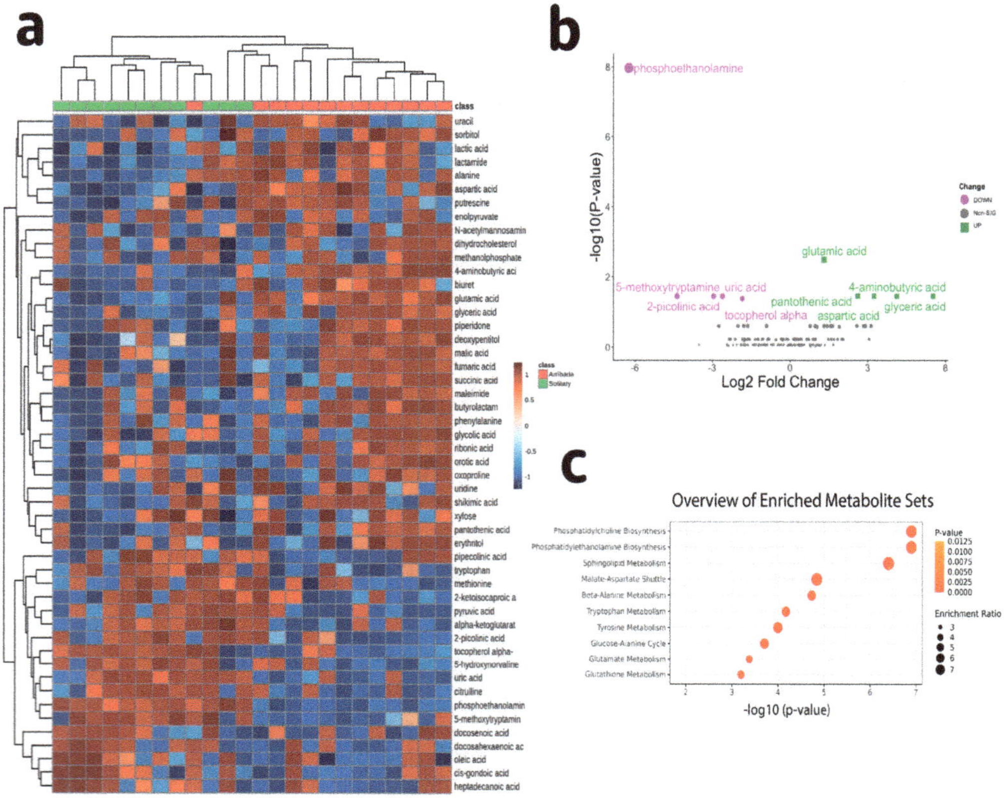

Figure 3. Metabolite profiling. (**a**) Heat map showing the 50 most abundant metabolites in olive ridley plasma, (**b**) Volcano plot showing differentially expressed metabolites in arribada compared to solitary nesters (5% FDR), and (**c**) Enrichment pathways showing significantly enriched pathways in arribada compared to solitary nesters.

3.3. Stable Isotopes

We conducted stable isotope analysis to detect potential differences in resource acquisition between nesting modes. We found significant differences in %C (t = 2.74, df = 18, n = 20, p = 0.013) and %N (t = 2.73, df = 18, n = 20, p = 0.014) between nesting modes, but not in C:N ratios (Figure 4a). When comparing stable isotope signatures, we also found small overlaps between nesting modes on 40% ellipses. Our results suggest that before coming to nest, individuals nesting in arribada are likely feeding at lower trophic levels and in more productive benthic feeding grounds than individuals nesting solitarily (Figure 4b).

3.4. Oxidative Damage

We evaluated whether oxidative damage represents a potential proximate cost for nesting in arribada by comparing four circulating markers of oxidative damage among nesting modes. Plasma levels of the lipid peroxidation products 4-HNE (arribada 100% positive vs. solitary 25% positive for 4-HNE, $\chi^2(2) = 10.01$, n = 17, p = 0.0016), and MDA (0.0025 ± 0.0013 vs. 0.0011 ± 0.00034 pmol/µg of protein, t = 3.77, df = 15, n = 17, p = 0.0019), along with protein carbonyls (0.00246 ± 0.0013 vs. 0.00103 ± 0.00032 nmol/µg of protein, t = 3.64, df = 14, n = 16, p = 0.0027) were higher in arribada than in solitary nesters (Figure 5a–c). In contrast, protein nitration (3-Nitrotyrosine) did not vary between nesting modes (solitary 0.039 ± 0.026 vs. arribada 0.058 ± 0.031 nM/mg, t = −1.29, df = 15, n = 17, p = 0.22;

Figure 5d). These results suggest that oxidative stress is a potential cost of arribada nesting in olive ridley sea turtles.

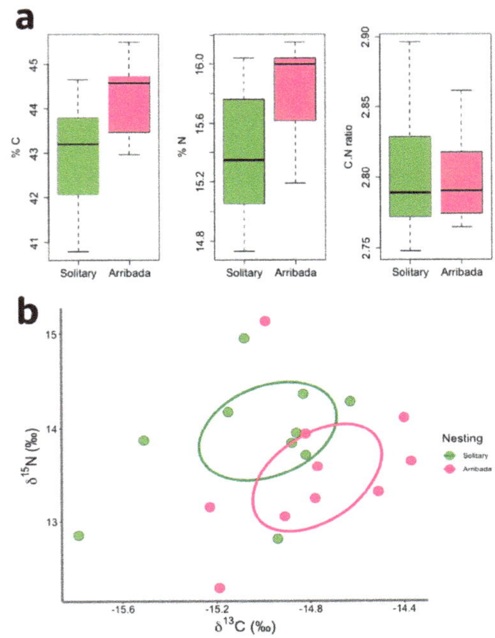

Figure 4. Stable isotope analysis. (**a**) Comparison of %C ($p = 0.013$) and %N ($p = 0.014$) and C:N ($p > 0.05$) ratio between solitary and arribada nesters, (**b**) δ^{13}C and δ^{15}N ratios. Standard ellipses represent a 40% overlap between nesting modes.

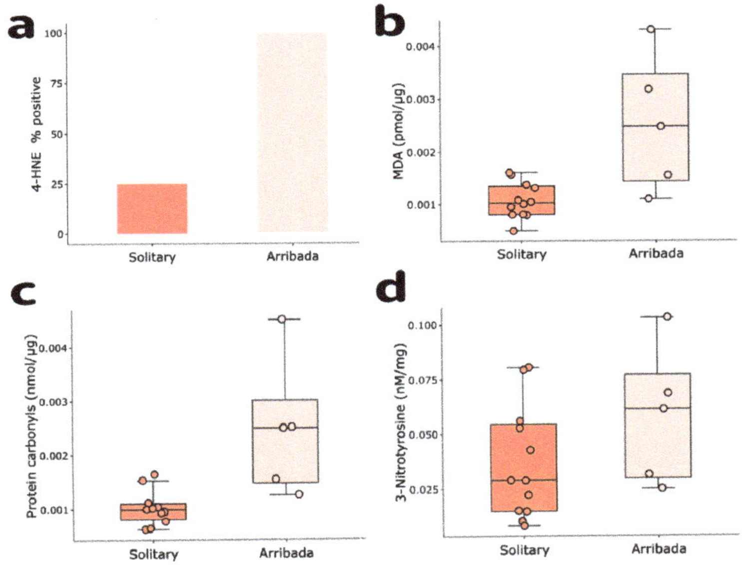

Figure 5. Oxidative damage. Lipid peroxidation products: (**a**) 4-HNE-protein adducts (binary test, percent positive arribada, 100% vs. solitary, 25%, $p = 0.0009$, $\chi^2(2) = 10.01$, $p = 0.0016$), and (**b**) MDA-protein adducts ($p = 0.0019$); (**c**) protein carbonyls ($p = 0.0027$) and (**d**) 3-Nitrotyrosine ($p = 0.22$).

4. Discussion

Potential fitness benefits associated with arribada nesting include mate finding during nearshore aggregations [14], predator satiation at the time of hatching [20], multiple paternity and increased genetic exchange [19]. Whether this specialized nesting mode carries a physiological cost was previously unknown. In this study, we found that arribada and solitary nesters have distinct circulating metabolic profiles and that arribada nesters are heavier and have higher P4, T3, T4, lipid peroxidation, and protein oxidation levels than solitary nesters. We also found differences in stable isotope signatures between nesting modes and enrichment for catabolic and antioxidant pathways in arribada compared to solitary nesters. Therefore, our results suggest that nesting in arribada might be energetically more expensive than nesting solitarily and that oxidative damage is a potential trade-off for the fitness benefits associated with arribada nesting in olive ridley sea turtles.

Little is known about the endogenous adjustments that allow individuals to synchronize to join an arribada. *Lepidochelys* spp. can retain their eggs for longer periods compared to other species of sea turtles [34]. Similarly, tolerance to hypoxia-induced pre-ovipositional embryonic arrest is higher in eggs from arribada than from solitary nesters [23]. Both increased egg retention capacity and embryonic arrest appear to be important to synchronize to join the arribada. Similarly, gonadosteroids might also influence whether an individual joins the arribada. We found higher P4 but not E2 or testosterone levels in arribada than in solitary nesters. A P4 surge during ovulation induces a rapid albumen release into the oviduct, which activates previously stored sperm [35]. Therefore, it is possible that higher P4 levels in arribada than in solitary nesters are related to increased mating opportunities during the arribadas. Of note, P4 levels do not differ between nesting modes in Kemp's ridleys [36]. Thus, there is no conclusive evidence about the role of gonadosteroids in promoting arribada nesting in *Lepidochelys* spp.

In a previous study we showed that arribada nesters have higher circulating corticosterone and glucose levels than solitary nesters [28]. Here, we found that arribada nesters are bigger and have higher thyroid hormone levels than solitary nesters. Moreover, we found strong positive correlations between corticosterone and glucose, corticosterone and lactate, glucose and lactate, and TG and NEFA in arribada but not in solitary nesters. Similarly, metabolite profiling shows differences in major pathways related to cell metabolism and antioxidant defense. Phosphoethanolamine (PETh) was the most downregulated metabolite in arribada compared to solitary nesters. Besides being important for cell membrane composition, PETh stimulates tolerance to nutrient starvation, and its levels increase in glutamine-deprived cells [37]. Consistent with this observation, levels of glutamic acid, 4-aminobutyric acid, pantothenic acid, aspartic acid, and glyceric acid were increased in arribada compared to solitary nesters, while levels of antioxidants α-tocopherol, uric acid, picolinic acid and, 5-methoxytryptamine were decreased. These metabolites are important for insulin production, glycolysis, and redox balance [38].

We also found enrichment for catabolic processes, including the glucose-alanine cycle and the malate-aspartate cycle. Of note, both of these pathways produce oxidants as byproducts either directly or through downstream effectors [39,40]. Thus, our results support the idea that arribada nesters have higher metabolic activity than solitary nesters. Similarly, our combined results also suggest that arribada nesters have larger energy reserves than solitary nesters. Our results are consistent with those reported in animals nesting in the Rushikulya Rookery of Orissa, India, where arribada nesters are also larger than solitary nesters [41]. As discussed earlier, the same individual can nest interchangeably in arribadas or in solitary, but the factors that determine whether an individual joins the arribada or nests solitarily remain unknown [14,16]. Our results suggest that arribadas are potentially more energetically costly than solitary nesting and that bigger animals with larger energetic reserves join the arribadas.

As capital breeders [42–44], sea turtles feed and build their energy reserves prior to migrating, mating, and nesting. Hence, if arribada nesting is more energetically costly than solitary nesting, individuals lacking appropriate energy reserves might choose solitary nesting over joining the arribadas despite losing the fitness benefits associated with arribada nesting. Solitary nesting requires shorter inter-nesting intervals. Thus, solitary nesting likely results in less time spent away from the feeding grounds [45], potentially allowing solitary nesters to build their energy reserves and join the arribadas during subsequent reproductive bouts. Our stable isotope analysis shows that solitary and arribada nesters likely feed in different grounds before arrival to the nesting beach. Both $\delta^{13}C$ and $\delta^{15}N$ values are consistent with reported data for this East Pacific population [46]. Although we were not able to measure prey items, the ellipse distances in $\delta^{13}C$ and $\delta^{15}N$ suggest that solitary and arribada nesters feed at different trophic levels and benthic zones. More specifically, the higher $\delta^{15}N$ values seen in solitary as opposed to arribada nesters suggest that solitary nesters feed at a higher trophic level than arribada nesters [47]. Moreover, the less negative $\delta^{13}C$ values observed in arribada than in solitary nesters suggest that arribada nesters feed in benthic and more productive feeding grounds than solitary nesters [47]. Therefore, differences in resource availability and allocation might ultimately affect whether an individual joins the arribada. According to life history theory, reproduction is a costly life history trait, and our results suggest that nesting in arribada is more energetically expensive than nesting solitarily, though it carries fitness benefits such as increased genetic exchange [19].

The cost of reproduction represents one of the most fundamental life history trade-offs [3], but there is inconclusive evidence about whether oxidative stress is a proximal cost of reproductive investment [1]. Work with laboratory versus wild animals often yields conflicting results [48]. Similarly, differences in reproductive strategies within wild vertebrates likely result in differential susceptibility to oxidative stress [7–9]. Whether capital breeders have higher susceptibility to oxidative stress than income breeders remains understudied. In macaques, North American red squirrels, aspic vipers and northern elephant seals, reproductive activities increase oxidative damage despite concurrent increases in antioxidant defenses [10–12,49]. Here we found higher lipid peroxidation and protein oxidation levels in arribada compared to solitary nesters, and reduced levels of α-tocopherol and uric acid, a primary circulating antioxidant which varies with stress levels [50]. We also found that arribada nesting is likely more energetically costly than solitary nesting. These results suggest that higher energy expenditure in olive ridley turtles nesting in arribadas is associated with increased oxidative stress. In northern elephant seals, breeding increases circulating lipid peroxidation in males but not in females [11]. Elephant seals are polygynous, sexually dimorphic capital breeders [51,52]. Males compete for position in a dominance hierarchy used to control access to females [51–54]. Thus, energy expenditure associated with breeding is higher in male than in female elephant seals [55]. Hence, it is possible that in capital breeders with unique reproductive behaviors that result in increased energy expenditure, such as male elephant seal males or female olive ridleys nesting in arribada, oxidative stress is a trade-off for the fitness benefits associated with such behaviors.

5. Conclusions

We found that arribada nesting in olive ridley sea turtles is associated with increased levels of circulating oxidative damage and reduced antioxidant levels compared to solitary nesting. We also found that biochemical, endocrine and metabolomic profiles and stable isotope signatures differ between nesting modes, with arribada nesters likely having increased energy reserves and metabolic activity compared to solitary nesters. These results suggest that oxidative damage is a potential cost of synchronous nesting in olive ridley sea turtles. As such, oxidative stress may be a trade-off for a reproductive mode that carries increased fitness benefits in a capital breeder. Whether these trade-offs are present in other capital breeders remains unknown and warrants further investigation.

Author Contributions: Conceptualization, B.G.A. and J.P.V.-M.; methodology, B.G.A., D.C.E., D.D.M.-S., H.M.-L., E.M.L.-R. and D.E.C.; validation, D.D.M.-S.; formal analysis, B.G.A.; investigation, B.G.A., D.C.E., M.H.-M. and E.M.L.-R.; resources, M.H.-M., E.M.L.-R., H.M.-L., D.E.C. and J.P.V.-M.; data curation, B.G.A., D.C.E., D.D.M.-S. and D.E.C.; writing—original draft preparation, B.G.A. and

J.P.V.-M.; writing—review and editing, J.P.V.-M.; visualization, B.G.A. and D.D.M.-S.; supervision, J.P.V.-M.; project administration: J.A.M.-V. and H.M.-L.; funding acquisition, H.M.-L., D.E.C. and J.P.V.-M. All authors have read and agreed to the published version of the manuscript.

Funding: B.G.A. was supported by the UC Berkeley Chancellor's Fellowship and a Ford Foundation Fellowship. José Alejandro Marmolejo-Valencia and Horacio Merchant-Larios were funded by FORDECYT-PRONACES 137721-2020. Research funded by Sonoma State University and UC Berkeley.

Institutional Review Board Statement: The animal study protocol was approved by Sonoma State University's IACUC (protocol number: 2016-53, approval date: 11 September 2016).

Informed Consent Statement: Not applicable.

Data Availability Statement: The datasets generated for this study can be found in the figshare https://figshare.com/articles/dataset/Arango_etal_OR_DATA_xlsx/14864427 (accessed on 6 September 2022).

Acknowledgments: We thank Centro Mexicano de la Tortuga and Campamento Tortuguero Palmarito personnel, especially Ernesto Albavera-Padilla for his assistance with sampling logistics and useful discussions, and Allison Raymundo, Carmelo Ambrosio, Alberto Jarquín and Antonio Santiago for their assistance during fieldwork. We thank Cooperativa la Tortuga Feliz, Heradio-Santillán and Sóstenes Rodríguez-Reyes from La Escobilla Campamento Tortuguero, and Javier Beltrán-Robledo for their assistance with fieldwork and sample processing. We thank Stefania Mambelli for her help and guidance with sample preparation for stable isotope analysis and Jackie Lebenzon for advice on data analysis for metabolomics.

Conflicts of Interest: The authors declare no conflict of interest.

Appendix A

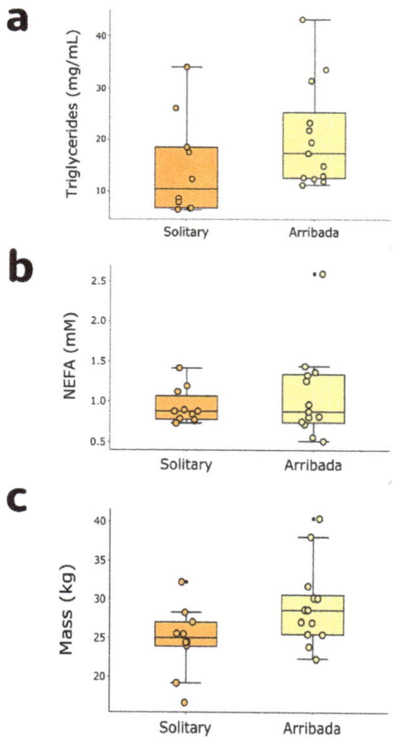

Figure A1. (**a**) Triglycerides ($p = 0.07$), (**b**) NEFA ($p = 0.66$), and (**c**) Mass ($p = 0.042$) between solitary and arribada nesters.

References

1. Metcalfe, N.B.; Monaghan, P. Does Reproduction Cause Oxidative Stress? An Open Question. *Trends Ecol. Evol.* **2013**, *28*, 347–350. [CrossRef] [PubMed]
2. Alonso-Alvarez, C.; Bertrand, S.; Devevey, G.; Prost, J.; Faivre, B.; Sorci, G. Increased Susceptibility to Oxidative Stress as a Proximate Cost of Reproduction. *Ecol. Lett.* **2004**, *7*, 363–368. [CrossRef]
3. Dowling, D.K.; Simmons, L.W. Reactive Oxygen Species as Universal Constraints in Life-History Evolution. *Proc. R. Soc. B Biol. Sci.* **2009**, *276*, 1737–1745. [CrossRef] [PubMed]
4. Monaghan, P.; Metcalfe, N.B.; Torres, R. Oxidative Stress as a Mediator of Life History Trade-Offs: Mechanisms, Measurements and Interpretation. *Ecol. Lett.* **2009**, *12*, 75–92. [CrossRef] [PubMed]
5. Garratt, M.; Vasilaki, A.; Stockley, P.; McArdle, F.; Jackson, M.; Hurst, J.L. Is Oxidative Stress a Physiological Cost of Reproduction? An Experimental Test in House Mice. *Proc. R. Soc. B Biol. Sci.* **2011**, *278*, 1098–1106. [CrossRef] [PubMed]
6. Ołdakowski, Ł.; Piotrowska, Z.; Chrząścik, K.M.; Sadowska, E.T.; Koteja, P.; Taylor, J.R.E. Is Reproduction Costly? No Increase of Oxidative Damage in Breeding Bank Voles. *J. Exp. Biol.* **2012**, *215*, 1799–1805. [CrossRef] [PubMed]
7. Birnie-Gauvin, K.; Costantini, D.; Cooke, S.J.; Willmore, W.G. A Comparative and Evolutionary Approach to Oxidative Stress in Fish: A Review. *Fish Fish.* **2017**, *18*, 928–942. [CrossRef]
8. Costantini, D. Meta-Analysis Reveals That Reproductive Strategies Are Associated with Sexual Differences in Oxidative Balance across Vertebrates. *Curr. Zool.* **2018**, *64*, 1–11. [CrossRef]
9. Godoy, R.S.; Lanés, L.E.K.; Castro, B.D.; Weber, V.; Wingen, N.; Pires, M.M.; Oliveira, G.T.; Maltchik, L. Oxidative Stress Resistance in a Short-Lived Neotropical Annual Killifish. *Biogerontology* **2020**, *21*, 217–229. [CrossRef]
10. Georgiev, A.V.; Thompson, M.E.; Mandalaywala, T.M.; Maestripieri, D. Oxidative Stress as an Indicator of the Costs of Reproduction among Free-Ranging Rhesus Macaques. *J. Exp. Biol.* **2015**, *218*, 1981–1985. [CrossRef]
11. Sharick, J.T.; Vazquez-Medina, J.P.; Ortiz, R.M.; Crocker, D.E. Oxidative Stress Is a Potential Cost of Breeding in Male and Female Northern Elephant Seals. *Funct. Ecol.* **2015**, *29*, 367–376. [CrossRef]
12. Fletcher, Q.E.; Selman, C.; Boutin, S.; McAdam, A.G.; Woods, S.B.; Seo, A.Y.; Leeuwenburgh, C.; Speakman, J.R.; Humphries, M.M. Oxidative Damage Increases with Reproductive Energy Expenditure and Is Reduced by Food-Supplementation. *Evolution* **2013**, *67*, 1527–1536. [CrossRef]
13. Dornfeld, T.C.; Robinson, N.J.; Tomillo, P.S.; Paladino, F.V. Ecology of Solitary Nesting Olive Ridley Sea Turtles at Playa Grande, Costa Rica. *Mar. Biol.* **2015**, *162*, 123–139. [CrossRef]
14. Bernardo, J.; Plotkin, P.T. An Evolutionary Perspective on the Arribada Phenomenon and Reproductive Behavioral Polymorphism of Olive Ridley Sea Turtles (Lepidochelys Olivacea). In *Biology and Conservation of Ridley Sea Turtles*; Plotkin, P.T., Ed.; The Johns Hopkins University Press: Baltimore, MD, USA, 2007; pp. 59–87.
15. Ocana, M.; Harfush-Melendez, M.; Heppell, S. Mass Nesting of Olive Ridley Sea Turtles Lepidochelys Olivacea at La Escobilla, Mexico: Linking Nest Density and Rates of Destruction. *Endanger. Species Res.* **2012**, *16*, 45–54. [CrossRef]
16. Hirth, H.F. Some Aspects of the Nesting Behavior and Reproductive Biology of Sea Turtles. *Integr. Comp. Biol.* **1980**, *20*, 507–523. [CrossRef]
17. Plotkin, P.T.; Byles, R.A.; Rostal, D.C.; Owens, D.W. Independent versus Socially Facilitated Oceanic Migrations of the Olive Ridley, Lepidochelys Olivacea. *Mar. Biol.* **1995**, *122*, 137–143. [CrossRef]
18. Plotkin, P.T.; Owens, D.W.; Byles, R.A.; Patterson, R. Departure of Male Olive Ridley Turtles (*Lepidochelys olivacea*) from a Nearshore Breeding Ground. *Herpetologica* **1996**, *52*, 1–7.
19. Jensen, M.P.; Abreu-Grobois, F.A.; Frydenberg, J.; Loeschcke, V. Microsatellites Provide Insight into Contrasting Mating Patterns in Arribada vs. Non-Arribada Olive Ridley Sea Turtle Rookeries. *Mol. Ecol.* **2006**, *15*, 2567–2575. [CrossRef]
20. Eckrich, C.E.; Owen, D.W. Solitary versus Arribada Nesting in the Olive Ridley Sea Turtles (*Lepidochelys olivacea*): A Test of the Predator-Satiation Hypothesis. *Herpetologica* **1995**, *51*, 349–354.
21. Honarvar, S.; O'Connor, M.P.; Spotila, J.R. Density-Dependent Effects on Hatching Success of the Olive Ridley Turtle, Lepidochelys Olivacea. *Oecologia* **2008**, *157*, 221–230. [CrossRef]
22. Bézy, V.S.; Valverde, R.A.; Plante, C.J. Olive Ridley Sea Turtle Hatching Success as a Function of Microbial Abundance and the Microenvironment of In Situ Nest Sand at Ostional, Costa Rica. *J. Mar. Biol.* **2014**, *2014*, 351921. [CrossRef]
23. Williamson, S.A.; Evans, R.G.; Robinson, N.J.; Reina, R.D. Synchronised Nesting Aggregations Are Associated with Enhanced Capacity for Extended Embryonic Arrest in Olive Ridley Sea Turtles. *Sci. Rep.* **2019**, *9*, 9783. [CrossRef]
24. Dutton, P.H. Methods for Collection and Preservation of Samples for Sea Turtle Genetic Studies. In Proceedings of the Proceedings of the International Symposium, Miami, FL, USA, 12–14 September 1995; pp. 17–24.
25. Espinoza-Romo, B.A.; Sainz-Hernández, J.C.; Ley-Quiñónez, C.P.; Hart, C.E.; Leal-Moreno, R.; Aguirre, A.A.; Zavala-Norzagaray, A.A. Blood Biochemistry of Olive Ridley (*Lepidochelys olivacea*) Sea Turtles Foraging in Northern Sinaloa, Mexico. *PLoS ONE* **2018**, *13*, e0199825. [CrossRef]
26. Mettee, N. Sample Collection Techniques. Available online: http://www.seaturtleguardian.org/sample-collection-techniques (accessed on 6 September 2022).
27. Allen, C.D.; Robbins, M.N.; Eguchi, T.; Owens, D.W.; Meylan, A.B.; Meylan, P.A.; Kellar, N.M.; Schwenter, J.A.; Nollens, H.H.; LeRoux, R.A.; et al. First Assessment of the Sex Ratio for an East Pacific Green Sea Turtle Foraging Aggregation: Validation and Application of a Testosterone ELISA. *PLoS ONE* **2015**, *10*, e0138861. [CrossRef]

28. Arango, B.G.; Harfush-Meléndez, M.; Marmolejo-Valencia, J.A.; Merchant-Larios, H.; Crocker, D.E. Blood Oxygen Stores of Olive Ridley Sea Turtles, Lepidochelys Olivacea Are Highly Variable among Individuals during Arribada Nesting. *J. Comp. Physiol. B Biochem. Syst. Environ. Physiol.* **2020**, *19*, 185–194. [CrossRef]
29. Vazquez-Medina, J.P.; Zenteno-Savin, T.; Tift, M.S.; Forman, H.J.; Crocker, D.E.; Ortiz, R.M. Apnea Stimulates the Adaptive Response to Oxidative Stress in Elephant Seal Pups. *J. Exp. Biol.* **2011**, *214*, 4193–4200. [CrossRef]
30. Fiehn, O.; Wohlgemuth, G.; Scholz, M.; Kind, T.; Lee, D.Y.; Lu, Y.; Moon, S.; Nikolau, B. Quality Control for Plant Metabolomics: Reporting MSI-Compliant Studies. *Plant J.* **2008**, *53*, 691–704. [CrossRef]
31. Xia, J.; Wishart, D.S. Using Metaboanalyst 3.0 for Comprehensive Metabolomics Data Analysis. *Curr. Protoc. Bioinforma.* **2016**, *2016*, 14.10.1–14.10.91. [CrossRef] [PubMed]
32. Tiessen, A.; Cubedo-Ruiz, E.A.; Winkler, R. Improved Representation of Biological Information by Using Correlation as Distance Function for Heatmap Cluster Analysis. *Am. J. Plant Sci.* **2017**, *08*, 502–516. [CrossRef]
33. Frolkis, A.; Knox, C.; Lim, E.; Jewison, T.; Law, V.; Hau, D.D.; Liu, P.; Gautam, B.; Ly, S.; Guo, A.C.; et al. SMPDB: The Small Molecule Pathway Database. *Nucleic Acids Res.* **2009**, *38*, 480–487. [CrossRef]
34. Plotkin, P.T.; Rostal, D.C.; Byles, R.A.; Owens, D.W. Reproductive and Developmental Synchrony in Female Lepidochelys Olivacea. *J. Herpetol.* **1997**, *31*, 17–22. [CrossRef]
35. Owens, D.W. The Comparative Reproductive Physiology of Sea Turtles. *Integr. Comp. Biol.* **1980**, *20*, 549–563. [CrossRef]
36. Rostal, D.C.; Grumbles, J.S.; Byles, R.A.; Marquez-M, R.; Owens, D.W. Nesting Physiology of Kemp's Ridley Sea Turtles, Lepidochelys Kempi, at Rancho Nuevo, Tamaulipas, Mexico, with Observations on Population Estimates. *Chelonian Conserv. Biol.* **1997**, *2*, 538–547.
37. Osawa, T.; Shimamura, T.; Saito, K.; Hasegawa, Y.; Ishii, N.; Nishida, M.; Ando, R.; Kondo, A.; Anwar, M.; Tsuchida, R.; et al. Phosphoethanolamine Accumulation Protects Cancer Cells under Glutamine Starvation through Downregulation of PCYT2. *Cell Rep.* **2019**, *29*, 89–103.e7. [CrossRef]
38. Wang, X.; Sun, H.; Zhang, A. *Metabolic Profiling and Biomarkers Analysis of XinQiXu Syndrome*; Elsevier Inc.: Amsterdam, The Netherlands, 2015; ISBN 9780128031186.
39. Felig, P. The Glucose-Alanine Cycle. *Metabolism* **1973**, *22*, 179–207. [CrossRef]
40. Palmieri, L.; Pardo, B.; Lasorsa, F.M.; Del Arco, A.; Kobayashi, K.; Iijima, M.; Runswick, M.J.; Walker, J.E.; Saheki, T.; Satrústegui, J.; et al. Citrin and Aralar1 Are Ca2+-Stimulated Aspartate/Glutamate Transporters in Mitochondria. *EMBO J.* **2001**, *20*, 5060–5069. [CrossRef]
41. Tripathy, B. An Assessment of Solitary and Arribada Nesting of Olive Ridley Sea Turtles (*Lepidochelys olivacea*) at the Rushikulya Rookery of Orissa, India. *Asiat. Herpetol. Res.* **2008**, *11*, 136–142.
42. Hamann, M.; Limpus, C.J.; Whittier, J.M. Patterns of Lipid Storage and Mobilisation in the Female Green Sea Turtle (*Chelonia mydas*). *J. Comp. Physiol. B* **2002**, *172*, 485–493. [CrossRef]
43. Goldberg, D.W.; Leitao, S.A.T.; Godfrey, M.H.; Lopez, G.G.; Santos, A.J.B.; Neves, F.A.; de Souza, E.P.G.; Moura, A.S.; Bastos, J.d.C.; Bastos, V.L.F.d.C. Ghrelin and Leptin Modulate the Feeding Behaviour of the Hawksbill Turtle *Eretmochelys imbricata* during Nesting Season. *Conserv. Physiol.* **2013**, *1*, 1–13. [CrossRef]
44. Plot, V.; Jenkins, T.; Robin, J.-P.; Fossette, S.; Georges, J.-Y. Leatherback Turtles Are Capital Breeders: Morphometric and Physiological Evidence from Longitudinal Monitoring. *Physiol. Biochem. Zool.* **2013**, *86*, 385–397. [CrossRef]
45. Owens, D.W.; Morris, Y.A. The Comparative Endocrinology of Sea Turtles. *Copeia* **1985**, *3*, 723–735. [CrossRef]
46. Peavey, L.E.; Popp, B.N.; Pitman, R.L.; Gaines, S.D.; Arthur, K.E.; Kelez, S.; Seminoff, J.A. Opportunism on the High Seas: Foraging Ecology of Olive Ridley Turtles in the Eastern Pacific Ocean. *Front. Mar. Sci.* **2017**, *4*, 348. [CrossRef]
47. Fleming, A.H.; Kellar, N.M.; Allen, C.D.; Kurle, C.M. The Utility of Combining Stable Isotope and Hormone Analyses for Marine Megafauna Research. *Front. Mar. Sci.* **2018**, *5*, 338. [CrossRef]
48. Ensminger, D.C.; Salvador-pascual, A.; Arango, B.G.; Allen, K.N.; Vazquez-Medina, J.P. Fasting Ameliorates Oxidative Stress: A Review of Physiological Strategies across Life History Events in Wild Vertebrates. *Comp. Biochem. Physiol. -A Mol. Integr. Physiol.* **2021**, *256*. [CrossRef]
49. Stier, A.; Dupoué, A.; Picard, D.; Angelier, F.; Brischoux, F.; Lourdais, O. Oxidative Stress in a Capital Breeder (*Vipera aspis*) Facing Pregnancy and Water Constraints. *J. Exp. Biol.* **2017**, *220*, 1792–1796. [CrossRef]
50. Cohen, A.; Klasing, K.; Ricklefs, R. Measuring Circulating Antioxidants in Wild Birds. *Comp. Biochem. Physiol. -B Biochem. Mol. Biol.* **2007**, *147*, 110–121. [CrossRef]
51. Deutsch, C.J.; Haley, M.P.; Le Boeuf, B.J. Reproductive Effort of Male Northern Elephant Seals: Estimates from Mass Loss. *Can. J. Zool.* **1990**, *68*, 2580–2593. [CrossRef]
52. Deutsch, C.J.; Crocker, D.E.; Costa, D.P.; Le Boeuf, B.J. Sex- and Age-Related Variation in Reproductive Effort of Northern Elephant Seals. In *Elephant seals: Population Ecology, Behavior, and Physiologypopulation Ecology, Behavior, and Physiology*; Le Boeuf, B.J., Laws, R.M., Eds.; University of California Press: Berkeley, CA, USA, 1994; pp. 169–210, ISBN 9780520328150.
53. Le Boeuf, B.J. Male-Male Competition and Reproductive Success in Elephant Seals. *Am. Zool.* **1974**, *14*, 163–176. [CrossRef]
54. Haley, M.P.; Deutsch, C.J.; Le Boeuf, B.J. Size, Dominance and Copulatory Success in Male Northern Elephant Seals, Mirounga Angustirostris. *Anim. Behav.* **1994**, *48*, 1249–1260. [CrossRef]
55. Crocker, D.E.; Houser, D.S.; Webb, P.M. Impact of Body Reserves on Energy Expenditure, Water Flux, and Mating Success in Breeding Male Northern Elephant Seals. *Physiol. Biochem. Zool.* **2012**, *85*, 11–20. [CrossRef]

antioxidants

MDPI

Review

The Glutathione System: A Journey from Cyanobacteria to Higher Eukaryotes

Corinne Cassier-Chauvat, Fanny Marceau, Sandrine Farci, Soufian Ouchane and Franck Chauvat *

Université Paris-Saclay, CEA, CNRS, Institute for Integrative Biology of the Cell (I2BC), F-91190 Gif-sur-Yvette, France; corinne.cassier-chauvat@cea.fr (C.C.-C.); fanny.marceau@cea.fr (F.M.); soufian.ouchane@i2bc.paris-saclay.fr (S.O.)

* Correspondence: franck.chauvat@cea.fr; Tel.: +33-1-69-08-78-11

Abstract: From bacteria to plants and humans, the glutathione system plays a pleiotropic role in cell defense against metabolic, oxidative and metal stresses. Glutathione (GSH), the γ-L-glutamyl-L-cysteinyl-glycine nucleophile tri-peptide, is the central player of this system that acts in redox homeostasis, detoxification and iron metabolism in most living organisms. GSH directly scavenges diverse reactive oxygen species (ROS), such as singlet oxygen, superoxide anion, hydrogen peroxide, hydroxyl radical, nitric oxide and carbon radicals. It also serves as a cofactor for various enzymes, such as glutaredoxins (Grxs), glutathione peroxidases (Gpxs), glutathione reductase (GR) and glutathione-S-transferases (GSTs), which play crucial roles in cell detoxication. This review summarizes what is known concerning the GSH-system (GSH, GSH-derived metabolites and GSH-dependent enzymes) in selected model organisms (*Escherichia coli*, *Saccharomyces cerevisiae*, *Arabidopsis thaliana* and human), emphasizing cyanobacteria for the following reasons. Cyanobacteria are environmentally crucial and biotechnologically important organisms that are regarded as having evolved photosynthesis and the GSH system to protect themselves against the ROS produced by their active photoautotrophic metabolism. Furthermore, cyanobacteria synthesize the GSH-derived metabolites, ergothioneine and phytochelatin, that play crucial roles in cell detoxication in humans and plants, respectively. Cyanobacteria also synthesize the thiol-less GSH homologs ophthalmate and norophthalmate that serve as biomarkers of various diseases in humans. Hence, cyanobacteria are well-suited to thoroughly analyze the role/specificity/redundancy of the players of the GSH-system using a genetic approach (deletion/overproduction) that is hardly feasible with other model organisms (*E. coli* and *S. cerevisiae* do not synthesize ergothioneine, while plants and humans acquire it from their soil and their diet, respectively).

Keywords: cyanobacteria; human; plants; glutathione; glutaredoxins; glutathione-S-transferases; iron-sulfur cluster; methylglyoxal; ergothioneine; ophthalmate; norophthalmate

Citation: Cassier-Chauvat, C.; Marceau, F.; Farci, S.; Ouchane, S.; Chauvat, F. The Glutathione System: A Journey from Cyanobacteria to Higher Eukaryotes. *Antioxidants* **2023**, *12*, 1199. https://doi.org/10.3390/antiox12061199

Academic Editors: Marcelo Hermes-Lima, Daniel Carneiro Moreira and Tania Zenteno-Savín

Received: 5 May 2023
Revised: 25 May 2023
Accepted: 29 May 2023
Published: 31 May 2023

1. Introduction

Most life forms are continuously challenged with toxic reactive oxygen species (ROS) present in our oxygenic atmosphere (ozone, O_3), and/or generated by respiration and cell metabolism [1–3] and photosynthesis in cyanobacteria [4–8], algae and plants [9–11]. In addition, photosynthetic organisms are exposed to solar UV that also generate ROS [12,13].

ROS molecules encompass singlet oxygens ($^1O_2$), superoxide anions ($O_2^{\bullet-}$), hydrogen peroxides (H_2O_2), and hydroxyl radicals ($^{\bullet}OH$) that cause damages to target molecules, namely: lipids, nucleic acids and proteins [2,10], thereby generating cell death in microorganisms and multiple disorders and diseases in humans [14–16] that reduce longevity [17].

Superoxide anions and hydrogen peroxides can both react with proteins containing iron-sulfur [Fe-S] clusters, liberating their Fe ions. Free or complexed Fe^{2+} ions reduce H_2O_2, yielding hydroxyl radicals that modify all kinds of biomolecules at a diffusion-limited rate. Hence, radicals, sulfenic acids, disulfides and (hydro)peroxides are directly or indirectly formed by ROS [3,18]. ROS also oxidize cysteines to form thiyl (sulfenyl) radical (-S•) by one-electron transition; sulfenic acid (-SOH) and disulfide (-S-S-) by a two-electrons transition; sulfinic acid (-SO_2H) by a four-electrons transition; and eventually sulfonic acid (-SO_3H) by a six-electrons transition [19]. Concerning disulfides, two types can be distinguished considering whether they link two cysteinyl residues from either the same or different proteins (intra- or inter-molecular disulfide bridges), or from a protein and a molecule of glutathione (glutathionylation). Glutathione is the γ-L-glutamyl-L-cysteinyl-glycine tri-peptide (hereafter designated as GSH) that plays a prominent role in ROS detoxification from bacteria to higher eukaryotes [3,11,16,20,21]. It directly scavenges ROS and also serves as a redox cofactor for various antioxidant enzymes, such as glutaredoxins (Grxs), glutathione peroxidases (Gpxs), glutathione reductase (GR) and glutathione S-transferases (GSTs). The above-mentioned glutathionylation can protect cysteinyl residues against irreversible oxidation (generation of sulfinic and sulfonic acids), and/or act in regulation [9,11,18,22], as shown in Figure 1.

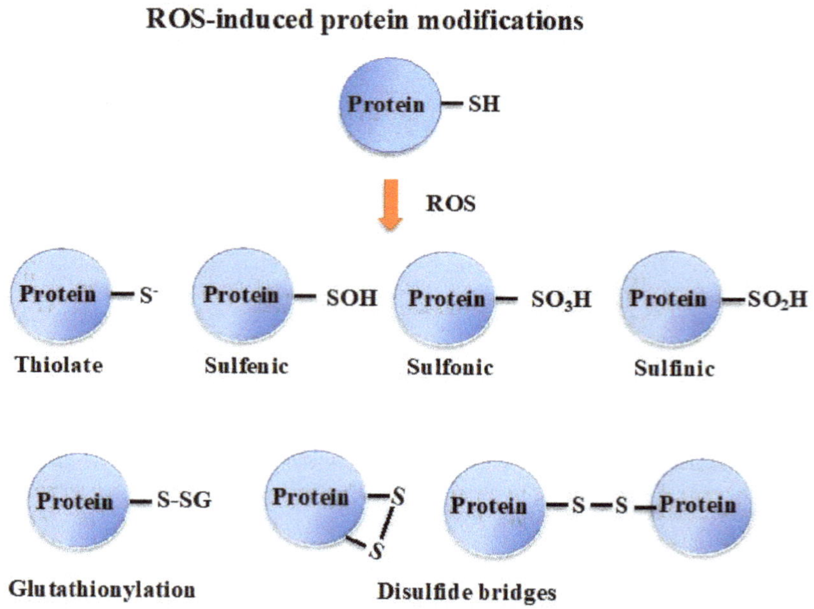

Figure 1. Schematic representation of the oxidation of the cysteinyl residue of protein to sulfenic (-SOH), sulfinic (-SO_2H) and sulfonic (-SO_3H); and disulfide (-S-S-) with another cysteinyl residue from the same or another protein, or a molecule of glutathione.

ROS can also be detoxified by various metabolites (ascorbate, carotenoids, vitamins, etc.) and several enzymes [23]. The superoxide dismutase (SOD) converts $O_2^{\bullet-}$ to H_2O_2, which is then detoxified to H_2O by the catalase and peroxidase enzymes [18]. H_2O_2 can also be detoxified by the hydroperoxide activity of some glutaredoxins [18]. The protein disulfides and glutathione-protein mixed disulfides are repaired by thioredoxins [20,24], glutaredoxins [3,11,25,26] and glutathione-S-transferases [14,16,25,27–29].

ROS-removing systems are usually viewed as beneficial antioxidants that maintain damaging ROS below dangerous levels [3,20,30]. However, ROS are also a necessary part of subcellular and intercellular communication in living organisms [3,18]. Indeed ROS species can serve as signal mediators in the redox regulation of cell metabolism [19], as they are enzymatically produced and degraded by NADPH-oxidases, which generate superoxide anions [3], SOD, which generates H_2O_2, and catalase and peroxidase, which detoxify H_2O_2 into H_2O. Furthermore, H_2O_2 oxidizes protein thiols in disulfides or sulfenic acids, which can be reduced back to thiols, and are thereby good thiol redox switches for signaling [10,18]. Consequently, it has been proposed that "redox biology" [31] or "ROS processing systems" [10] would be a more accurate term than "(anti)oxidative systems" to describe cellular components that interact with ROS.

This review presents what is known concerning the evolutionary-conserved glutathione-system in selected model organisms, *E. coli*, *S. cerevisiae*, *A. thaliana* and human, emphasizing cyanobacteria for several reasons (See the next paragraphs for details and references). Cyanobacteria are environmentally crucial prokaryotes regarded as having evolved the oxygenic photosynthesis process, the chloroplast of algae and plants, and the glutathione-system to protect themselves against the ROS produced by their active photoautotrophic metabolism. Furthermore, cyanobacteria synthesize the thiol-less GSH homologs, ophthalmate and norophthalmate, and the ergothioneine antioxidant that operates in signaling and/or detoxication in humans. Moreover, cyanobacteria combine several important properties, such as (i) a simple nutritional requirement, (ii) a great physiological robustness, (iii) an important metabolic plasticity and (iv) the powerful genetics of some model strains. Hence, they are regarded as promising "low-cost" microbial factories for (i) the sustainable production of food and high-value chemicals for health and energy, (ii) the bioremediation of polluted waters and (iii) the fertilization of cultures.

2. Biological Importance and Biotechnological Interests of Cyanobacteria

Cyanobacteria are primordial prokaryotes regarded as the "inventor" of oxygenic photosynthesis [32], which played an important role in the evolution of Early Earth and the biosphere by absorbing a huge amount of the greenhouse gas carbon dioxide (CO_2), and evolving a huge amount of dioxygen (O_2) [33–37]. Indeed, cyanobacteria are regarded as responsible for the oxygenation (and oxidation) of the atmosphere since the Great Oxidation Event around 2.4 Ga [32,33,38–42].

As a consequence, cyanobacteria have long been challenged by ROS $^1O_2$, $O_2^{\bullet -}$, H_2O_2 and OH [2,3,18], which are generated by their active photosynthesis and, sometimes, their respiration [8,10,11]. Singlet oxygens are unavoidably produced by the interaction of sunlight with photosynthetic pigments (chlorophyll a, carotenoids and phycobiliproteins) while superoxide anions, hydrogen peroxides and hydroxyl radicals are generated when the light-driven electron transport exceeds what is needed for nutrients assimilation [8,10,43]. Cyanobacteria are also strongly exposed to solar UV radiations (UVR) that also generate ROS [20,44]. Consequently, cyanobacteria represent a major source of ROS in aquatic environments [45]. Furthermore, cyanobacteria were also the first organisms to cope with the oxygen-promoted changes in metal availability: decrease of iron (Fe), cobalt (Co), nickel (Ni) and manganese (Mn), and increase of zinc (Zn), molybdenum (Mo) and copper (Cu). This constitutes a real challenge because approximately one-quarter to one-third of all cellular proteins require metals [46].

Cyanobacteria are also frequently challenged by heavy metals (cadmium, cesium, chromate, mercury, lead, uranium, etc.) which are released by natural sources (volcanoes and forest fires) and anthropogenic activities (mining, fossil fuel burning, etc.). The toxicity of heavy metals is based on their chemical properties, which allow them to promote the production of ROS and the inactivation of enzymes [47–49], basically by reaction with SH groups, including that of GSH [50,51]. The presence of heavy metals in soils and waters is especially problematic because metals are persistent in the environment and they accumulate throughout the food chain, thereby threatening human health [46,52,53]. Cyanobacteria are important organisms to investigate the relations between metals and oxidative stress as they constitute the first biological barrier against entry of heavy metals into the food chain. Furthermore, cyanobacteria perform the two metal-requiring ROS-generating processes, photosynthesis and respiration, in the same membrane system [54,55]. Moreover, cyanobacteria are regarded as promising organisms for bioremediation of metal pollutants thanks to their robust photoautotrophic metabolism [55,56] and their multifarious mechanisms (biosorption, bioaccumulation and biotransformation) to sequester and minimize the toxic effects of heavy metals [57,58].

To cope with ROS and environmental stresses, cyanobacteria have evolved the glutathione system [5,59,60], which is crucial to their photoautotrophic lifestyle [4,44,61] and has been conserved during evolution [5,6,59,60,62]. The glutathione system comprises the glutathione tripeptide itself (γ-L-glutamyl-L-cysteinyl-glycine, GSH) and its cysteine-less homologs (ophthalmate and norophthalmate); in humans, these serve as biomarkers of diseases (See below), as well as numerous GSH-dependent enzymes, such as glutaredoxins and glutathione-S-transferases [18,63], which have been conserved during evolution [11,26,27,62,64,65]. In addition, cyanobacteria possess other promiscuous antioxidant enzymes, such as superoxide dismutases, catalases and peroxidases [8].

Contemporary cyanobacteria continue to play a key role in the global ecosystem. They fix enormous amounts of atmospheric CO_2 and N_2 to produce huge amounts of O_2 and biomass for our food chain [32,33,36,37,56,66–69]. They have been consumed by humans and used as soil fertilizers for over a thousand years [70–73]. Furthermore, they produce a wealth of metabolites, such as vitamins, antioxidants (such as ergothioneine mentioned below), antibiotics, antifreezes, drugs, osmoprotectants and toxins [56,74–77] that can influence human health and/or improve plant growth and/or resistance to stress (drought, salt, heavy metals and pathogens) [78]. Currently, several cyanobacteria are being tested as a way to replenish O_2, provide food, and recycle CO_2 and urea wastes during long-term space missions [79,80]. Moreover, cyanobacteria are viewed as promising cell factories for the production of chemicals (biofuels, biodegradable bioplastics, drugs, solvents, etc.) from highly abundant natural resources: solar energy, water (fresh/marine), CO_2 and minerals [74,75,81], thanks to their active and robust photoautotrophic metabolism [55,56] and the synthetic biology tools of model species [82], as shown in Figure 2.

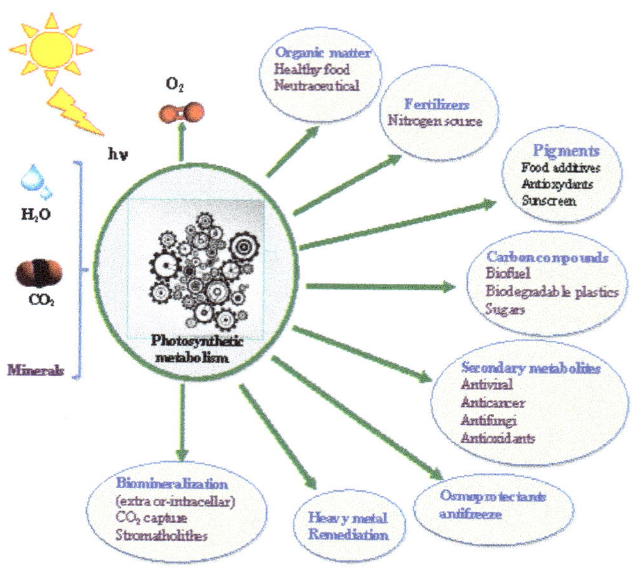

Figure 2. Schematic representation of the biotechnological interests of cyanobacteria. Minerals include calcium, nitrogen (N_2, ammonium, nitrate and urea), phosphate, sulfate etc.

Finally, in colonizing most aquatic ecosystems and soils of our planet, where they face various environmental challenges and interactions with competitors, predators or symbiotic hosts (angiosperms, bryophytes, fungi and gymnosperms), cyanobacteria have evolved as widely diverse organisms. They display various cell morphologies [83] and cellular differentiation [84], as well as widely diverse genome sizes (1.44–12.07 Mb), GC content (30–60%) and organization (presence of a circular chromosome with or without one to several linear chromosomes and circular plasmids) [82,85]. Hence, cyanobacteria are good model organisms to study the impact of environmental conditions and interactions with other organisms on the physiology, metabolism and morphology of microbial cells.

Together, the above-mentioned environmental importance of cyanobacteria and their interest for basic and applied science highlight the value of studying the glutathione system of cyanobacteria that has been conserved during evolution.

3. Synthesis and Importance of Glutathione in Living Organisms

Glutathione (GSH) was discovered in 1888 by J. de Rey-Pailhade, and its composition as γ-L-glutamyl-L-cysteinyl-glycine was established much later, in 1935 [86]. That the Cys and Glu of GSH are linked through the γ-carbonyl group of Glu instead of the typical α-carboxyl group confers a high stability to GSH since only very specific enzymes under particular conditions may operate on its degradation (See below). GSH is the most abundant non-protein thiol (concentration ranging from 0.1 mM to about 20 mM) in all three kingdoms of life: Bacteria (mostly Gram-negative, rarely Gram-positive), Archaea, and Eukarya, where it plays pleiotropic roles in cell life and resistance to stresses [3,5,11,15,18,20,21,87–89]. GSH is a nucleophilic metabolite that directly scavenges ROS, nitric oxides and carbon radicals [3,5,11,18]. GSH also serves as electron donor to various antioxidant enzymes, including glutaredoxins, glutathione peroxidases and glutathione-S-transferases (See below). Furthermore, GSH can also act in the synthesis of ergothioneine, another antioxidant catalyzed by the EgtB enzyme (See below). In addition, the cysteinyl thiols of GSH can complex metal [3,15,20], as shown in Figure 3.

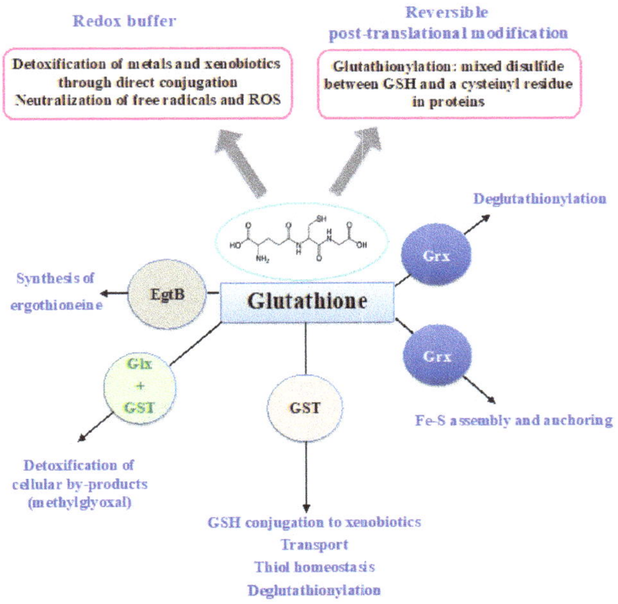

Figure 3. Schematic representation of the pleiotropic roles of glutathione.

GSH plays a prominent role in iron homeostasis in many prokaryotes [3,5] and most eukaryotes [18]. It is a key component of the cytoplasmic pool of labile iron, mostly occurring under the Fe(II)GSH complex [90], which likely supplies Fe for the synthesis of the Fe or [Fe-S] cluster cofactors of a wealth of enzymes involved in electron transfers (photosynthesis respiration) and central metabolism [15,18,88,91].

GSH is synthesized by two sequential ATP-requiring enzymes, namely the γ-glutamyl-cysteine (γ-GC) synthetase (γ-GCS, EC 6.3.2.2), which forms γ-GC from L-glutamic acid and L-cysteine, and the GSH synthetase (GS, EC 6.3.2.3), which forms GSH from γ-GC and L-glycine, as shown in Figure 4.

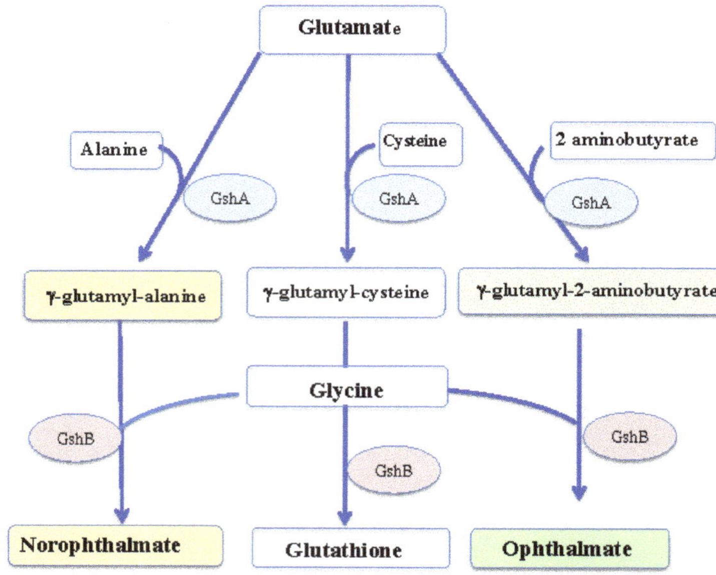

Figure 4. Schematic representation of the synthesis of glutathione and its thiol-less analogs. GshA: gamma-glutamylcysteine ligase; GshB: glutathione synthase.

In most cells, the two GSH-synthesis enzymes are encoded by separate genes, named *gshA* and *gshB* in prokaryotes and *gsh1* and *gsh2* in eukaryotes [9,20,92]. The evolutionary history of the GSH biosynthesis genes is complex in that the two genes in the pathway were acquired independently [60]. The gene encoding γ-glutamyl-cysteine ligase most probably arose in cyanobacteria [59] and was subsequently transferred to other bacteria and eukaryotes [5,6]. Then, eukaryotes and most bacteria apparently recruited a protein from the ATP-grasp superfamily of enzymes to synthesize glutathione from γ-glutamyl-cysteine and glycine [5,6,60]. In many organisms, the activity of γ-GCS (GshA), which is the rate-limiting step in the GSH biosynthesis pathway, is subjected to feedback inhibition by GSH to avoid over-accumulation of GSH [20,30,93].

Glutathione is vital in many organisms, including yeast [20,94,95], mice [96], plants [10,87,97] and cyanobacteria [4,61], but not in *Escherichia coli* [20,92,98,99]. *E. coli* mutants devoid of GSH do not exhibit enhanced sensitivity to oxidative stress (H_2O_2, cumene hydroperoxide, ionizing (gamma) radiations) in exponentially growing culture [92], but stationary-phase cultures are more sensitive to H_2O_2 than the wild-type [100]. However, an *E. coli* strain overproducing glutathione is more resistant to gamma-irradiations than the corresponding wild-type strain, not merely because it has a higher content of GSH per se, but because it has an increased capacity to synthesize GSH when irradiated [101]. It has also been shown that *E. coli* and *Salmonella typhimurium* accumulate reduced glutathione in the growth medium during the exponential phase [102] to protect cells against external toxic compounds such as H_2O_2, N-methyl-N′-nitro-N-nitrosoguanidine, iodoacetamide and heavy metals [103,104]. In microbial systems (bacteria and yeast) a specific glutathione uptake process exists to salvage glutathione from lysing cells [9].

In the yeast *S. cerevisiae gsh1*, mutants grow more slowly than the wild-type strain in rich medium and can grow in minimal medium only in the presence of exogenous GSH [20]. In contrast, *gsh2* mutants grow well in the absence of GSH and are not particularly sensitive to either H_2O_2 or *t*-butyl hydroperoxide. These results indicate that γ-glutamyl-cysteine, which accumulates in the *gsh2* strains, has some of the antioxidant activities of GSH [20].

In plants, where the complete absence of GSH causes death at the embryonic stage [97], mutants with a less severe decrease in GSH content are viable but more sensitive to many biotic and abiotic stresses [11], including Cd [105] and Zn [106]. The biosynthesis of γ-glutamyl-cysteine catalyzed by glutamate cysteine ligase takes place in chloroplasts [9], which likely originated from cyanobacteria [34,107,108], while the formation of GSH catalyzed by glutathione synthetase can occur both in the chloroplasts and in the cytosol. Then, GSH is transported to mitochondria and the nucleus [9,11,18,109]. In plants [109,110] and some cyanobacteria [111,112], GSH is also polymerized into phytochelatin to chelate metals that are coordinated by its numerous thiol groups. In addition, plants [113,114], cyanobacteria [115,116] and many other organisms [117] use the cysteine-rich protein metallothionein to chelate metals.

In humans, low levels of GSH and high levels of ROS are associated with HIV, diabetes mellitus and/or neurodegenerative diseases [16,118,119]. Cancer cells, with their high levels of GSH (and glutathione reductase activity, see below), are refractory to some of the therapies inducing oxidative stress. GSH can also react with the bioactive nitric oxide (NO) gas to produce S-nitrosoglutathione (GSNO), a storage form of this gaseous radical in tissues [89]. The reaction of NO with $O_2^{\bullet-}$ to form peroxynitrite ($ONOO^-$) conveys the GSNO-derived NO in the extracellular fluid. This process controls the physiological levels of this signaling molecule in tissues [120]. These properties likely explain the NO-like vasodilation properties of GSNO that are shared by other pharmacological NO donors [89]. In addition to GSH, human cells are protected from metal stress by synthesizing the metal-binding protein metallothionein [121], like many other organisms [117].

Some Gram-positive bacteria, including *Actinobacillus pleuropneumoniae* [122], *Listeria monocytogenes* [123], *Pasteurella multocida* [124] and both *Streptococcus agalactiae* and *Streptococcus thermophilus* [125,126], contain a newly discovered bifunctional enzyme, termed GshF, which possesses both GshA and GshB activities. The N-terminal sequence of GshF is similar to that of *E. coli* GshA, but the C-terminal sequence is more similar to the D-Ala, D-Ala ligase than to any known GshB [125]. Interestingly, GSH inhibits neither the GshA activity nor the GshB activity of the GshF, and the GshA activity of GshF is higher compared with that in other organisms, and it is not inhibited by GSH [127]. More than 20 bacteria, mostly Gram-positive, possess a *gshF*-like gene. Recently, the *S. thermophilus gshF* gene was overexpressed in tobacco plants, *E. coli* and yeast cells to increase their production of GSH, an objective of biotechnological interest [126–130].

Due to its critical role in antioxidation, xenobiotic detoxification, and immune regulation pathways, GSH has been widely used in the food, cosmetic, and pharmaceutical industries [131]. So far, GSH is commercially produced mainly by *Saccharomyces cerevisiae* strains, which have generally been recognized as safe. Previous studies have focused on overproducing GshA and GshB, but the production yield and titer of GSH in such *Saccharomyces cerevisiae* strains remain low due to the feedback inhibition on GshA [127]. To overcome this limitation, the GshF bifunctional enzyme from Gram-positive bacteria was produced in *S. cerevisiae*, as GshF is insensitive to feedback inhibition. The resulting strain produced 240 mg L^{-1} GSH with GSH content and yield of 4.3% and 25.6 $mg_{glutathione}/g_{glucose}$, respectively [127]. However, this production of GSH by *S. cerevisiae* competes with glucose demands of other industries and results in high production costs [131]. To save production cost, we think that this approach of using GshF for GSH production should be tested in cyanobacteria because they can produce high-value chemicals from sunlight and CO_2 instead of glucose.

Instead of GSH itself, some organisms employ its precursors or derivates [3,59,120], such as γ-glutamyl-cysteine in halobacteria and halophilic archaea or trypanothione in kinetoplastid parasites [132]. They may also use other thiols, e.g., bacillithiol in Gram-positive Firmicutes [133] or mycothiol in many actinobacteria, such as the human pathogen *Mycobacterium tuberculosis* [134].

In some organisms, such as mycobacteria [135], the γ-glutamyl-cysteine peptide is also used for the synthesis of ergothioneine (hereafter EGT), an unusual thio-histidine betaine amino acid (also known as 2-mercaptohistidine trimethylbetaine) that has potent antioxidant and cytoprotective activities [136–140]. Hence, in mycobacteria there is competition between EGT and glutathione biosynthesis. EGT has both a thiol (antioxidant) and a thione form [136,140], with the latter thione tautomer being predominant at physiological pH, thereby making EGT unusually resistant to oxidation by molecular O_2 [137]. Its midpoint potential, +0.06 V, is unusually high compared to typical thiols, including GST (−0.2 to −0.4 V) [137]. EGT can serve as a reductant via one-electron reaction or as a nucleophilic reagent via two-electrons exchange. In vitro studies have shown that EGT can scavenge ROS, such as singlet oxygen, hydroxyl radical and, more slowly, hydrogen peroxide H_2O_2. When EGT acts as a direct antioxidant (reductant), it is oxidized in EGT disulfide, EGT sulfenic or sulfinic acid, depending on the conditions (pH and/or the presence of thiols, the strength of oxidants, etc.) [136]. EGT-sulfinic is unstable and irreversibly degraded into L-hercynine, or oxidized to EGT sulfonic acid [139]. EGT can also participate in the chelation of divalent metals (Co, Cu, Hg, Ce, Pt) and in radiative reactions by physically deactivating high-energy molecules via energy transfer [137,139].

Few other organisms, such as cyanobacteria (see below) and certain fungi (*Neurospora crassa*, the fission yeast and mushroom fruiting bodies) are able to synthesize EGT [136,141,142], unlike plants and animals who acquire it via their soil and their diet, respectively [135,137]. EGT is stable in the body for a long time after ingestion, and is viewed as protecting the central nervous system against diseases [138]. Animals have evolved a highly selective transporter for it, originally known as a carnitine transporter (OCTN1) [138,140] and also called ergothioneine transporter ETT because of its 100-fold higher affinity for EGT [139]. Genetic analysis in mycobacteria [142] and EGT consumption in mammals has shown that EGT protects cells against oxidative [137,139], metal [143] and UV [137,139] stresses. Hence, EGT has been considered safe by regulatory agencies and may have value as a nutraceutical and antioxidant [137,139].

Returning to glutathione, some of its homologs have no cysteine residue, and therefore no reducing properties, such as ophthalmate (L-γ-glutamyl-L-α-aminobutyryl-L-glycine) and norophthalmate (L-γ-glutamyl-L-alanyl-L-glycine). Ophthalmate (hereafter OPH) and norophthalmate (NOPH) were initially discovered in various animal organs (lens, brain and liver) [144,145], where they are regarded as biomarkers of GSH depletion elicited by oxidative stress [146,147]. OPH was also found to have accumulated in stressed plants [148], yeasts [149], bacteria (*E. coli*) [150] and cyanobacteria (*Synechocystis* PCC 6803) [61]. Like GSH, both OPH and NOPH are synthesized by the GshA and GshB enzymes [61,146–149], as seen in Figure 4.

Evolutionary Interest of Glutathione Synthesis in Cyanobacteria

Cyanobacteria contain a high concentration of intracellular GSH (2–10 mM), mainly in the reduced form [7,151,152], and they can accumulate GSH when supplied with GSH precursor amino acids (Glu, Cys or Gly), especially Cys [152], similar to what was observed in the yeast *Saccharomyces cerevisiae* [153]. Such cyanobacterial cells with a higher GSH content were more tolerant to heat [154]. Similarly, the unicellular model *Synechococcus elongatus* PCC 7942 that contains more GSH than the other unicellular model *Synechocystis* PCC 6803 is more tolerant to chromate than the latter cyanobacterium [155].

The activity of GshA from the model cyanobacteria *Anabaena* PCC 7120 [156] and *Synechocystis* PCC 6803 [6], as well as GshB from *Synechococcus elongatus* PCC 7942 [157], have been verified following their production in *E. coli.* GshA was found to be inhibited by both GSH and 1-buthionine sulfoximine [6,156], like most GshA enzymes [20,30,93]. GshA most probably arose in cyanobacteria and was subsequently transferred to other bacteria and eukaryotes, which then recruited GshB to synthesize GSH, like cyanobacteria [5,6,59,60]. Supporting this hypothesis, both *gshA* and *gshB* genes appeared to be essential in the model cyanobacterium *Synechocystis* PCC 6803 [4,61]. Furthermore, the GshB-depleted mutant was showed to be sensitive to O_2 [61] and photo-oxidative stress [4,61], cadmium [158] and the antibiotic gentamicin [159].

Due to its importance for the food, cosmetic and pharmaceutical industries [131], GSH is commercially produced mainly by fermentation of *Saccharomyces cerevisiae* overproducing both GshA and GshB [127], or the more active GshF bifunctional enzyme [127]. However, this GSH production competes with glucose demands of other industries [131]. To save glucose costs, we propose to overproduce GshF in cyanobacteria, so as to produce GSH from sunlight and CO_2. For this purpose, *Arthrospira* species (commercial name *Spirulina*) are of special interest because they are generally recognized as safe.

In agreement with the conservation of the GSH system from cyanobacteria to higher eukaryotes, cyanobacteria can polymerize GSH into phytochelatin ((γ-Glu-Cys)$_{2\text{–}11}$-Gly) to chelate metals on its thiol groups [111,112,160], like plants [109]. Cyanobacteria also synthesize the cysteine-rich metal-chelating protein metallothionein [115,116], like many other organisms [117], including plants [113,114] and humans [121].

Cyanobacteria were also found to synthesize thiol-less γ-glutamyl peptides (γ-glutamyl-Ala, γ-Glu-2-aminobutyryl, γ- Glu-Leu, γ- Glu-iLeu, γ-Glu-Met, γ-Glu-Phe, γ-Glu-Thr) [61,161], in response to oxidative stress [61] or other slow-growth conditions [161], which are viewed as reservoirs of amino acids in cyanobacteria. Such γ-glutamyl dipeptides that differ from γ-Glu-Cys have been identified in eukaryotes, such as *Saccharomyces cerevisiae* [162] and mice [163], supporting the notion that the GSH system has been evolutionarily conserved from (cyano)bacteria to higher eukaryotes. In mammals, these γ-glutamyl peptides are regarded as being beneficial for human consumption [164], but also involved in inflammation, oxidative stress and/or glucose metabolism leading to cardio-metabolic diseases and/or diabetes [165,166].

Thiol-less analogues of glutathione, γ-Glu-Ala-Gly (NOPH) and γ-Glu-2-aminobutyryl-Gly (OPH), accumulated in cyanobacteria challenged by glucose-triggered metabolic and oxidative stress [61]. OPH was also found to be accumulated in stressed bacteria [150], yeasts [149] and plants [148]. The synthesis of OPH and NOPH in the cyanobacterium *Synechocystis* PCC 6803 was found to depend on GshA and GshB [61], as observed in mammals [18,147]. In mammals, OPH and NOPH are regarded as liver and heart markers of GSH depletion elicited by oxidative stress [146–149]. Glucose-stressed cyanobacterial cells accumulate not only OPH and NOPH, but also GSH [167]. Thus, OPH and NOPH are likely generated by the depletion of cysteine (not GSH per se) caused by its accelerated incorporation into GSH to cope with the increased GSH need for the detoxification of oxidants and glucose-catabolites (methylglyoxal, See below).

Many cyanobacteria synthesize large amounts of the EGT antioxidant [61,168–171], and fishes that feed, to some extent, on cyanobacteria are thus provided with plenty of EGT [137]. This finding is interesting because there is an increasing demand for EGT [135,140], and extracting it and chemically synthesizing it from edible fungi has a high cost and low yield [137,139]. The biosynthesis of EGT is well understood in *Mycobacterium* [135]. The SAM-dependent methyltransferase EgtD transforms histidine into histidine betaine. EgtA, the glutamate cysteine ligase, synthesizes γ-Glu-Cys. EgtB adds the thiol group of γ-Glu-Cys to the side chain of histidine betaine, which is transformed by EgtC into the histidine betaine cysteine sulfoxide metabolite. This intermediate is then converted to EGT by EgtE, the pyridoxal 5′-phosphate-dependent cysteine desulfurization enzyme [135,136]. All mycobacterial EGT-synthesis genes have been overexpressed in *E. coli*, leading to EGT production [135,172].

In EGT-producing cyanobacteria, which necessarily possess all EGT-synthesis genes, *egtB*, *egtC* and *egtD*, but not *egtE*, have been identified by sequence homology with their mycobacterial orthologs [171], and we have observed in the model species *Synechocystis* PCC 6803 that it is GSH itself, not its γ-Glu-Cys precursor, that serves for EGT synthesis [61]. We propose to use *Synechocystis* PCC 6803 to identify the possibly unknown EGT-synthesis genes as well as the molecular role and the selectivity/redundancy of EGT, GSH, OPH and NOPH, in signaling and/or detoxification of oxidative and metabolic stresses. Furthermore, we propose to increase the expression of all EGT-synthesis genes as an attempt to generate a cyanobacterial factory for cheap, high-level production of EGT from solar energy and CO_2.

4. Glutathione Degradation

Because of its γ-linkage between the carboxyl group of glutamate and the amine group of cysteine, GSH cannot be degraded by genuine protease. Dedicated peptidases, γ-glutamyl transpeptidase (GGTs; E.C. 2.3.2.2) and/or γ-glutamyl cyclotransferases (γ-GCT or GGCTs, EC 4.3.2.9), catabolize GSH in bacteria and/or eukaryotes [9,11,21,110].

The GGT enzyme is conserved throughout all three domains of life [11,16,173]. Bacterial GGTs are generally soluble and localized in the periplasmic space or secreted in the extracellular environment [173], whereas eukaryotic GGTs are embedded in plasma or vacuole membranes [9]. GGT enzymes release the cysteinyl-glycine dipeptide and 5-oxoproline, a cyclized form of glutamate (Zhang and Forman, 2012). Then, the cysteinyl-glycine dipeptide is broken down into cysteine and glycine by specific Cys-Gly peptidases (including the leucine aminopeptidase (EC 3.4.13.18)) while the 5-oxoproline is converted into glutamate by the ATP-dependent 5-oxoprolinase (EC 3.5.2.9). The released glutamate, cysteine and glycine can be ploughed back into the synthesis of reduced GSH [9,11]. In *E. coli* and a few other bacteria, the tripeptidase (PepT) also acts in GSH degradation, and both the GGT- and the PepT-encoding genes are dispensable for cell growth under favorable laboratory conditions [9,21].

In plants, GSH degradation seems to be as important as GSH synthesis for sulfur metabolism [110]. The apoplastic and vacuolar GSH pools are degraded by GGTs, which cleave the γ-glutamyl moiety of GSH, GSSG and GS-conjugates, or transfer the Glu residue is transferred to amino acids to produce γ-Glu amino acids [9,11,110], which are beneficial for human consumption [164]. *Arabidopsis* possesses four GGTs. GGT1 and GGT2 are localized in the apoplast and degrade extracellular GSSG into Glu and Cys–Gly [110], similar to mammalian GGTs. The Cys–Gly dipeptide is further broken down by a dipeptidase into Cys and Gly. They are then translocated to the cytosol where they serve in the synthesis of protein and GSH, followed by a novel round of export/degradation in the apoplast [9]. *GGT3* is considered a pseudogene since its transcript encodes a protein lacking a sequence important for GGT catalytic activity. The *GGT4*-encoded isoform is present in the vacuole where it supports detoxification processes by degrading GSH conjugates formed by glutathione S-transferase following exposure to toxic xenobiotics [11,28,110]. GSH degradation mainly occurs in the cytosol [11]. In *Arabidopsis*, it involves three γ-glutamyl cyclotransferases (GGCTs, EC 4.3.2.9). They specifically hydrolyze GSH into a Cys-Gly dipeptide and 5-oxoproline, which are further broken down into Glu, Cys and Gly by (i) the Leu aminopeptidase 1 (EC 3.4.13.18) encoded by AtLAP1 and (ii) by the 5-oxoprolinase (5-OPase, EC 3.5.2.9) [11,110]. The pathogenic bacterium *Ralstonia solanacearum* injects many proteinic virulence factors in plant host cells, including an active GGCT enzyme that degrades intracellular glutathione, as a strategy to subvert plant defenses systems [11].

In mammals, where it was first reported, the GGT enzyme is a cell-surface protein that contributes to the extracellular catabolism of GSH, but it has no role in either GSH or γ-glutamyl-Cys transport back into cells [9]. Interestingly, GGT is used as a diagnostic marker for many human diseases [16,174]. Mammals also have both ChaC1- and ChaC2-type GGCT enzymes, which are expressed only under endoplasmic reticulum stress or constitutively, respectively. ChaC1 and ChaC2 have high specificity and comparable K_m values for GSH, but ChaC1 has 10- to 20-fold higher catalytic activity than ChaC2 [110].

Glutathione Degradation in Cyanobacteria: An Overlooked Aspect of Their Metabolism

Little is known in cyanobacteria concerning the degradation of glutathione. Freshwater cyanobacteria possess one to four presumptive GGT-encoding genes that have not been studied yet, whereas marine species have no such genes. Furthermore, cyanobacteria have no *pepT* orthologous genes. In the cyanobacterium *Synechococcus elongatus* PCC 7942, the inactivation of a gene encoding a leucyl aminopeptidase (LAP) with cysteinyl-glycinase activity was found to decrease the tolerance to UV [44].

5. Glutathione Reductase

The function of GSH depends on the reactivity of its cysteinyl thiol group, which can complex metals, be alkylated to thioethers or oxidized to disulfides, thereby forming a glutathione disulfide dimer (GSSG) [3,15,20]. Four processes can remove GSSG after oxidative challenge: (i) ATP-driven export of GSSG [20]; (ii) degradation of GSSG by peptidase [9,11]; (iii) reduction of GSSG by the glutathione reductase (GR) enzyme (see below and Figure 2) or the thioredoxin reductase/thioredoxin (TrxR/Trx) couple [3,15,20] and (iv) the glutathionylation/deglutathionylation of proteins yielding RSSG and GSH [11,18,22]. Under normal conditions, GSH is about a 100-fold more abundant than GSSG [18,20,21]. For example, the GSH/GSSG molecular ratio is about 200 in *E. coli* cells growing in the rich standard-medium LB.

The flavoenzyme GR (EC 1.8.1.7) is a highly conserved enzyme across the tree of life, which converts GSSG to two molecules of GSH, by using NADPH (mostly) or NADH (rarely) as the reducing agents [11,15,18,20,21,175–178], as shown in Figure 5.

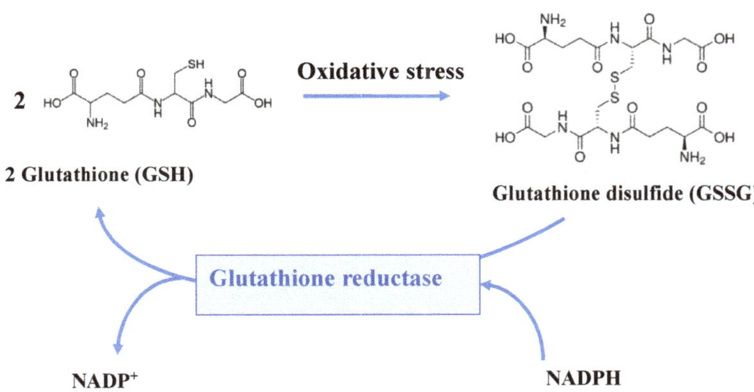

Figure 5. Schematic representation of the role of the glutathione reductase enzyme.

In vitro analyses showed that GR from humans, yeast and *E. coli* are inhibited by reduced glutathione [179].

GR belongs to the pyridine nucleotide-disulfide oxidoreductase family that includes the related enzymes dihydrolipoamide dehydrogenase, mercuric ion reductase trypanothione reductase and some TrxR-isoforms [3]. GR from pro- and eukaryotes share about 50% identity and form stable homodimers of ~110 kDa each comprising three domains [3,120,175]. The FAD- and NADP-(binding) domains are globular, whereas the interface dimerization domain is somewhat flat. It contains two regions, at the N-terminus 71–104 and C-terminus 372–482. The FAD-binding and the NADPH-binding domains are in residues 1–157 and 198–238, respectively. The catalytic site of GRs possesses two conserved cysteines (C61, C65) that can form a disulfide bond. In the *S. cerevisiae* enzyme, the cysteine C239 (not conserved) can bind excess GSH when required.

GR accumulate in cellular regions of high electron flux, where ROS are generated [120]. In prokaryotes, GR is localized in the periplasm, associated with the inner membrane facing the cytoplasm, and it can be secreted to the extracellular environment. In eukaryotes, GR is present in the cytoplasm, the endoplasmic reticulum lumen, the lysosomes and the organelles nucleus, mitochondria and chloroplasts, thanks to its transport from the cytosol [3,11,18,20,120,175,178].

In *E. coli*, GR deficient mutants, originally isolated in a screen for diamide sensitive mutants [180,181], have wild-type (WT) growth rates under standard conditions [20]. They are more sensitive than the WT strain to cumene hydroperoxide and the $O_2^{\bullet -}$-generating compound paraquat, but not to *t*-butyl hydroperoxide [182,183]. Furthermore, their increased sensitivity to H_2O_2 could be uncovered only in a catalase mutant background [184]. Despite their weak GR activity (0.45 units as compared to 35 units in WT cells), the ratio of GSH to GSSG is not altered significantly from that of WT. This finding indicates that GSSG can also be reduced by other enzymes, such as thioredoxin reductase/thioredoxin (TrxR/Trx) couples [3,20,181]. In addition, an increased GSH synthesis in GR mutants may also help to maintain their high GSH levels [182].

In *Saccharomyces cerevisiae* and human cells, the GR-encoding gene was cloned on the basis of its sequence homology with the *E. coli* gene [20]. This eukaryotic GR gene expresses both the cytosolic and mitochondrial forms of GR, which are synthesized using alternative in-frame start codons. Starting at the first AUG codon, the synthesis generates a long GR isoform marked for transport to the mitochondria. The translation starting at the second AUG codon generates a shorter GR isoform remaining in the cytosol. Usually, the pre-sequence of the mitochondrial form is cleaved off upon import by mitochondrial proteases so that the mitochondrial and cytosolic forms have a similar length [20,120]. Yeast mutants lacking GR show WT growth, but accumulate increased levels of GSSG [185,186] and increased export of GSSG into the vacuole to maintain the highly reducing environment of the cytosol [18]. These mutants are very sensitive to H_2O_2 and the thiol oxidant diamide and are partially sensitive to cumene hydroperoxide, t-butyl hydroperoxide and paraquat [20,187]. Interestingly, GR-less mutants, which also lack the genes for thioredoxin 1 and thioredoxin 2, are nonviable under aerobic conditions and grow poorly anaerobically [186]. Thus, yeast cells require the presence of either the GSH- or the thioredoxin-dependent reducing systems for growth [185], as observed in *E. coli* (See above).

In humans, GR activity is positively correlated to longevity, and centenarians have an increased level of GR [17], but cancer cells in having high levels of GSH and GR are refractory to some therapies that induce oxidative stress [120]. Low GR activity is correlated with a higher susceptibility of cataract development during early adulthood [3] and HIV-1 infection [118]. In mice, GR was shown to act in the defense against bacterial infections [188].

In plants, GR activity is increased in response to abiotic stresses triggered by heavy metals, salts, drought, UV radiation and chilling temperatures [10,11]. Transgenic approaches elevating GR activity increased resistance to oxidative stress, whereas mutants with lower GR activities were more sensitive to oxidative stress and were affected in their development [3,10]. In *Arabidopsis*, GRs are encoded by two nuclear genes, *GR1* (*At3g24170*) and *GR2* (*At3g54660*), which are more expressed in roots and in photosynthetic tissues, respectively [11,120]. GR1 is present in the cytosol, nucleus and peroxisomes, whereas GR2 is dual-targeted to mitochondria [189] and chloroplasts [190]). Similarly, the two GR enzymes of the alga *Chlamydomonas reinhardtii* were shown to act in the protection against photo-oxidative stress [178].

Also interestingly, in vitro analysis of the two GR enzymes of the marine diatom *Thalassiosira oceanica* produced as recombinant proteins from *E. coli* were found to couple the oxidation of NADPH to the reduction of not only GSSG, but also oxygen, thereby generating superoxides in the absence of GSSG. As *Thalassiosira oceanica* is abundant in oceans, its GR activity is likely to be a contributor to the production of the large quantity of superoxides observed in oceans. As these ToGR are similar to GRs from bacteria, yeast and humans, it has been proposed that the counterintuitive production of ROS by GR could be widespread [191].

Glutathione Reductase in Cyanobacteria

The GR enzyme of the filamentous cyanobacterium *Anabaena* PCC 7119 was analyzed in cell extracts [192] and purified as a dimeric flavin adenine dinucleotide-containing protein [193]. Its kinetic parameters are comparable to those of chloroplast enzymes, but its molecular weight is lower, similar to that of non-photosynthetic microorganisms [193]. However, it has three differences with respect to GRs from heterotrophic organisms: (i) a strong acidic character of the protein, (ii) an absolute specificity for NADPH and (iii) an optimum pH of 9.0 [193]. Furthermore, the activity of the *Anabaena* PCC 7119 GR is inhibited by sulfhydryl reagents, Zn^{2+} ions and heavy-metal ions, with GSSG behaving as a protective agent [193].

The GR enzyme of the other filamentous cyanobacterium *Anabaena* PCC 7120 has been purified directly or after production in a GR deficient *E. coli* strain, before or after addition of a hexa-histidine tag at the C-terminal of the protein to facilitate its purification [194,195]. Its amino acid sequence has 41 to 48% identity with GRs of *Escherichia coli, Pseudomonas aeruginosa*, pea, *Arabidopsis thaliana* and humans. Like most GRs, the *Anabaena* PCC 7120 GR uses NADPH as a cofactor, but its K_m values for NADPH and GSSG are higher than those of other GRs [194,195]. The *Anabaena* PCC 7120 GR also shows significant activity when NADH is used as a reductant, in agreement with the finding that it carries the GXGXXG "fingerprint" motif (amino acids 173–178) [196] typical of NADH-dependent enzymes [176,177] instead of the GXGXXA consensus motif of NADPH-dependent GR [176,177]. Furthermore, the *Anabaena* PCC 7120 GR harbors (i) a Lys residue (Lys203) in place of an Arg residue involved in NADPH binding by other GRs, and (ii) an insertion of 10 amino-acid residues that form an extra loop near the entrance of the pyridine-nucleotide-binding site [196]. Removal of this loop increased the catalytic efficiency of the *Anabaena* PCC 7120 GR with NADPH by reducing K_M, and with NADH by increasing k_{cat} [196].

Another GR has been purified from the other filamentous cyanobacterium *Spirulina maxima* [197]. Its amino acid composition was very similar to other GRs, and its optimum pH was 7.0. The *Spirulina maxima* GR is predominantly tetrameric, in equilibrium with a minor dimeric form. Its dissociation into dimers was observed at pH of 9.5 or in 6 mM urea. However, its equilibrium at neutral pH was altered by neither NADPH nor by disulfide reducing reagents [197]. Cyanobacterial GR activities were increased in response to unusual growth conditions (increasing the concentration of phosphate [197] or replacing nitrate by ammonium [194]) or stresses triggered by pesticides [198] and metals [199,200]. In addition to other players [201], GR was shown to protect the O_2-sensitive nitrogen-fixing nitrogenase enzyme from oxidative stress in the cyanobacterium *Gloeocapsa* LB795 [202].

As compared to filamentous cyanobacteria, GR has been poorly studied in unicellular species, likely because two of the best-studied species, including for biotechnological purposes (photosynthetic production of high-value chemicals and bioremediation), *Synechocystis* PCC 6803 and *Synechococcus* PCC 7002, have no GR enzymes [82,203]. Interestingly, the insect *Drosophila melanogaster* has no GR enzyme, and it employs a thioredoxin reductase/thioredoxin system to reduce GSSG back to GSH [204]. It will be very interesting to study and compare the GSH system of both *Synechocystis* PCC 6803 and *Synechococcus* PCC 7002 with that of the other well-studied unicellular model *Synechococcus elongatus* PCC 7942 that contains a genuine GR [203]. As these three cyanobacteria are robust, it is conceivable that *Synechocystis* PCC 6803 and *Synechococcus* PCC 7002 could compensate the absence of GR by using a thioredoxin reductase/thioredoxin system yet to be identified, or by having an increased GSH synthesis and GSSG turnover as compared to *Synechococcus elongatus* PCC 7942.

6. Importance of the Evolutionary-Conserved Glutathione-S-Transferase Enzymes

As the reactivity of glutathione (GSH) with proteins, small molecules and xenobiotics can be low in vivo [18,205], GSH-dependent reactions are accelerated by enzymes, such as glutaredoxins (See below) and glutathione-S-transferases (GSTs).

Glutathione-S-transferases (EC 2.5.1.18) constitute a superfamily of enzymes that play prominent roles in specialized secondary metabolism and detoxication. The presence of GSTs in most living organisms highlights their ancient origin and the preservation of their functions during evolution [3]. GSTs are widely studied in eukaryotes because of their great relevance to human health [14,64,206–209], plant growth [28,210] and responses to stresses [27,63,211,212] and pathogens [213]. GSTs are less well studied in prokaryotes even though they act in bacterial protection against metabolite by-product (methylglyoxal, see below) [214] and pollutants such as polychlorinated biphenyls (PCBs), dichloroacetate and polycyclic aromatic hydrocarbons (PHA) [215–217].

GSTs catalyze reactions where GSH is consumed (GSH-conjugation on metabolites, chemicals or metals) or not (isomerization and dehalogenation) and reactions where GSH is oxidized (GSH-dependent peroxidases, -thiol-transferase, -dehydro-ascorbate reductase) [3,18]. GSTs can also bind and transport molecules through their noncatalytic ligandin properties [212,218]. Finally, GSTs can also interact with proteins to modulate their activity by glutathionylation/deglutathionylation (formation/reduction of a disulfide bridge between the cysteinyl residue of GSH and a cysteinyl residue of a target protein) [14,22,25,29], as shown in Figure 6.

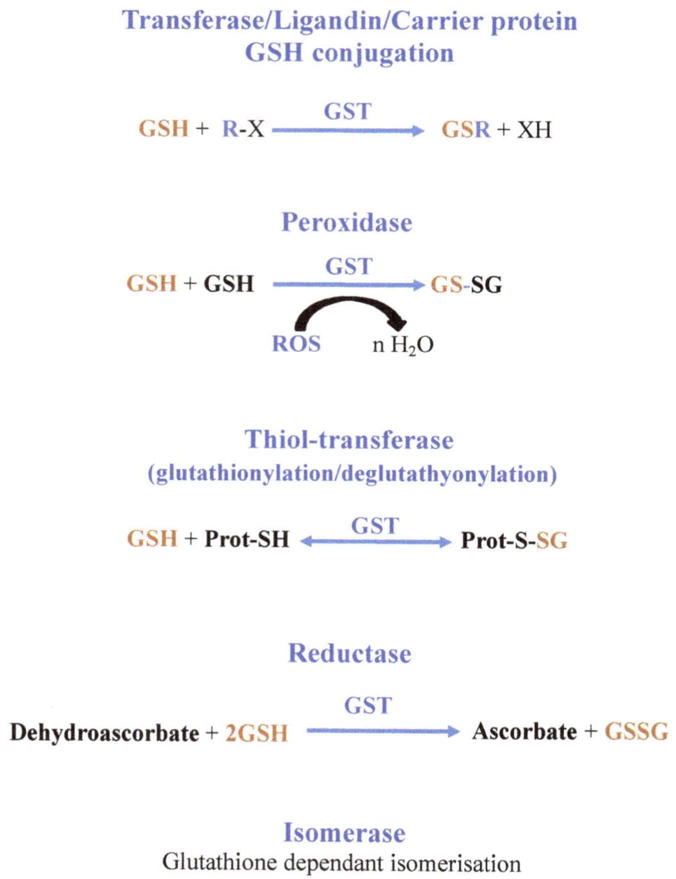

Figure 6. Schematic representation of the reactions catalyzed by glutathione-S-transferases.

GSTs are mainly homo- or heterodimeric enzymes, where each subunit contains an N-terminal thioredoxin (TRX) domain linked to an α-helical C-terminal domain [14,210,219]. The active site, located in a cleft between both domains, contains a GSH-binding site and a hydrophobic-substrate binding site. Based on their amino-acid sequence, GSTs were classified into various classes designated by a Greek letter [8,220]. GSTs having a sequence identity greater than 40% or lower than 25% belong to the same class or different classes, respectively [215]. GSTs are also grouped based on their localization in the cell, namely cytosolic, mitochondrial and microsomal, commonly referred as membrane-associated proteins in eicosanoid and glutathione metabolism (MAPEG) [3,18,62,215,217,219]. Finally, GSTs were further distinguished into four catalytic types, depending on an assumed important residue for catalysis, namely: tyrosine (TyrGSTs), serine (SerGSTs), cysteine (CysGSTs) and atypical (AtyGSTs) [8,14,210,217,221–223].

Glutathione-S-Transferase in Cyanobacteria

The multiplicity of GSTs in plants (55 in *Arabidopsis thaliana* [28]) and humans (18 GSTs [206]) and their localization in one or several compartments (cytosol, chloroplast and/or mitochondria) [3,28,208] makes their analysis difficult. By contrast, cyanobacteria, the basic organisms that possess fewer GSTs, are attractive models for studying the selectivity/redundancy of these enzymes at the level of a whole organism. Furthermore, cyanobacteria are the primordial photosynthetic organisms and served as hosts for the evolution of GSTs with diversity in their structures, substrate recognition and catalytic functions [224].

We have recently started an analysis of all six evolutionarily conserved GSTs (designated as Sll0067, Sll1147, Sll1545, Sll1902, Slr0236 and Slr0605 [7]) of the unicellular model *Synechocystis* PCC 6803 that is well-studied for basic and applied research (bioremediation, bioproduction of high-value chemicals) [56,82]. Only Sll1545 appeared to be crucial to cell life [7,62,214], and our unpublished data. The rho class GST Sll1545 was found to have a good GSH-transferase activity with cumene hydroperoxide CuOOH [225], and to catalyze the detoxification of the water pollutant dichloroacetate [216], a toxic by-product of water chlorination and a metabolite of drugs [226]. We showed that Sll1545 and Slr0236 act defensively against photo-oxidative stress triggered by high light or H_2O_2 [7]. We also reported that the MAPEG-type GST, Sll1147 and its human orthologs, play a prominent role in the tolerance to membrane stresses triggered by heat, cold and lipid peroxidation [62]. That human orthologs of Sll1147 could rescue the stress tolerance of a *Synechocystis* PCC 6803 mutant lacking Sll1147 showed that the function of this MAPEG-type GST has been conserved during evolution from cyanobacteria to humans. The chi-class GST Sll0067 and its orthologs have been characterized biochemically from *Thermosynechococcus elongatus* BP-1, *Synechococcus elongatus* PCC 6301 and *Synechocystis* PCC 6803 [214,220,223,227]. Sll0067 purified from a recombinant *E. coli* strain as homo-dimer composed of about 20 kDa subunit appeared to have a good GSH-transferase activity with isothiocyanates (especially phenethyl isothiocyanate) and a low activity with the model substrate CDNB (1-chloro-2, 4-dinitrobenzene) [220,223]. Very interestingly, Sll0067 was found to play a crucial role in the detoxication of methylglyoxal ([214] and see below).

The halotolerant cyanobacterium *Halothece* PCC 7418 contains four putative GST [228]. One of them, H3557, is encoded by a salt-stress inducible gene and it exhibited the classical GST activity toward GSH and CDNB, which appeared to be salt tolerant [228]. Furthermore, recombinant *E. coli* cells producing H3557 exhibited an increased tolerance to H_2O_2 [228], in agreement with the amino-acids sequence similarity of H3557 with the rho-class GST Sll1545 of the model cyanobacterium *Synechocystis* PCC 6803 that protects it against H_2O_2 [7]. In *Anabaena cylindrica* and *Anabaena laxa* the GST activities were induced in response to the herbicide bentazone [229] and the fungicide R-metalaxyl, respectively [198]

7. Glutathione Acts in the Detoxification of Methylglyoxal, a By-Product of Cell Metabolism from Cyanobacteria to Humans

Methylglyoxal (MG) is a very dangerous dicarbonyl compound that strongly interacts with lipids, nucleic acids and the arginine and lysine residues of proteins (Figure 7), generating advanced glycation end products (AGEs) that disturb cell metabolism in prokaryotes [230,231] and eukaryotes [230,232,233].

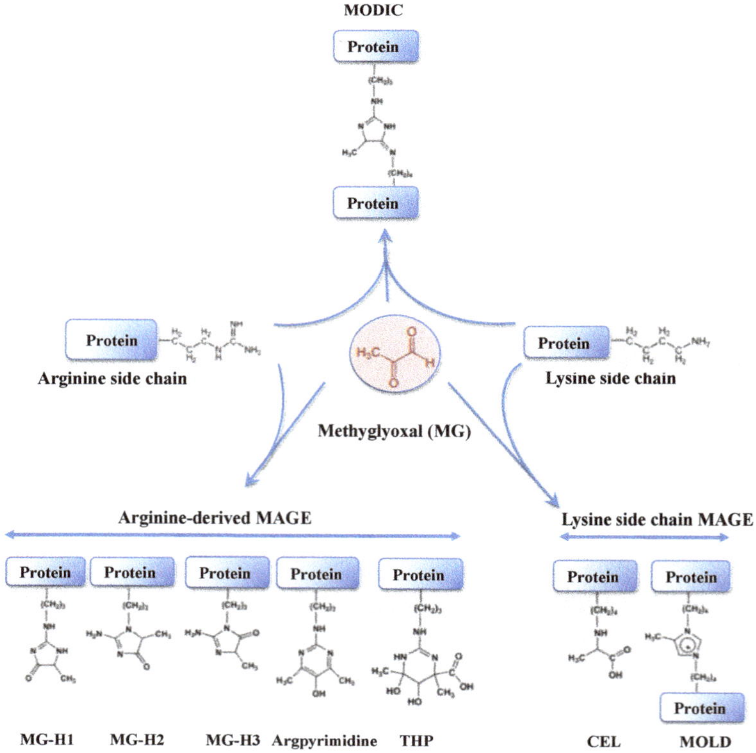

Figure 7. Schematic representation of protein modification resulting from the crosslinking of methylglyoxal onto arginine or lysine amino-acid residues. The abbreviations are as follows: MODIC, 2-ammonio-6-({2-[4-ammonio-5-oxido-5-oxopently)amino]-4-methyl-4,5-dihydro-1H-imidazol-5-ylidene}amino)hexanoate; MG-H1, N(delta)-(5-methyl-4-imidazolon-2-yl)-L-ornithine; MG-H2, 2-amino-5-(2-amino-5-hydro-5-methyl-4-imidazolon-1-yl)pentanoic acid; MG-H3, 2-amino-5-(2-amino-4-hydro-4-methyl-5-imidazolon-1-yl)pentanoic acid; THP, N(delta)-(4-carboxy-4,6-dimethyl-5,6-dihydroxy-1,4,5,6-tetrahydropyrimidine-2-yl)-L-ornithine; CEL, N^{ε}-(carboxyethyl)lysine and MOLD: 1,3-di(N^{ε}-lysino)-4-methyl-imidazolium.

Like ROS, MG has a dual nature depending on its concentrations within the cells; it acts in signaling at low concentrations, but provokes detrimental effects, such as glutathionylation [234] (see below), at high concentrations [232,235]. In humans, MG is implicated in diabetes [236] and age-related disorders [230], such as retinopathy, nephropathy, cancer, and Parkinson's and Alzheimer's diseases [233,234,237]. Hence, MG is increasingly regarded as a marker of diabetes-related diseases. The calculated MG concentrations in mammalian cells were reported to vary between 0.5 and 5 μM, similar to what was observed in yeast (4 μM) [18]. In plants, elevated MG levels are a general response to abiotic and biotic stresses, such as salinity, heavy metals and drought [11]. Furthermore, MG is viewed as acting in signaling via abscisic acid, Ca^{2+}, K^{+} and ROS, and these processes are regarded as useful for the development of stress-resilient crops [11,232,238].

MG is mainly formed in all cells both under normal and pathological conditions by the non-enzymatic breakdown of the triose phosphate isomers dihydroxyacetone phosphate (DHAP) and glyceraldehyde-3-phosphate (G3P) [230,231,233,234], which rapidly lose α-carbonyl protons and their phosphate groups, generating MG. MG is also generated by the spontaneous auto-oxidation of ketone bodies and sugars, lipid peroxidation and the Maillard reaction between reducing sugars and amino acids [239,240]. In addition, various enzymes generate MG from: (i) the fatty-acid derived acetone (cytochrome P450); (ii) the aminoacetone produced by the metabolism of glycine and threonine (monoamine oxidases); and (iii) the elimination of an inorganic phosphate from DHAP by the MG synthase in bacteria [236,241] and mammals [3,236].

MG is predominantly detoxified by the glyoxalase pathway, which starts by the supposedly "spontaneous" (non-enzymatic) conjugation of MG with GSH to form a hemithioacetal (HTA). HTA is then isomerized by glyoxalase I (GlxI, *S*-d-lactoylglutathione lyase; EC 4.4.1.5) to S-d-lactoylglutathione that is hydrolyzed by glyoxalase II (GlxII, *S*-2-hydroxyacylglutathione hydrolase; EC 3.1.2.6) to produce D-lactate and release GSH [3,11,18,236,237,242]. That GSH is required for MG detoxication explains why MG is accumulated in response to GSH depletion caused by oxidative stress [234]. GlxI belongs to vicinal oxygen chelate enzymes, which contain an ancient βαβββ-motif required for binding metal ions [3,234]. GlxI (also called Glo1) from humans, yeast and the parasite *Plasmodium falciparum*, is a dimeric protein that prefers Zn^{2+} (or Fe^{2+}) at its active site, whereas the enzymes from *E. coli* and the pathogens *Yersinia pestis*, *Pseudomonas aeruginosa* and *Neisseria meningitidis* and the protist *Leishmania major* are optimally activated in the presence of Ni^{2+} (or Co^{2+}) [237]. In *Arabidopsis*, overexpression of GlxI increased tolerance against salinity and maintained lower levels of MG as compared to the wild type (WT) plant [11]. GlxII (also called Glo2) is a monomeric protein composed of an N-terminal β-lactamase domain with a conserved Fe(II)Zn(II) center at the active site and a C-terminal domain with five α-helices [3,237]. In plants, GlxI and GlxII enzymes are encoded by multi-gene families, unlike what was observed in most prokaryotes and animals, which have single such genes [11]. In *E. coli* and yeast, both the *glxI* and *glxII* genes are non-essential to life, and growth phenotypes only became obvious upon challenge with exogenous MG [3,242].

MG can also be detoxified by other enzymes that do not require GSH, such as MG dehydrogenase, aldehyde dehydrogenases, aldo-keto reductases (AKR), α-dicarbonyl/L-xylulose reductase and the MG reductase [8,11,233,235,237]. AKRs, which exist across all phyla, are primarily NADP(H)-dependent monomeric oxidoreductases with a molecular weight ranging from 34 to 37 kDa [8].

Methylglyoxal Detoxication: Lessons from a Cyanobacterium

Little attention has been paid to MG production, signaling and detoxification systems in photosynthetic organisms, although they produce MG by the catabolism of sugars, amino acids and lipids, such as heterotrophic organisms, but also by their active assimilation of CO_2 driven by photosynthesis [54,232,243]. This issue is even more acute in cyanobacteria, the environmentally important prokaryotes [244], because they perform the two MG and/or ROS-producing pathways, photosynthesis (CO_2 fixation and gluconeogenesis) and respiration (glucose catabolism), in the same cell compartment [54,55]. Furthermore, cyanobacteria are regarded as the inventor of oxygenic photosynthesis [33–35,37] and GSH and GSH-utilizing enzymes, such as glyoxalase (Glx) and glutathione transferases (GST), to cope with the ROS often produced by their active photosynthesis system [5,59,60].

In the model of cyanobacterium *Synechocystis* PCC 6803, we showed that (i) MG is toxic, (ii) both the *glxI* and *glxII* genes are non-essential to cell life but are required for the protection against MG and (iii) the MG synthase (EC.4.2.99.11.) gene is also non-crucial to cell growth [61,214]. We also reported that Sll0067 operates in the protection against MG, unlike the other five GSTs [7,62]. The Δ*sll0067* deletion mutant, which grew as fit the WT strain under standard (photoautotrophic) conditions, was found to be hypersensitive to MG. Furthermore, the MG-sensitive Δ*sll0067* mutant exposed to exogenous MG (or glucose) accumulated not only MG but also GSH, indicating that Sll0067 acts in an MG elimination process that requires GSH, similar to the GSH-dependent detoxification of MG catalyzed by the glyoxalase system. This interpretation was confirmed by in vitro tests showing that Sll0067 indeed catalyzes the conjugation of GSH with MG to form the hemithioacetal metabolite, which is known to be subsequently isomerized by GlxI and hydrolyzed by GlxII to release d-lactate and GSH [233,235]. Furthermore, the fixation of one

MG molecule on the first subunit of the Sll0067 dimeric protein was found to stimulate the fixation of another MG molecule on the other Sll0067 monomer, thereby increasing Sll0067 activity [214]. The fixation of MG on Sll0067 also enhanced its affinity for GSH, and Sll0067 was also found to be activated by S-d-lactoylGSH, the intermediate metabolite in MG-detoxification. Collectively, these findings showed that MG enhances the Sll0067-driven conjugation of GSH and MG to promote MG detoxification by the glyoxalase pathway [214]. These data are important because, so far, the conjugation of GSH with MG was considered as spontaneous (non-enzymatic) in all organisms [232,233,235,237]. The finding that Sll0067 acts in the detoxification of MG, which causes diabetes in humans [231,233,234,237], is consistent with the correlation between the occurrence of diabetes and the poor activity of a human GST orthologous to Sll0067 [245]. These studies will certainly stimulate research on MG detoxification in mammals (possibly leading to the identification of biomarkers and/or drugs); plants (with interest for agriculture) and cyanobacteria (with interest for the sustainable production of valuable chemicals, such as lactate [246]).

8. Glutathione Maintains the Redox Homeostasis of Protein Thiols via Glutathionylation/Deglutathionylation Catalyzed by Glutathione-S-Transferases and Some Glutaredoxins

Oxidative stress promotes the covalent modification of proteins by GSH, i.e., formation of a disulfide bridge between the thiol group of a cysteinyl residue of a protein and a molecule of GSH [3,18,205]. This post-translational modification called S-glutathionylation is regarded as a transient protection of critical cysteines against irreversible oxidation towards sulfinic and sulfonic acid forms during oxidative stress [25,247]. It occurs only at specific cysteinyl residues of proteins, in response to ROS and MG [234,248], and not randomly. A basic environment or the proximity of a metal cation are key determinants for the tendency of thiol groups to become deprotonated and consequently be affected by oxidation and spontaneous S-glutathionylation [11]. As GSH is a bulky molecule, its ligation to proteins can have an impact on their structure, function, catalytic capacity and/or subcellular localization [11,22,25,29,207]. For example, three main glycolytic enzymes, GAPDH (glyceraldehyde-3-P-dehydrogenase) [249], aldolase and TPI (triose-P-isomerase) [250,251], are inhibited by S-glutathionylation under oxidative conditions in plants and other organisms, probably to redirect the glycolytic carbon flux towards the oxidative pentose phosphate pathway (OPPP) to generate reductive power in the form of NADPH [11].

GSTs [14,22,25,29,252] and glutaredoxins (Grxs) [22,25,26,248] catalyze both S-glutathionyla tion and deglutathionylation, while thioredoxins (Trxs) catalyze deglutathionylation [11,25,26].

Grx are small thiol proteins found in all kingdoms of life [3,10,18,22,26,29,253]. The first identified function of Grx was described as an electron donor for the ribonucleotide reductase enzyme (RNR) in a *E. coli* mutant lacking Trx [254], the classic hydrogen donor for RNR [255]. Grx can detoxify hydroperoxide thanks to their hydroperoxidase activity [18]. Bacterial Grxs have the most basic form of the Trx-fold, consisting of a four to five central β-sheet surrounded by three α-helices. Grxs of higher organisms frequently display additional N- and C-terminal helices. Interestingly, GSTs, the other glutathione-dependent enzymes, have similar architectures, supporting the theory of a common ancestor for Grxs and GSTs [3,256]. In the last two decades, the Grx family has impressively grown, and it has become clear that Grx is much more than a back-up system for Trx [257]. For example, in mammals at the physiological concentration of GSH, the GSH-Grx system sustains the RNR activity more efficiently than the Trxs [258].

Grx-isoforms can be structurally categorized as monomeric or dimeric proteins, which possess an active site with the sequence motif CXXC (dithiol Grxs) or CXXS (monothiol Grxs), with or without an Fe/S-cluster [3,18,26,88,91]. Grx can be furthermore grouped based on enzymatic activities, subcellular localizations or (putative) physiological functions (ROS detoxication, iron metabolism, etc.) [15,91]. In plants, Grx are involved in the regulation of development through interaction with distinct transcription regulators [10]. In humans, Grx functions have been implied in various physiological and pathological conditions, from immune defense to neurodegeneration and cancer development [26].

In general, CxxC-type Grxs function primarily in redox regulation and electron supply to metabolic enzymes. They catalyze the formation and reduction of disulfides, i.e., inter- and intra-molecular protein disulfides, including glutathionylation/deglutathionylation [15,22,25,26,29,234,252]. These redox-active Grxs often contain a consensus Cys-Pro-Tyr-Cys active site motif (CPYC-type of Grxs). The formation and reduction of protein disulfides require both their active site cysteinyl residues (dithiol reaction), while glutathionylation/deglutathionylation requires only their more N-terminal cysteinyl residue (monothiol mechanism). Both reactions start with a nucleophilic attack of the more N-terminal cysteinyl residue, which has a particularly low pK_a value, on the target disulfide. In the dithiol mechanism, the intermediate disulfide between the target protein and the Grx is reduced by its more C-terminal cysteinyl residue. The monothiol mechanism results in a reduced target protein, and a disulfide between the Grx and GSH. This disulfide can be reduced by another molecule of GSH, generating Grx and GSSG. Both reactions are fully reversible, as Grxs catalyze both the oxidation and reduction of their targets [15,18].

Glutaredoxins and Glutathionylation in Cyanobacteria

Grxs are mostly studied in the model cyanobacterium *Synechocystis* PCC 6803 that possess only three Grxs: two CxxC-type Grxs, Grx1 and Grx2, and one CGFS-type Grx (Grx3), which are all dispensable to cell growth under standard conditions [203,259,260]. Both Grx1 and Grx2 act defensively against oxidative and metal stresses [49,203,260–262]. Interestingly, from the point of view of the selectivity/redundancy of these Grx, both Grx2 and Grx3, but not Grx1, were found to protect cells against H_2O_2, heat and high light [260]. Grx1, but neither Grx2 nor Grx3, was shown to physically interact with the mercuric/uranyl reductase enzyme MerA, which can be inhibited by glutathionylation, and subsequently reactivated by Grx1 [49]. Furthermore, Grx1, but not Grx2, was found to interact with NTR (NTR stands for NADPH-thioredoxin reductase). Interestingly, Grx1 and Grx2 were shown to act in an integrative redox pathway, NTR–Grx1–Grx2–Fed7 (Fed7 stands for ferredoxin 7) that protects *Synechocystis* PCC 6803 against selenate toxicity [203].

In *Synechocystis* PCC 6803, about 380 proteins involved in carbon and nitrogen metabolisms, photosynthesis, cell division and tolerance to stresses (GshB, Grx3 and GST sll1145) can be glutathionylated [263]. For four of these *Synechocystis* PCC 6803 proteins (the AbrB2 transcription regulator, the mercuric reductase, a peroxiredoxin and a 3-phosphoglycerate dehydrogenase), we have verified that their activity was indeed controlled by glutathionylation [49,263,264]. These data, together with similar findings obtained with other prokaryotes [26,247] and higher eukaryotes [11,14,22,25,26,249–251,265], showed that the glutathionylation/deglutathionylation regulatory process has been conserved during evolution.

9. Glutathione, Glutaredoxins and the Biogenesis of the Iron-Sulfur Cluster of Proteins

Glutathione plays a crucial role in cellular iron metabolism [15,90,266–268] and the synthesis and repair of iron-sulfur cluster [Fe-S] of a wealth of enzymes (See below). Iron (Fe) and sulfur (S) are crucial elements in all kingdoms of life [267–269]. Iron, the fourth most plentiful element in the Earth's crust [270], is frequently a growth-limiting factor because ancient cyanobacteria raised the oxygen levels that oxidized the soluble ferrous ions (Fe^{2+}) to insoluble ferric ions (Fe^{3+}) [33,35,38,39,41,271,272]. Fe atoms are associated with many proteins as part of hemes, mono- or di-iron non-heme centers, or iron–sulfur [Fe-S] clusters [273,274]. Sulfur (S) is the fifteenth and the sixth most abundant chemical elements in Earth's crust and aquatic environment, respectively [270]. Sulfur is essential in living organisms and is notably required for the synthesis of cysteine, which also serves for the synthesis of GSH, [Fe-S] and methionine [275]. Biological organisms absorb and assimilate sulfate from their environment via a reductive pathway involving a series of transporters, enzymes and GSH. This process is tightly controlled because it consumes energy and produces toxic compounds, notably sulfite and sulfide. In particular, it provides electrons to adenosine 5′-phosphosulfate reductases but also regulates the activity of glutamate-cysteine ligase by reducing a regulatory disulfide [275].

[Fe-S] clusters are critical cofactors in all categories of life. They participate in the transfer of electrons (photosynthesis and respiration), transcriptional and translation regulation, DNA repair and replication [15,18,26,88,91,266,273,276–278]. The chemically simplest Fe-S clusters are the rhombic [2Fe-2S] and the cubane [4Fe-4S] types, which contain iron ($Fe^{2+/3+}$) and sulfide (S^{2-}) [273]. [Fe-S] clusters were discovered in the early 1960s by purifying enzymes, including plant and bacterial ferredoxins, with characteristic electron paramagnetic resonance signals [273]. Later, chemists and biochemists devised in vitro protocols to assemble [Fe-S] clusters into apoproteins, and thereby assumed that these co-factors can assemble spontaneously on proteins in living cells. However, genetic, biochemical and cell-biological studies in the 1990s proved that the assembly and maturation of [Fe-S] centers on proteins is a catalyzed process rather than a spontaneous one [273].

[2Fe-2S] centers are synthesized from iron and cysteine-derived sulfur by highly conserved multi-protein machineries, including the iron-sulfur cluster (ISC) synthesis machinery [273,277,278] and many types of Grx [15]. The first [Fe-S]-Grx were isolated from humans (CSYC-type Grx, [279]) and the poplar tree (CGYC-type Grx, [280]). In both enzymes, the apo form (monomer) is a regular redox active Grx, while the holo form (dimer) has a bridging [2Fe-2S] cluster but no oxidoreductase activity. This [2Fe-2S] center lies at the interface of a dimeric complex of two Grxs ligated by the two N-terminal active site thiols and the thiols of two non-covalently bound GSH molecules [91]. In both CSYC- and CGYC-type Grxs, the exchange of the seryl or glycyl residue, respectively, for a prolyl residue abolished cluster ligation, in agreement with the absence of [Fe-S] cluster in natural CPYC-type Grx [15]. These CPYC-type Grx catalyze the GSH-dependent reduction of protein disulfides and deglutathionylation [91].

Following the discovery of these C(non-P)YC-type Grxs, CGFS-type Grxs have been characterized as [2Fe-2S]-proteins [15]. They occur as homodimer containing a subunit-bridging [2Fe-2S] cluster ligated by the catalytic cysteine of the CGFS motif of each monomer and the cysteines of two molecules of GSH from bacteria to higher eukaryotes [15,65,281]. With few exceptions, CGFS-type Grxs are inactive as oxidoreductases; they act in iron metabolism and [Fe-S] cluster formation and transfer [15]. Interestingly, the engineering of a CxxC-type Grx with a CGFS-type loop switched its function from oxidoreductase to [Fe-S] transferase, and the introduction of a CxxC-type loop into a CGFS-type Grx abolished its [Fe-S] transferase activity and activated the oxidative half-reaction of the oxidoreductase [91].

Glutaredoxin and the Biogenesis of Iron-Sulfur Clusters: Lessons from Cyanobacteria

Cyanobacteria have a very high requirement for Fe (~10-fold more than non-photosynthetic prokaryotes) [271]. The electron flow associated with the operation of photosystems I and II (PS I and PS II) requires ~22 Fe atoms [270]. The unicellular model *Synechocystis* PCC 6803 contains the conserved gene clusters involved in Fe-S cluster biosynthesis: the *suf* (sulfur utilization factor) operon, some of the *isc* ([Fe-S] cluster) genes, and a single *nif* (nitrogen fixation) gene [278]. Only the gene products from the *nif* and *suf* operons are required for growth under standard photoautotrophic conditions [278]. The small NifU protein (SyNfu, 76 amino-acids) with a conserved CXXC motif harboring a [2Fe-2S] cluster serves as the principal scaffold protein required for iron-sulfur cluster biosynthesis and transfer to apo-ferredoxin [282]. SyNfu is also capable to deliver cluster to both the yeast monothiol Grx3 and the human dithiol Grx2, albeit at a lower rate [278]. Cluster exchange experiments showed that GSH can extract the cluster from holo-SyNfu, but the transfer is unidirectional [278].

We previously reported that the third Grx of *Synechocystis* PCC 6803, Grx3, exists as a monomeric apoprotein or a dimeric holoprotein. The dimer contains a subunit-bridging [2Fe-2S] cluster ligated by the cysteinyl residue of the CGFS motif of each Grx3 monomer and the cysteinyl residue of two GSH molecules [65,281]. Very interestingly, this feature has been conserved in Grx3 orthologs from bacteria [65,281,283], yeasts [65,284,285], plants [15,65,284,286–289] and humans [15,65,91,284,290], supporting the notion that Grx functions have been conserved throughout evolution.

10. Discussion

The glutathione system (GSH, GSH-derived metabolites and GSH-dependent enzymes), which plays pleiotropic roles in cell detoxication in most living organisms, is crucial to human food, health and longevity. However, the in vivo analysis of the multiple players of the GSH-system is difficult in plants and mammals because of their physiological and genetic complexities (slow development, aging, tissue specificity, cellular differentiation, sub-cellular compartmentation, traffic between organelles, multiple genes families, etc.). In contrast, the analysis of the selectivity/redundancy of each player of the GSH-system is easier in basic organisms, such as unicellular cyanobacteria, which are endowed with a small genome that is easy to manipulate. It is indeed important to thoroughly analyze the GSH system of cyanobacteria for the following reasons. Cyanobacteria are environmentally crucial and biotechnologically important organisms. They are regarded as having evolved the GSH system to protect themselves against the ROS produced by their active photosynthetic metabolism and solar UV. They synthesize the thiol-less GSH homologs ophthalmate and norophthalmate as well as the toxic metabolite by-product methylglyoxal (MG) that serve as biomarkers of several diseases in humans. In addition, cyanobacteria synthesize the GSH-derived metabolites ergothioneine (EGT) and phytochelatin, which play crucial roles in cell detoxication in humans and plants, respectively. Hence, cyanobacteria are well-suited organisms to thoroughly analyze the roles of the GSH system and its crosstalk with the ergothioneine player, using a genetic approach (deletion/overproduction) hardly feasible with other model organisms (*E. coli* and *S. cerevisiae* do not synthesize EGT, and plants and humans acquire EGT from their soil and their diet, respectively). Attesting the value of studying the GSH system of the model cyanobacterium *Synechocystis* PCC 6803, it was recently shown that (i) the CGFS-type glutaredoxins (Grxs) of *Synechocystis* PCC 6803, *E. coli, S. cerevisiae, A. thaliana* and humans harbor a GSH-ligated [2Fe-2S] cluster and (ii) the Sll1147 MAPEG-type glutathione-S-transferase of *Synechocystis* PCC 6803 and its human orthologs play a crucial role in the tolerance to oxidative stress. Furthermore, a *Synechocystis* PCC 6803 GST having orthologs in higher eukaryotes was shown to catalyze the conjugation of GSH on MG, the first step of MG detoxication that is always presented as spontaneous (not catalyzed by an enzyme) from bacteria to humans. These results show that cyanobacteria can indeed be used to characterize the evolutionarily conserved functions or features of actors of the GSH system. Therefore, we argue in favor of a comparative analysis of the GSH system of the three robust model cyanobacteria *Synechocystis* PCC 6803, *Synechococcus* PCC 7002 and *Synechococcus elongatus* PCC 7942 because they have interesting differences: *Synechococcus elongatus* PCC 7942 possesses a glutathione reductase (GR) enzyme but does not synthesize EGT, whereas the contrary is true for *Synechocystis* PCC 6803 and *Synechococcus* PCC 7002. As these three cyanobacteria are robust, it is conceivable that *Synechocystis* PCC 6803 and *Synechococcus* PCC 7002 compensate the absence of GR in using an as yet unknown thioredoxin reductase/thioredoxin system, similar to what was observed in *D. melanogaster*. Alternatively, *Synechocystis* PCC 6803 and *Synechococcus* PCC 7002 could compensate their lack of GR by having a very active GSH synthesis and/or GSSG turnover as compared to *Synechococcus elongatus* PCC 7942.

11. Conclusions

We have shown in this study that cyanobacteria, the basic organisms regarded as having invented the glutathione system for cell detoxication, are well-suited organisms to study the selectivity/redundancy of the evolutionarily conserved players of this system, which are important for agriculture and medicine.

Author Contributions: Conceptualization, C.C.-C. and F.C.; writing—original draft preparation, F.C., C.C.-C., F.M., S.O. and S.F.; writing—review and editing, F.C., C.C.-C., F.M., S.O. and S.F.; supervision, F.C. and C.C.-C.; project administration, F.C. and C.C.-C.; funding acquisition, F.C. and C.C.-C. All authors have read and agreed to the published version of the manuscript.

Funding: This research was funded in part by the FocusDem program of CEA (A-BIOEN-04-AP-02) that paid the phD fellowship of FM.

Conflicts of Interest: The authors declare no conflict of interest. The funders had no role in the design of the study; in the collection, analyses, or interpretation of data; in the writing of the manuscript; or in the decision to publish the results.

References

1. Vaish, S.; Gupta, D.; Mehrotra, R.; Mehrotra, S.; Basantani, M.K. Glutathione S-transferase: A versatile protein family. *3 Biotech* **2020**, *10*, 321. [CrossRef] [PubMed]
2. Imlay, J.A. The molecular mechanisms and physiological consequences of oxidative stress: Lessons from a model bacterium. *Nat. Rev. Microbiol.* **2013**, *11*, 443–454. [CrossRef] [PubMed]
3. Deponte, M. Glutathione catalysis and the reaction mechanisms of glutathione-dependent enzymes. *Biochim. Biophys. Acta Gen. Subj.* **2013**, *1830*, 3217–3266. [CrossRef] [PubMed]
4. Cameron, J.C.; Pakrasi, H.B. Essential role of glutathione in acclimation to environmental and redox perturbations in the cyanobacterium *Synechocystis* sp. PCC 6803. *Plant Physiol.* **2010**, *154*, 1672–1685. [CrossRef]
5. Fahey, R.C. Glutathione analogs in prokaryotes. *Biochim. Biophys. Acta Gen. Subj.* **2013**, *1830*, 3182–3198. [CrossRef]
6. Musgrave, W.B.; Hankuil, Y.; Kline, D.; Cameron, J.C.; Wignes, J.; Dey, S.; Pakrasi, H.B.; Jez, J.M. Probing the origins of glutathione biosynthesis through biochemical analysis of glutamate-cysteine ligase and glutathione synthetase from a model photosynthetic prokaryote. *Biochem. J.* **2013**, *450*, 63–72. [CrossRef]
7. Kammerscheit, X.; Chauvat, F.; Cassier-Chauvat, C. First in vivo evidence that glutathione-S-transferase operates in photo-oxidative stress in cyanobacteria. *Front. Microbiol.* **2019**, *10*, 1899. [CrossRef]
8. Rai, R.; Singh, S.; Rai, K.K.; Ral, A.; Sriwastaw, S.; Rai, L.C. Regulation of antioxidant defense and glyoxalase systems in cyanobacteria. *Plant Physiol. Biochem.* **2021**, *168*, 353–372. [CrossRef]
9. Bachhawat, A.K.; Yadav, S. The glutathione cycle: Glutathione metabolism beyond the γ-glutamyl cycle. *IUBMB Life* **2018**, *70*, 585–592. [CrossRef]
10. Noctor, G.; Reichheld, J.-P.; Foyer, C.H. ROS-related redox regulation and signaling in plants. *Semin. Cell Dev. Biol.* **2018**, *80*, 3–12. [CrossRef]
11. Dorion, S.; Ouellet, J.C.; Rivoal, J. Glutathione Metabolism in Plants under Stress: Beyond Reactive Oxygen Species Detoxification. *Metabolites* **2021**, *11*, 641. [CrossRef] [PubMed]
12. He, Y.-Y.; Hader, D.-P. Reactive oxygen species and UV-B: Effect on cyanobacteria. *Photochem. Photobiol. Sci.* **2002**, *1*, 729–736. [CrossRef] [PubMed]
13. Nawkar, G.N.; Maiban, P.; Park, J.H.; Sahi, V.P.; Lee, S.Y.; Kang, C.H. UV-Induced cell death in plants. *Int. J. Mol. Sci.* **2013**, *14*, 1608–1628. [CrossRef] [PubMed]
14. Sing, R.R.; Reindi, K.M. Glutathione S-Transferases in Cancer. *Antioxidants* **2021**, *10*, 701. [CrossRef]
15. Trnka, D.; Hossain, M.F.; Magdalena, J.L.; Gellert, M.; Lillig, C.H. Role of GSH and Iron-Sulfur Glutaredoxins in Iron Metabolism-Review. *Molecules* **2020**, *27*, 3860. [CrossRef]
16. Hristov, B.D. The Role of Glutathione Metabolism in Chronic Illness Development and Its Potential Use as a Novel Therapeutic Target. *Cureus* **2022**, *14*, e29696. [CrossRef]
17. Belenguer-Varea, A.; Tarazona-Santabalbina, F.J.; Avellana-Zaragoza, J.A.; Martinez-Reig, M.; Mas-Bargues, C.; Ingles, M. Oxidative stress and exceptional human longevity: Systematic review. *Free Radic. Biol. Med.* **2020**, *149*, 51–63. [CrossRef]
18. Deponte, M. The Incomplete Glutathione Puzzle: Just Guessing at Numbers and Figures? *Antioxid. Redox Signal* **2017**, *27*, 1130–1161. [CrossRef]
19. Forman, H.J.; Maiorino, M.; Ursini, F. Signaling Functions of Reactive Oxygen Species. *Biochemistry* **2010**, *49*, 835–842. [CrossRef]
20. Carmel-Harnel, O.; Storz, G. Roles of the Glutathione- and Thioredoxin-Dependent Reduction Systems in the *Escherichia Coli* and *Saccharomyces cerevisiae* Responses to Oxidative Stress. *Annu. Rev. Microbiol.* **2000**, *54*, 439–451. [CrossRef]
21. Masip, L.; Veeravalli, K.; Georgiou, G. The many faces of glutathione in bacteria. *Antioxid. Redox Signal.* **2006**, *8*, 753–762. [CrossRef] [PubMed]
22. Mailloux, R.J. Protein S-glutathionylation reactions as a global inhibitor of cell metabolism for the desensitization of hydrogen peroxide signals. *Redox Biol.* **2020**, *32*, 101472. [CrossRef] [PubMed]
23. Jena, A.B.; Samal, R.R.; Bhol, N.K.; Duttaroy, A.K. Cellular Red-Ox system in health and disease: The latest update. *Biomed. Pharmacother.* **2023**, *162*, 114606. [CrossRef] [PubMed]
24. Mallén-Ponce, M.J.; Huertas, M.J.; Florencio, F.J. Exploring the Diversity of the Thioredoxin Systems in Cyanobacteria. *Antioxidants* **2022**, *11*, 654. [CrossRef]

25. Kalinina, E.; Novichkova, M. Glutathione in Protein Redox Modulation through S-Glutathionylation and S-Nitrosylation. *Molecules* **2021**, *26*, 435. [CrossRef]
26. Ogata, F.T.; Branco, V.; Vaie, F.F.; Coppo, L. Glutaredoxin: Discovery, redox defense and much more. *Redox Biol.* **2021**, *43*, 101975. [CrossRef]
27. Nianiou-Obeidat, I.; Madesis, P.; Kissoudis, C.; Voulgari, G.; Chronopoulou, E.; Tsaftaris, A.; Labrou, N.E. Plant glutathione transferase-mediated stress tolerance: Functions and biotechnological applications. *Plant Cell Rep.* **2017**, *36*, 791–805. [CrossRef]
28. Gallé, Á.; Czékus, Z.; Bela, K.; Horváth, E.; Ördög, A.; Csiszár, J.; Poór, P. Plant glutathione transferases and light. *Front. Plant Sci.* **2019**, *9*, 1944. [CrossRef]
29. Musaogullari, A.; Chai, Y.C. Redox regulation by protein s-glutathionylation: From molecular mechanisms to implications in health and disease. *Int. J. Mol. Sci.* **2020**, *21*, 9113. [CrossRef]
30. Murata, K.; Kimura, A. Overproduction of glutathione and its derivatives by genetically engineered microbial cells. *Biotechnol. Adv.* **1990**, *8*, 59–96. [CrossRef]
31. Schieber, M.; Chandel, N.S. ROS Function in Redox Signaling and Oxidative Stress. *Curr. Biol.* **2014**, *24*, R453–R462. [CrossRef]
32. Sanchez-Baracaldo, P.; Bianchini, G.; Wilson, J.D.; Knoll, A.H. Cyanobacteria and biogeochemical cycles through Earth history. *Trends Microbiol.* **2022**, *30*, 143–157. [CrossRef] [PubMed]
33. Berman-Frank, I.; Lundgren, P.; Falkowski, P. Nitrogen fixation and photosynthetic oxygen evolution in cyanobacteria. *Res. Microbiol.* **2003**, *154*, 157–164. [CrossRef] [PubMed]
34. Archibald, J.M. The Puzzle of Plastid Evolution. *Curr. Biol.* **2009**, *19*, 81–88. [CrossRef]
35. Schopf, J.W. The paleobiological record of photosynthesis. *Photosynth. Res.* **2011**, *107*, 87–101. [CrossRef] [PubMed]
36. Hamilton, T.L.; Bryant, D.A.; Macalady, J.L. The role of biology in planetary evolution: Cyanobacterial primary production in low-oxygen Proterozoic oceans. *Environ. Microbiol.* **2016**, *18*, 325–340. [CrossRef]
37. West, J.B. The strange history of atmospheric oxygen. *Physiol. Rep.* **2022**, *10*, e15214. [CrossRef]
38. Rasmussen, B.; Fletcher, I.R.; Brocks, J.J.; Kilburn, M.R. Reassessing the first appearance of eukaryotes and cyanobacteria. *Nature* **2008**, *455*, 1101–1104. [CrossRef]
39. Fisher, W.W.; Hemp, J.; Valentine, J.S. How did life survive Earth's great oxygenation? *Curr. Opin. Chem. Biol.* **2016**, *31*, 166–178. [CrossRef]
40. Demoulin, C.F.; Lara, Y.J.; Cornet, L.; François, C.; Baurain, D.; Wilmotte, Y.; Javaux, E.J. Cyanobacteria evolution: Insight from the fossil record. *Free Radic. Biol. Med.* **2019**, *140*, 206–223. [CrossRef]
41. Schad, M.; Konhauser, K.O.; Sanchez-Baracaldo, P.; Kappler, A.; Bryce, C. How did the evolution of oxygenic photosynthesis influence the temporal and spatial development of the microbial iron cycle on ancient Earth? *Free Radic. Biol. Med.* **2019**, *140*, 154–188. [CrossRef] [PubMed]
42. Fourmier, G.P.; Morre, K.R.; Rangel, L.T.; Fayette, J.G.; Momper, L.; Bosak, T. The Archean origin of oxygenic photosynthesis and extant cyanobacterial lineages. *Proc. R. Soc. B* **2021**, *288*, 20210675. [CrossRef] [PubMed]
43. Shimakawa, G.; Matsuda, Y.; Nakajima, K.; Tamoi, M.; Shigeoka, S.; Miyake, C. Diverse strategies of O2 usage for preventing photo-oxidative damage under CO2 limitation during algal photosynthesis. *Sci. Rep.* **2017**, *7*, 41022. [CrossRef] [PubMed]
44. Weiss, E.L.; Fang, M.; Taton, A.; Szubin, R.; Palsson, B.O.; Mitchell, B.G.; Golden, S.S. An unexpected role for leucyl aminopeptidase in UV tolerance revealed by a genome-wide fitness assessment in a model cyanobacterium. *Proc. Natl. Acad. Sci. USA* **2022**, *119*, e2211789119. [CrossRef]
45. Diaz, J.M.; Plummer, S. Production of extracellular reactive oxygen species by phytoplankton: Past and future directions. *J. Plankton Res.* **2018**, *40*, 655–666. [CrossRef]
46. Waldron, K.J.; Robinson, N.J. How do bacterial cells ensure thatmetalloproteins get the correct metal. *Nat. Rev. Microbiol.* **2009**, *7*, 25–35. [CrossRef]
47. Faisal, M. Chromium-resistant bacteria and cyanobacteria: Impact on Cr(VI) reduction potential and plant growth. *J. Ind. Microbiol. Biotechnol.* **2005**, *32*, 615–621. [CrossRef]
48. Houot, L.; Floutier, M.; Marteyn, B.; Michaut, M.; Picciocchi, A.; Legrain, P.; Aude, J.C.; Cassier-Chauvat, C.; Chauvat, F. Cadmium triggers an integrated reprogramming of the metabolism of *Synechocystis* PCC6803, under the control of the Slr1738 regulator. *BMC Genom.* **2007**, *8*, 350. [CrossRef]
49. Marteyn, B.; Sakr, S.; Farci, S.; Bedhomme, M.; Chardonnet, S.; Decottignies, P.; Lemaire, S.D.; Cassier-Chauvat, C.; Chauvat, F. The *Synechocystis* PCC6803 MerA-like enzyme operates in the reduction of both mercury and uranium under the control of the glutaredoxin 1 enzyme. *J. Bacteriol.* **2013**, *195*, 4138–4145. [CrossRef]
50. Garcia-Calleja, J.; Cossart, T.; Pedrero, Z.; Santos, J.P.; Ouerdane, L.; Tessier, E.; Slaveykova, V.I.; Arnoux, D. Determination of the Intracellular Complexation of Inorganic and Methylmercury in Cyanobacterium *Synechocystis* sp. PCC 6803. *Environ. Sci. Technol.* **2021**, *55*, 13971–13979. [CrossRef]
51. Ali, S.; Mir, R.A.; Tyagi, A.; Manzar, N.; Kashyap, A.S.; Mushtaq, M.; Raina, A.; Park, S.; Sharma, S.; Mir, Z.A.; et al. Chromium Toxicity in Plants: Signaling, Mitigation, and Future Perspectives. *Plants* **2023**, *12*, 1502. [CrossRef] [PubMed]
52. Collado-Lopez, S.; Betanzos-Robledo, L.; Tellez-Rojo, M.M.; Lamadrid-Figueroa, H.; Reyes, M.; Rios, C.; Cantoral, A. Heavy Metals in Unprocessed or Minimally Processed Foods Consumed by Humans Worldwide: A Scoping Review. *Int. J. Environ. Res. Public Health* **2022**, *19*, 8651. [CrossRef] [PubMed]

53. Wang, R.; Sang, P.; Guo, Y.; Jin, P.; Cheng, Y.; Yu, H.; Xie, Y.; Yao, W.; Qian, H. Cadmium in food: Source, distribution and removal. *Food Chem.* **2023**, *405*, 134666. [CrossRef]
54. Shimakawa, G.; Suzuki, M.; Yamamoto, E.; Saito, R.; Iwamoto, T.; Nishi, A.; Miyake, C. Why don't plants have diabetes? Systems for scavenging reactive carbonyls in photosynthetic organisms. *Biochem. Soc. Trans.* **2014**, *42*, 543–547. [CrossRef]
55. Lea-Smith, D.J.; Bombelli, P.; Vasudevan, R.; Howe, C.J. Photosynthetic, respiratory and extracellular electron transport pathways in cyanobacteria. *Biochim. Biophys. Acta Bioenerg.* **2016**, *1857*, 247–255. [CrossRef] [PubMed]
56. Veaudor, T.; Blanc-Garin, V.; Chenebault, C.; Diaz-Santos, E.; Sassi, J.F.; Cassier-Chauvat, C.; Chauvat, F. Recent advances in the photoautotrophic metabolism of cyanobacteria: Biotechnological implications. *Life* **2020**, *10*, 71. [CrossRef]
57. Cui, J.; Xie, Y.; Sun, T.; Chen, L.; Zhang, W. Deciphering and engineering photosynthetic cyanobacteria for heavy metal bioremediation. *Sci. Total Environ.* **2021**, *761*, 144111. [CrossRef]
58. Chakdar, H.; Thapa, S.; Srivastava, A.; Shukla, P. Genomic and proteomic insights into the heavy metal bioremediation by cyanobacteria. *J. Hazard. Mater.* **2022**, *424*, 127609. [CrossRef]
59. Fahey, R.C.; Buschbacher, R.M.; Newton, G.L. The evolution of glutathione metabolism in phototrophic microorganisms. *J. Mol. Evol.* **1987**, *25*, 81–88. [CrossRef]
60. Copley, S.D.; Dhillon, J. Lateral gene transfer and parallel evolution in the history of glutathione biosynthesis genes. *Genome Biol.* **2002**, *3*, research0025. [CrossRef]
61. Narainsamy, K.; Farci, S.; Braun, E.; Junot, C.; Cassier-Chauvat, C.; Chauvat, F. Oxidative-stress detoxification and signalling in cyanobacteria: The crucial glutathione synthesis pathway supports the production of ergothioneine and ophthalmate. *Mol. Microbiol.* **2016**, *100*, 15–24. [CrossRef] [PubMed]
62. Kammerscheit, X.; Chauvat, F.; Cassier-Chauvat, C. From cyanobacteria to human, MAPEG-type glutathione-S-transferases operate in cell tolerance to heat, cold, and lipid peroxidation. *Front. Microbiol.* **2019**, *10*, 2248. [CrossRef] [PubMed]
63. Hasanuzzaman, M.; Nahar, K.; Anee, T.I.; Fujita, M. Glutathione in plants: Biosynthesis and physiological role in environmental stress tolerance. *Physiol. Mol. Biol. Plants* **2017**, *23*, 249–268. [CrossRef] [PubMed]
64. Hayes, J.D.; Flanagan, J.U.; Jowsey, I.R. Glutathione transferases. *Annu. Rev. Pharmacol. Toxicol.* **2005**, *45*, 51–88. [CrossRef]
65. Picciocchi, A.; Saguez, C.; Boussac, A.; Cassier-Chauvat, C.; Chauvat, F. CGFS-type monothiol glutaredoxins from the cyanobacterium *Synechocystis* PCC6803 and other evolutionary distant model organisms possess a glutathione-ligated [2Fe-2S] cluster. *Biochemistry* **2007**, *46*, 15018–15026. [CrossRef] [PubMed]
66. Partensky, F.; Hess, W.R.; Vaulot, D. *Prochlorococcus*, a Marine Photosynthetic Prokaryote of Global Significance. *Microbiol. Mol. Biol. Rev.* **1999**, *63*, 106–127. [CrossRef]
67. Zwirglmaier, K.; Jardillier, L.; Ostrowski, M.; Mazard, S.; Garczarek, L.; Vaulot, D.; Not, F.; Massana, R.; Ulloa, O.; Scanlan, D.J. Global phylogeography of marine *Synechococcus* and *Prochlorococcus* reveals a distinct partitioning of lineages among oceanic biomes. *Environ. Microbiol.* **2008**, *10*, 147–161. [CrossRef]
68. Dai, W.; Chen, M.; Myers, C.; Ludtke, S.J.; Pettitt, B.M.; King, J.A.; Schmid, M.F.; Chiu, W. Visualizing Individual RuBisCO and Its Assembly into Carboxysomes in Marine Cyanobacteria by Cryo-Electron Tomography. *J. Mol. Biol.* **2018**, *430*, 4156–4167. [CrossRef]
69. Nelson, C.; Garcia-Pichel, F. Beneficial Cyanosphere Heterotrophs Accelerate Establishment of Cyanobacterial Biocrust. *Appl. Environ. Microbiol.* **2021**, *87*, e0123621. [CrossRef]
70. Singh, J.S.; Kumar, A.; Rai, A.N.; Singh, D.P. Cyanobacteria: A precious bio-resource in agriculture, ecosystem, and environmental sustainability. *Front. Microbiol.* **2016**, *7*, 529. [CrossRef]
71. Panjiar, N.; Mishra, S.; Yadav, A.N.; Verma, P. Functional Foods from Cyanobacteria. In *Microbial Functional Foods and Nutraceuticals*; Gupta, V.K., Treichel, H., Shapaval, V., de Oliveira, L.A., Tuohy, M.G., Eds.; Wiley: Hoboken, NY, USA, 2017.
72. Sheha, G.R.; Yadav, R.K.; Chatrath, A.; Gerard, M.; Tripathi, K.; Govindsamy, V.; Abraham, G. Perspectives on the potential application of cyanobacteria in the alleviation of drought and salinity stress in crop plants. *J. Appl. Phycol.* **2021**, *33*, 3761–3778. [CrossRef]
73. Liu, D.; Liberton, M.; Hendry, J.I.; Aminian-Dehkordi, J.; Maranas, C.D.; Pakrasi, H.B. Engineering biology approaches for food and nutrient production by cyanobacteria. *Curr. Opin. Biotechnol.* **2021**, *67*, 1–6. [CrossRef] [PubMed]
74. Cassier-Chauvat, C.; Dive, V.; Chauvat, F. Cyanobacteria: Photosynthetic factories combining biodiversity, radiation resistance, and genetics to facilitate drug discovery. *Appl. Microbiol. Biotechnol.* **2017**, *101*, 1359–1364. [CrossRef] [PubMed]
75. Demay, J.; Bernard, C.; Reinhardt, A.; Marie, B. Natural products from cyanobacteria: Focus on beneficial activities. *Mar. Drugs* **2019**, *17*, 320. [CrossRef] [PubMed]
76. Kirsch, F.; Klähn, S.; Hagemann, M. Salt-Regulated Accumulation of the Compatible Solutes Sucrose and Glucosylglycerol in Cyanobacteria and Its Biotechnological Potential. *Front. Microbiol.* **2019**, *10*, 2139. [CrossRef]
77. Pade, N.; Hagemann, M. Salt Acclimation of Cyanobacteria and Their Application in Biotechnology. *Life* **2015**, *5*, 25–49. [CrossRef]
78. Kollmen, J.; Strieth, D. The Beneficial Effects of Cyanobacterial Co-Culture on Plant Growth. *Life* **2022**, *12*, 223. [CrossRef]
79. Verseux, C.; Baqué, M.; Lehto, K.; De Vera, J.P.P.; Rothschild, L.J.; Billi, D. Sustainable life support on Mars—The potential roles of cyanobacteria. *Int. J. Astrobiol.* **2016**, *15*, 65–92. [CrossRef]
80. Keller, R.; Goli, K.; Porter, W.; Alrabaa, A.; Jones, J.A. Cyanobacteria and Algal-Based Biological Life Support System (BLSS) and Planetary Surface Atmospheric Revitalizing Bioreactor Brief Concept Review. *Life* **2023**, *13*, 816. [CrossRef]

81. Rautela, A.; Kumar, S. Engineering plant family TPS into cyanobacterial host for terpenoids production. *Plant Cell Rep.* **2022**, *41*, 1791–1803. [CrossRef]
82. Cassier-Chauvat, C.; Blanc-Garin, V.; Chauvat, F. Genetic, Genomics, and Responses to Stresses in Cyanobacteria: Biotechnological Implications. *Genes* **2021**, *12*, 500. [CrossRef] [PubMed]
83. Cassier-Chauvat, C.; Chauvat, F. Cell division in cyanobacteria. In *The Cell Biology of Cyanobacteria*; Flores, E., Herrero, A., Eds.; Caister Academic Press: Poole, UK, 2014.
84. Herrero, A.; Stavans, J.; Flores, E. The multicellular nature of filamentous heterocyst-forming cyanobacteria. *FEMS Microbiol. Rev.* **2016**, *40*, 831–854. [CrossRef] [PubMed]
85. Cassier-Chauvat, C.; Veaudor, T.; Chauvat, F. Comparative genomics of DNA recombination and repair in cyanobacteria: Biotechnological implications. *Front. Microbiol.* **2016**, *7*, 1809. [CrossRef] [PubMed]
86. Meister, A. On the discovery of glutathione. *Trends Biochem. Sci.* **1988**, *13*, 185–188. [CrossRef] [PubMed]
87. Noctor, G.; Mhamdi, A.; Chaouch, S.; Han, Y.; Neukermans, J.; Marquez-Garcia, B.; Queval, G.; Foyer, C.H. Glutathione in plants: An integrated overview. *Plant Cell Environ.* **2012**, *35*, 454–484. [CrossRef]
88. Berndt, C.; Lillig, C.H. Glutathione, Glutaredoxins, and Iron. *Antioxid. Redox Signal.* **2017**, *27*, 1235–1251. [CrossRef]
89. Scire, A.; Cianfruglia, L.; Minnelli, C.; Bartolinin, D.; Torquato, P.; Principato, G.; Galli, F.; Armeni, T. Glutathione compartmentalization and its role in glutathionylation and other regulatory processes of cellular pathways. *BioFactors* **2018**, *45*, 152–168. [CrossRef]
90. Hider, R.C.; Kong, X.L. Glutathione: A key component of the cytoplasmic labile iron pool. *BioMetals* **2011**, *24*, 1179–1187. [CrossRef]
91. Trnka, D.; Engelke, A.D.; Gellert, M.; Moseler, A.; Hossain, M.F.; Lindenberg, T.T.; Pedroletti, L.; Odermatt, B.; de Souza, J.V.; Bronowska, A.K.; et al. Molecular basis for the distinct functions of redox-active and FeS-transfering glutaredoxins. *Nat. Commun.* **2020**, *11*, 3445. [CrossRef]
92. Greenberg, J.T.; Demple, B. Glutathione in *Escherichia coli* is dispensable for resistance to H_2O_2 and gamma radiation. *J. Bacteriol.* **1986**, *168*, 1026–1029. [CrossRef]
93. Richman, P.G.; Meister, A. Regulation of gamma-glutamyl-cysteine synthetase by nonallosteric feedback inhibition by glutathione. *J. Biol. Chem.* **1975**, *250*, 1422–1426. [CrossRef] [PubMed]
94. Grant, C.M.; Maciver, F.H.; Dawes, I.W. Glutathione is an essential metabolite required for resistance to oxidative stress in the yeast *Saccharomyces cerevisiae*. *Curr. Genet.* **1996**, *29*, 511–515. [CrossRef]
95. Spector, D.; Labarre, J.; Toledano, M.B. A genetic investigation of the essential role of glutathione: Mutations in the proline biosynthesis pathway are the only suppressors of glutathione auxotrophy in yeast. *J. Biol. Chem.* **2001**, *276*, 7011–7016. [CrossRef] [PubMed]
96. Shi, Z.-Z.; Osei-Frimpong, J.; Kala, G.; Lieberman, M.W. Glutathione synthesis is essential for mouse development but not for cell growth in culture. *Proc. Natl. Acad. Sci. USA* **2000**, *97*, 5101–5106. [CrossRef] [PubMed]
97. Cairns, N.G.; Pasternak, M.; Wachter, A.; Cobbett, C.S.; Meyer, A.J. Maturation of Arabidopsis seeds is dependent on glutathione biosynthesis within the embryo. *Plant Physiol.* **2006**, *141*, 446–455. [CrossRef] [PubMed]
98. Fuchs, J.A.; Warner, H.R. Isolation of an *Escherichia coli* mutant deficient in glutathione synthesis. *J. Bacteriol.* **1975**, *124*, 140–148. [CrossRef]
99. Daws, T.; Lim, C.J.; Fuchs, J.A. In vitro construction of gshB::kan in *Escherichia coli* and use of gshB::kan in mapping the gshB locus. *J. Bacteriol.* **1989**, *171*, 5218–5221. [CrossRef]
100. Chesney, J.A.; Eaton, J.W.; Mahoney, J.R. Bacterial Glutathione: A Sacrificial Defense against Chlorine Compounds. *J. Bacteriol.* **1996**, *178*, 2131–2135. [CrossRef]
101. Moore, W.R.; Anderson, M.E.; Meister, A.; Murata, K.; Kimura, A. Increased capacity for glutathione synthesis enhances resistance to radiation in *Escherichia coli*: A possible model for mammalian cell protection. *Proc. Natl. Acad. Sci. USA* **1989**, *86*, 1461–1464. [CrossRef]
102. Owens, R.A.; Hartman, P.E. Export of glutathione by some widely used Salmonella typhimurium and *Escherichia coli* strains. *J. Bacteriol.* **1986**, *168*, 109–114. [CrossRef]
103. Owens, R.A.; Hartman, P.E. Glutathione: A protective agent in Salmonella typhimurium and *Escherichia coli* as measured by mutagenicity and by growth delay assays. *Environ. Mutagen.* **1986**, *8*, 659–673. [CrossRef] [PubMed]
104. Smirnova, G.V.; Muzyka, N.G.; Ushakov, V.Y.; Tyulenev, A.V.; Oktyabrsky, O.N. Extracellular superoxide provokes glutathione efflux from *Escherichia coli* cells. *Res. Microbiol.* **2015**, *166*, 609–617. [CrossRef] [PubMed]
105. Howden, R.; Andersen, C.R.; Coldsbrough, P.; Cobbett, C.S. A cadmium-sensitive, glutathione-deficient mutant of *Arabidopsis thaliana*. *PLANT Physiol.* **1995**, *107*, 1067–1073. [CrossRef]
106. Shanmugam, V.; Tsednee, L.; Yeh, K.-C. Zinc Tolerance Induced by Iron 1 reveals the importance of glutathione in the cross-homeostasis between zinc and iron in *Arabidopsis thaliana*. *Plant J.* **2012**, *69*, 1006–1017. [CrossRef]
107. Ponce-Toledo, R.I.; Deschamps, P.; López-García, P.; Zivanovic, Y.; Benzerara, K.; Moreira, D. An Early-Branching Freshwater Cyanobacterium at the Origin of Plastids. *Curr. Biol.* **2017**, *27*, 386–391. [CrossRef] [PubMed]
108. Pinevich, A.V. Chloroplast history clarified by the criterion of light-harvesting complex. *Biosystem* **2020**, *196*, 104176. [CrossRef] [PubMed]

109. Seregin, I.V.; Kozhevnikova, A.D. Phytochelatins: Sulfur-Containing Metal(loid)-Chelating Ligands in Plants. *Int. J. Mol. Sci.* **2023**, *24*, 2430. [CrossRef]
110. Ito, T.; Ohkama-Ohtsu, N. Degradation of glutathione and glutathione conjugates in plants. *J. Exp. Bot.* **2023**, erad018. [CrossRef]
111. Bhargawa, P.; Srivastava, A.K.; Urmil, S.; Rai, L.C. Phytochelatin plays a role in UV-B tolerance in N2-fixing cyanobacterium *Anabaena doliolum*. *J. Plant Physiol.* **2005**, *162*, 1220–1225. [CrossRef]
112. Pandey, S.; Rai, R.; Rai, L.C. Proteomics combines morphological, physiological and biochemical attributes to unravel the survival strategy of *Anabaena* sp. PCC7120 under arsenic stress. *J. Proteom.* **2012**, *75*, 921–937. [CrossRef]
113. Freisinger, E. Plant MTs-long neglected members of the metallothionein superfamily. *Dalton Trans.* **2008**, *21*, 6663–6675. [CrossRef] [PubMed]
114. Gao, C.; Gao, K.; Yang, H.; Ju, T.; Zhu, J.; Tang, Z.; Zhao, L.; Chen, Q. Genome-wide analysis of metallothionein gene family in maize to reveal its role in development and stress resistance to heavy metal. *Biol. Res.* **2022**, *55*, 1. [CrossRef] [PubMed]
115. Xu, J.; Tian, Y.-S.; Peng, R.-H.; Xiong, A.-S.; Zhu, B.; Hou, X.-L.; Yao, Q.-H. Cyanobacteria MT gene SmtA enhance zinc tolerance in Arabidopsis. *Mol. Biol. Rep.* **2010**, *37*, 1105–1110. [CrossRef] [PubMed]
116. Mikhaylina, A.; Scott, L.; Scanlan, D.J.; Blindauer, C.A. A metallothionein from an open ocean cyanobacterium removes zinc from the sensor protein controlling its transcription. *J. Inorg. Biochem.* **2022**, *230*, 111755. [CrossRef] [PubMed]
117. Ziller, A.; Fraissinet-Tachet, L. Metallothionein diversity and distribution in the tree of life: A multifunctional protein. *Metallomics* **2018**, *10*, 1549–1559. [CrossRef] [PubMed]
118. Saing, T.; Lagman, M.; Castrillon, J.; Gutierrez, E.; Guilford, F.T.; Venketaraman, V. Analysis of glutathione levels in the brain tissue samples from HIV-1-positive individuals and subject with Alzheimer's disease and its implication in the pathophysiology of the disease process. *Biochim. Biophys. Acta Clin.* **2016**, *29*, 38–44. [CrossRef]
119. Dwivedi, D.; Megha, K.; Mishra, R.; Mandal, P.K. Glutathione in Brain: Overview of Its Conformations, Functions, Biochemical Characteristics, Quantitation and Potential Therapeutic Role in Brain Disorders. *Neurochem. Res.* **2020**, *45*, 1461–1480. [CrossRef]
120. Couto, N.; Wood, J.; Barber, J. The role of glutathione reductase and related enzymes on cellular redox homoeostasis network. *Free Radic. Biol. Med.* **2016**, *95*, 27–42. [CrossRef]
121. Dai, H.; Wang, L.; Li, L.; Huang, Z.; Ye, L. Metallothionein 1: A New Spotlight on Inflammatory Diseases. *Front. Immunol.* **2021**, *5*, 739918. [CrossRef]
122. Yang, J.; Li, W.; Wang, D.; Wu, H.; Li, Z.; Ye, Q. Characterization of bifunctional L-glutathione synthetases from *Actinobacillus pleuropneumoniae* and *Actinobacillus succinogenes* for efficient glutathione biosynthesis. *Appl. Microbiol. Biotechnol.* **2016**, *100*, 6279–6289. [CrossRef]
123. Gopal, S.; Borovok, A.; Ofer, M.; Yanku, G.; Cohen, W.; Kreft, J.; Aharonowitz, Y. A multidomain fusion protein in *Listeria monocytogenes* catalyzes the two primary activities for glutathione biosynthesis. *J. Bacteriol.* **2005**, *187*, 3839–3847. [CrossRef] [PubMed]
124. Vergauwen, B.; De Vos, D.; Van Beeumen, J.J. Characterization of the bifunctional gamma-glutamate-cysteine ligase/glutathione synthetase (GshF) of *Pasteurella multocida*. *J. Biol. Chem.* **2006**, *281*, 4380–4394. [CrossRef]
125. Janowiak, B.; Griffith, O.W. Glutathione synthesis in *Streptococcus agalactiae* One protein accounts for gamma-glutamylcysteine synthetase and glutathione synthetase activities. *J. Biol. Chem.* **2005**, *280*, 11829–11839. [CrossRef] [PubMed]
126. Li, W.; Li, Z.; Yang, J.; Ye, Q. Production of glutathione using a bifunctional enzyme encoded by *gshF* from *Streptococcus thermophilus* expressed in *Escherichia coli*. *J. Bacteriol.* **2011**, *154*, 261–268. [CrossRef]
127. Jeon, G.-B.; Lee, H.-J.; Park, J.P.; Park, K.; Choi, C.-H.; Kim, S.-K. Efficient production of glutathione in *Saccharomyces cerevisiae* via a synthetic isozyme system. *Biotechnol. J.* **2022**, *8*, 2200398. [CrossRef] [PubMed]
128. Liedschulte, V.; Wachter, A.; Zhigang, A.; Rausch, T. Exploiting plants for glutathione (GSH) production: Uncoupling GSH synthesis from cellular controls results in unprecedented GSH accumulation. *Plant Biotechnol. J.* **2010**, *8*, 807–820. [CrossRef] [PubMed]
129. Wang, C.; Zhang, J.; Wu, H.; Li, Z.; Ye, Q. Heterologous *gshF* gene expression in various vector systems in *Escherichia coli* for enhanced glutathione production. *J. Biotechnol.* **2015**, *214*, 63–68. [CrossRef]
130. Ge, S.; Zhu, T.; Li, Y. Expression of bacterial GshF in *Pichia pastoris* for glutathione production. *Appl. Environ. Microbiol.* **2012**, *78*, 5435–5439. [CrossRef]
131. Santos, L.O.; Silva, P.G.P.; Lemos, W.J.F.L.; de Oliveira, V.S.; Anschau, A. Glutathione production by *Saccharomyces cerevisiae*: Current state and perspectives. *Appl. Environ. Microbiol.* **2022**, *106*, 1879–1894. [CrossRef]
132. Ali, V.; Behera, S.; Nawaz, A.; Equbal, A.; Pandey, K. Unique thiol metabolism in trypanosomatids: Redox homeostasis and drug resistance. *Adv. Parasitol.* **2022**, *117*, 75–155. [CrossRef]
133. Chandrangsu, P.; Van Loi, W.; Antelman, H.; Helman, J.D. The Role of Bacillithiol in Gram-Positive Firmicutes. *Antioxid. Redox Signal.* **2018**, *28*, 445–462. [CrossRef]
134. Sao Emani, C.; Gallant, J.L.; Wild, I.J.; Baker, B. The role of low molecular weight thiols in *Mycobacterium tuberculosis*. *Tuberculosis* **2019**, *116*, 44–55. [CrossRef] [PubMed]
135. Qiu, Y.; Chen, Z.; Su, E.; Wang, L.; Sun, L.; Lei, P.; Xu, H.; Li, S. Recent Strategies for the Biosynthesis of Ergothioneine. *J. Agric. Food Chem.* **2021**, *69*, 13682–13690. [CrossRef]
136. Cumming, B.M.; Chinta, K.C.; Reddy, V.P.; Steyn, A.J.C. Role of Ergothioneine in Microbial Physiology and Pathogenesis. *Antioxid. Redox Signal.* **2018**, *28*, 431–444. [CrossRef] [PubMed]

137. Borodina, I.; Kenny, L.C.; McCarthy, C.M.; Paramasivan, K.; Pretorius, E.; Roberts, T.J.; van der Hoek, S.A.; Kell, D.B. The biology of ergothioneine, an antioxidant nutraceutical. *Nutr. Res. Rev.* **2020**, *33*, 190–217. [CrossRef] [PubMed]
138. Nakamichi, N.; Tsuzuku, S.; Shibagaki, F. Ergothioneine and central nervous system diseases. *Neurochem. Res.* **2022**, *47*, 2513–2521. [CrossRef]
139. Yadan, J.-C. Matching chemical properties to molecular biological activities opens a new perspective on l-ergothioneine. *FEBS Lett.* **2022**, *596*, 1299–1312. [CrossRef]
140. Liu, H.-M.; Tang, W.; Wang, X.-Y.; Jiang, J.-J.; Zhang, W.; Wang, W. Safe and Effective Antioxidant: The Biological Mechanism and Potential Pathways of Ergothioneine in the Skin. *Molecules* **2023**, *28*, 1648. [CrossRef]
141. Fahey, R.C. Novel Thiols of Prokaryotes. *Annu. Rev. Microbiol.* **2001**, *55*, 333–356. [CrossRef]
142. Stampfli, A.R.; Blankenfeldt, W.; Seebeck, F.P. Structural basis of ergothioneine biosynthesis. *Curr. Opin. Struct. Biol.* **2020**, *65*, 1–8. [CrossRef]
143. Gokçe, G.; Zuhuri, M.; Erfuna, E. Ergothioneine prevents endothelial dysfunction induced by mercury chloride. *Exp. Ther. Med.* **2018**, *15*, 4697–4702. [CrossRef]
144. Waley, S.G. Acidic peptides of the lens. 3. The structure of ophthalmic acid. *Biochem. J.* **1958**, *68*, 189–192. [CrossRef] [PubMed]
145. Tsuboii, S.; Hirota, K.; Ogata, K.; Ohmori, S. Ophthalmic and Norophthalmic Acid in Lens, Liver, and Brain of Higher Animals. *Anal. Biochem.* **1984**, *136*, 520–524. [CrossRef] [PubMed]
146. Soga, T.; Baran, R.; Suematsu, M.; Ueno, Y.; Ikeda, S.; Sakurakawa, T.; Kakazu, Y.; Ishikawa, T.; Robert, M.; Nishioka, T.; et al. Differential metabolomics reveals ophthalmic acid as an oxidative stress biomarker indicating hepatic glutathione consumption. *J. Biol. Chem.* **2006**, *281*, 16768–16776. [CrossRef] [PubMed]
147. Dello, S.A.W.G.; Neis, E.P.J.G.; de Jong, M.C.; van Eijk, H.M.H.; Kicken, C.H.; Olde Damink, S.W.M.; Dejong, C.H.C. Systematic review of ophthalmate as a novel biomarker of hepatic glutathione depletion. *Clin. Nutr.* **2013**, *32*, 325–330. [CrossRef] [PubMed]
148. Servillo, L.; Castaldo, D.; Giovane, A.; Casale, R.; D'Onofrio, N.; Cautela, D.; Balestrieri, M.L. Ophthalmic acid is a marker of oxidative stress in plants as in animals. *Biochim. Biophys. Acta Gen. Subj.* **2018**, *1862*, 991–998. [CrossRef]
149. Sajiki, K.; Pluskai, T.; Shimanuki, M.; Yanagida, M. Metabolomic analysis of fission yeast at the onset of nitrogen starvation. *Metabolites* **2013**, *3*, 1118–1129. [CrossRef]
150. Ito, T.; Yamauchi, A.; Hemmi, H.; Yoshimura, T. Ophthalmic acid accumulation in an *Escherichia coli* mutant lacking the conserved pyridoxal 5′-phosphate-binding protein YggS. *J. Biosci. Bioeng.* **2016**, *122*, 689–693. [CrossRef]
151. Tel-Or, E.; Uhflejt, M.; Packer, L. The role of glutathione and ascorbate in hydroperoxide removal in cyanobacteria. *Biochem. Biophys. Res. Commun.* **1985**, *132*, 533–539. [CrossRef]
152. Suginata, K.; Yamamoto, K.; Ashiida, H.; Kono, Y.; Saw, Y.; Shibata, H. Cysteine uptake for accumulation of glutathione by the cyanobacterium *Synechocystis* strain PCC 6803. *Biosci. Biotechnol. Biochem.* **1998**, *62*, 424–428. [CrossRef]
153. Izawa, S.; Inoue, Y.; Kimura, A. Oxidative stress response in yeast: Effect of glutathione on adaptation to hydrogen peroxide stress in *Saccharomyces cerevisiae*. *FEBS Lett.* **1995**, *368*, 73–76. [CrossRef] [PubMed]
154. Suginata, K.; Yamamoto, K.; Ashida, H.; Sawa, Y.; Shibata, H. Effect of Intracellular Glutathione on Heat-induced Cell Death in the Cyanobacterium, *Synechocystis* PCC 6803. *Biosci. Biotechnol. Biochem.* **1999**, *63*, 1112–1115. [CrossRef] [PubMed]
155. Gupta, A.; Ballal, A. Unraveling the mechanism responsible for the contrasting tolerance of *Synechocystis* and *Synechococcus* to Cr(VI): Enzymatic and non-enzymatic antioxidants. *Aquat. Toxicol.* **2015**, *164*, 118–125. [CrossRef]
156. Ashida, H.; Sawa, Y.; Shibata, H. Cloning, biochemical and phylogenetic characterizations of gamma-glutamylcysteine synthetase from *Anabaena* sp. PCC 7120. *Plant Cell Physiol.* **2005**, *46*, 557–562. [CrossRef] [PubMed]
157. Okumura, N.; Masamoto, K.; Wada, H. The *gshB* gene in the cyanobacterium *Synechococcus* sp. PCC 7942 encodes a functional glutathione synthetase. *Microbiology* **1997**, *143*, 2883–2890. [CrossRef] [PubMed]
158. Cameron, J.C.; Pakrasi, H.B. Glutathione in *Synechocystis* 6803: A closer look into the physiology of a ΔgshB mutant. *Plant Signal. Behav.* **2011**, *6*, 89–92. [CrossRef] [PubMed]
159. Cameron, J.C.; Pakrasi, H.B. Glutathione Facilitates Antibiotic Resistance and Photosystem I Stability during Exposure to Gentamicin in Cyanobacteria. *Appl. Environ. Microbiol.* **2011**, *77*, 3547–3550. [CrossRef]
160. Tsuji, N.; Nishikori, S.; Iwake, O.; Shiraki, K.; Miyasaka, H.; Takagi, M.; Hirata, K.; Miyamoto, K. Characterization of phytochelatin synthase-like protein encoded by alr0975 from a prokaryote. *Biochem. Biophys. Res. Commun.* **2004**, *315*, 751–755. [CrossRef] [PubMed]
161. Jaiswal, D.; Nenwani, M.; Mishra, V.; Wangikar, P.P. Probing the metabolism of γ-glutamyl peptides in cyanobacteria via metabolite profiling and 13C labeling. *Plant J.* **2022**, *109*, 708–726. [CrossRef]
162. Sofyanovich, O.A.; Nishiuchi, H.; Yamagishi, K.; Matrosova, E.V.; Serebrianyl, V.A. Multiple pathways for the formation of the γ-glutamyl peptides γ-glutamyl-valine and γ-glutamyl-valyl-glycine in *Saccharomyces cerevisiae*. *PLoS ONE* **2019**, *14*, e0216622. [CrossRef]
163. Kobayashi, S.; Ikeda, Y.; Shigeno, Y.; Konno, H.; Fujii, J. γ-Glutamylcysteine synthetase and γ-glutamyl transferase as differential enzymatic sources of γ-glutamylpeptides in mice. *Amino Acids* **2020**, *52*, 555–566. [CrossRef]
164. Yang, J.; Bai, W.; Zeng, X.; Cui, C. Gamma glutamyl peptides: The food source, enzymatic synthesis, kokumi-active and the potential functional properties—A review. *Trends Food Sci. Technol.* **2019**, *91*, 339–346. [CrossRef]
165. Kang, Y.P.; Mockabee-Maclas, A.; Jiang, C.; Falzone, A.; Prieto-Farigua, N.; Stone, E.; Harris, I.S.; DeNicolas, G.M. Non-canonical glutamate-cysteine ligase activity protects against ferroptosis. *Cell Metab.* **2021**, *33*, 174–189. [CrossRef]

166. Wu, Q.; Li, J.; Zhu, J.; Sun, X.; He, D.; Li, J.; Cheng, Z.; Zhang, X.; Xu, Y.; Chen, Q.; et al. Gamma-glutamyl-leucine levels are causally associated with elevated cardio-metabolic risks. *Front. Nutr.* **2022**, *9*, 936220. [CrossRef] [PubMed]
167. Narainsamy, K.; Cassier-Chauvat, C.; Junot, C.; Chauvat, F. High performance analysis of the cyanobacterial metabolism via liquid chromatography coupled to a LTQ-Orbitrap mass spectrometer: Evidence that glucose reprograms the whole carbon metabolism and triggers oxidative stress. *Metabolomics* **2013**, *9*, 21–32. [CrossRef]
168. Baran, R.; Bowen, B.P.; Bouskill, N.J.; Brodie, E.L.; Yannone, S.M.; Northern, T.R. Metabolite identification in *Synechococcus* sp. PCC 7002 using untargeted stable isotope assisted metabolite profiling. *Anal. Chem.* **2010**, *82*, 9034–9042. [CrossRef] [PubMed]
169. Pfeiffer, C.; Bauer, T.; Surek, B.; Schoming, E.; Grundermann, D. Cyanobacteria produce high levels of ergothioneine. *Food Chem.* **2011**, *129*, 1766–1769. [CrossRef]
170. Baran, R.; Ivanova, N.N.; Jose, N.; Garcia-Pichel, F.; Kyrpides, N.G.; Gigger, M.; Northern, T.R. Functional Genomics of Novel Secondary Metabolites from Diverse Cyanobacteria Using Untargeted Metabolomics. *Mar. Drugs* **2013**, *11*, 3617–3631. [CrossRef] [PubMed]
171. Liao, C.; Seebeck, F.P. Convergent Evolution of Ergothioneine Biosynthesis in Cyanobacteria. *ChemBioChem* **2017**, *18*, 2115–2118. [CrossRef]
172. Tanaka, N.; Kawano, Y.; Satoh, Y.; Dairi, T.; Ohtsu, I. Gram-scale fermentative production of ergothioneine driven by overproduction of cysteine in *Escherichia coli*. *Sci. Rep.* **2019**, *9*, 1895. [CrossRef]
173. Saini, M.; Kashyap, A.; Bindal, S.; Saini, K.; Gupta, R. Bacterial Gamma-Glutamyl Transpeptidase, an Emerging Biocatalyst: Insights into Structure–Function Relationship and Its Biotechnological Applications. *Front. Microbiol.* **2021**, *12*, 641251. [CrossRef] [PubMed]
174. Lee, D.S.; Evans, J.C.; Robins, S.J.; Wilson, P.W.; Albano, I.; Fox, C.S.; Wang, T.J.; Benjamin, E.J.; D'Agostino, R.B.; Vasan, R.S. Gamma Glutamyl Transferase and Metabolic Syndrome, Cardiovascular Disease, and Mortality Risk. *Arterioscler. Thromb. Vasc. Biol.* **2007**, *27*, 127–133. [CrossRef] [PubMed]
175. Schulz, G.; Schirmer, S.H.; Sachsenheimer, W.; Pai, E.F. The structure of the flavoenzyme glutathione reductase. *Nature* **1978**, *273*, 120–124. [CrossRef]
176. Scrutton, N.S.; Berry, A.; Perham, R.N. Redesign of the coenzyme specificity of a dehydrogenase by protein engineering. *Nature* **1990**, *343*, 38–43. [CrossRef]
177. Perham, R.N.; Scrutton, N.S.; Berry, A. New enzymes for old: Redesigning the coenzyme and substrate specificities of glutathione reductase. *Bioessays* **1991**, *10*, 515–525. [CrossRef] [PubMed]
178. Lin, T.H.; Rao, M.Y.; Lu, H.W.; Chiou, C.W.; Lin, S.T.; Chao, H.W.; Zheng, Z.L.; Cheng, H.C.; Lee, T.M. A role for glutathione reductase and glutathione in the tolerance of *Chlamydomonas reinhardtii* to photo-oxidative stress. *Physiol. Plant.* **2018**, *162*, 35–48. [CrossRef] [PubMed]
179. Sexton, D.J.; Mutus, B. Glutathione reductases from a variety of sources are inhibited by physiological levels of glutathione. *Comp. Biochem. Physiol.* **1992**, *103*, 897–901. [CrossRef]
180. Davis, N.K.; Greer, S.; Jones-Mortimer, M.C.; Perham, R.N. Isolation and mapping of glutathione reductase-negative mutants of *Escherichia coli* K12. *J. Gen Microbiol.* **1982**, *128*, 1631–1634. [CrossRef]
181. Tuggle, C.K.; Fuchs, J.A. Glutathione reductase is not required for maintenance of reduced glutathione in *Escherichia coli* K-12. *J. Bacteriol.* **1985**, *162*, 448–450. [CrossRef]
182. Alonso-Moraga, A.; Bocanegra, A.; Torres, J.M.; Pueyo, C. Glutathione status and sensitivity to GSH-reacting compounds of *Escherichia coli* strains deficient in glutathione metabolism and/or catalase activity. *Mol. Cell. Biochem.* **1987**, *73*, 61–68. [CrossRef]
183. Kunert, K.J.; Cresswell, C.F.; Schmidt, A.; Mullineaux, P.M.; Foyer, C.H. Variations in the activity of glutathione reductase and the cellular glutathione content in relation to sensitivity to methylviologen in *Escherichia coli*. *Arch. Biochem. Biophys.* **1990**, *282*, 233–238. [CrossRef]
184. Barbado, C.; Ramirez, M.; Blanco, M.A.; Lopez-Barea, J.; Pueyo, C. Mutants of *Escherichia coli* sensitive to hydrogen peroxide. *Curr. Microbiol.* **1983**, *8*, 251–253. [CrossRef]
185. Collison, L.P.; Dawes, I.W. Isolation, characterization and overexpression of the yeast gene, GLRI, encoding glutathione reductase. *Gene* **1995**, *156*, 123–127. [CrossRef]
186. Muller, E.G. A glutathione reductase mutant of yeast accumulates high levels of oxidized glutathione and requires thioredoxin for growth. *Mol. Biol. Cell* **1996**, *7*, 1805–1813. [CrossRef] [PubMed]
187. Grant, C.M.; Collison, L.P.; Roe, J.H.; Dawes, I.W. Yeast glutathione reductase is required for protection against oxidative stress and is a target gene for yAP-1 transcriptional regulation. *Mol. Microbiol.* **1996**, *21*, 171–179. [CrossRef]
188. Yan, J.; Meng, X.; Wancket, L.M.; Lintner, K.; Nelin, L.D.; Chen, B.; Francis, K.P.; Smith, C.V.; Rogers, L.K.; Liu, Y. Glutathione reductase facilitates host defense by sustaining phagocytic oxidative burst and promoting the development of neutrophil extracellular traps. *J. Immunol.* **2012**, *1888*, 2316–2327. [CrossRef]
189. Chew, O.; Rudhe, C.; Glaser, E.; Whelan, J. Characterization of the targeting signal of dual-targeted pea glutathione reductase. *Plant Mol. Biol.* **2003**, *53*, 341–356. [CrossRef] [PubMed]
190. Muller-Schussele, S.J.; Wang, R.; Gutle, D.; Romer, J.; Rodriguez-Franco, M.; Scholz, M.; Buchert, F.; Luth, V.M.; Kopriva, S.; Dormann, P.; et al. Chloroplasts require glutathione reductase to balance reactive oxygen species and maintain efficient photosynthesis. *Plant J.* **2020**, *103*, 1140–1154. [CrossRef]
191. Diaz, J.M.; Shi, X. NOX-like ROS production by glutathione reductase. *iScience* **2022**, *25*, 105094. [CrossRef]

192. Karni, L.; Moss, S.J.; Tel-Or, E. Glutathione reductase activity in heterocysts and vegetative cells of the cyanobacterium *Nostoc muscorum*. *Arch. Microbiol.* **1984**, *140*, 215–217. [CrossRef]
193. Serrano, A.; Rivas, J.; Losada, M. Purification and properties of glutathione reductase from the cyanobacterium *Anabaena* sp. strain 7119. *J. Bacteriol.* **1984**, *158*, 317–324. [CrossRef] [PubMed]
194. Jiang, F.; Hellman, U.; Sroga, G.E.; Bergman, B.; Mannervik, B. Cloning, sequencing, and regulation of the glutathione reductase gene from the cyanobacterium *Anabaena* PCC 7120. *J. Biol. Chem.* **1995**, *270*, 22882–22889. [CrossRef] [PubMed]
195. Jiang, F.; Mannervik, B. Optimized Heterologous Expression of Glutathione Reductase from Cyanobacterium *Anabaena* PCC 7120 and Characterization of the Recombinant Protein. *Protein Expr. Purif.* **1999**, *15*, 92–98. [CrossRef]
196. Danielson, U.H.; Jiang, F.; Hansson, L.O.; Mannervik, B. Probing the kinetic mechanism and coenzyme specificity of glutathione reductase from the cyanobacterium *Anabaena* PCC 7120 by redesign of the pyridine-nucleotide-binding site. *Biochemistry* **1999**, *38*, 9254–9263. [CrossRef] [PubMed]
197. Rendon, J.L.; Calgano, M.; Mendeza-Hernandez, G.; Ondarza, R.N. Purification, properties, and oligomeric structure of glutathione reductase from the cyanobacterium Spirulina maxima. *Arch. Biochem. Biophys.* **1986**, *248*, 215–223. [CrossRef] [PubMed]
198. Hamed, S.M.; Hassan, S.H.; Selim, S.; Wadaan, M.A.M.; Mohany, M.; Hozzein, W.N.; AbdElgawad, H. Differential responses of two cyanobacterial species to R-metalaxyl toxicity: Growth. *Environ. Pollut.* **2020**, *258*, 113681. [CrossRef]
199. Prajapati, R.; Yadav, S.; Atri, N. Nickel and arsenite-induced differential oxidative stress and antioxidant responses in two *Anabaena* species. *J. Basic Microbiol.* **2018**, *12*, 1061–1070. [CrossRef]
200. Hamed, S.M.; Hassan, S.H.; Selim, S.; Kumar, A.; Khalaf, S.M.H.; Wadaan, M.A.M.; Hozzein, W.N.; AbdElgawad, H. Physiological and biochemical responses to aluminum-induced oxidative stress in two cyanobacterial species. *Environ. Pollut.* **2019**, *251*, 961–969. [CrossRef]
201. Pernil, R.; Schleiff, E. Metalloproteins in the biology of heterocysts. *Life* **2019**, *9*, 32. [CrossRef]
202. Tözüm, S.R.D.; Gallon, J.R. The effects of methyl viologen on *Gloeocapsa* sp. LB795 and their relationship to the inhibition of acetylene reduction (nitrogen fixation) by oxygen. *J. Gen. Microbiol.* **1979**, *111*, 313–326. [CrossRef]
203. Marteyn, B.; Domain, F.; Legrain, P.; Chauvat, F.; Cassier-Chauvat, C. The thioredoxin reductase-glutaredoxins-ferredoxin crossroad pathway for selenate tolerance in *Synechocystis* PCC6803. *Mol. Microbiol.* **2009**, *71*, 520–532. [CrossRef]
204. Kanzok, S.M.; Fechner, A.; Bauer, H.; Ulschmidt, J.K.; Müller, H.-M.; Botella-Munoz, J.; Schneuwly, S.; Schirmer, R.H.; Becker, K. Substitution of the Thioredoxin System for Glutathione Reductase in Drosophila melanogaster. *Science* **2001**, *291*, 643–646. [CrossRef] [PubMed]
205. Berndt, C.; Lillig, C.H.; Flohé, L. Redox regulation by glutathione needs enzymes. *Front. Pharmacol.* **2014**, *5*, 168. [CrossRef] [PubMed]
206. Josephy, P.D. Genetic variations in human glutathione transferase enzymes: Significance for pharmacology and toxicology. *Hum. Genom. Proteom.* **2010**, *2010*, 876940. [CrossRef]
207. Carvalho, A.N.; Marques, C.; Guedes, R.C.; Castro-Caldas, M.; Rodrigues, E.; Van Horssen, J.; Gama, M.J. S-Glutathionylation of Keap1: A new role for glutathione S-transferase pi in neuronal protection. *FEBS Lett.* **2016**, *590*, 1455–1466. [CrossRef] [PubMed]
208. Allocati, N.; Masulli, M.; Di Ilio, C.; Federici, L. Glutathione transferases: Substrates, inihibitors and pro-drugs in cancer and neurodegenerative diseases. *Oncogenesis* **2018**, *7*, 8. [CrossRef] [PubMed]
209. Bocedi, A.; Noce, A.; Marrone, G.; Noce, G.; Cattani, G.; Gambardella, G.; Di Lauro, M.; Di Daniele, N.; Ricci, G. Glutathione Transferase P1-1 an Enzyme Useful in Biomedicine and as Biomarker in Clinical Practice and in Environmental Pollution. *Nutrients* **2019**, *11*, 1741. [CrossRef] [PubMed]
210. Sylvestre-Gonon, E.; Law, S.R.; Schwartz, M.; Robe, K.; Keech, O.; Didierjean, C.; Dubos, C.; Rouhier, N.; Hecker, A. Functional, structural and biochemical features of plant serinyl-glutathione transferases. *Front. Plant Sci.* **2019**, *10*, 608. [CrossRef]
211. Kumar, S.; Trivedi, P.K. Glutathione S-Transferases: Role in Combating Abiotic Stresses Including Arsenic Detoxification in Plants. *Front. Plant Sci.* **2018**, *9*, 751. [CrossRef]
212. Schwartz, M.; Perrot, T.; Aubert, E.; Dumarçay, S.; Favier, F.; Gérardin, P.; Morel-Rouhier, M.; Mulliert, G.; Saiag, F.; Didierjean, C.; et al. Molecular recognition of wood polyphenols by phase II detoxification enzymes of the white rot Trametes versicolor. *Sci. Rep.* **2018**, *8*, 8472. [CrossRef]
213. Gullner, G.; Komives, T.; Kiraly, L.; Schröder, P. Glutathione S-Transferase Enzymes in Plant-Pathogen Interactions. *Front. Plant Sci.* **2018**, *9*, 1836. [CrossRef] [PubMed]
214. Kammerscheit, X.; Hecker, A.; Rouhier, N.; Chauvat, F.; Cassier-Chauvat, C. Methylglyoxal Detoxification Revisited: Role of Glutathione Transferase in Model Cyanobacterium *Synechocystis* sp. Strain PCC 6803. *MBio* **2020**, *11*. [CrossRef] [PubMed]
215. Allocati, N.; Federici, L.; Masulli, M.; Di Ilio, C. Glutathione transferases in bacteria. *FEBS J.* **2009**, *276*, 58–75. [CrossRef] [PubMed]
216. Pandey, T.; Chhetri, G.; Chinta, R.; Kumar, B.; Singh, D.B.; Tripathi, T.; Singh, A.K. Functional classification and biochemical characterization of a novel rho class glutathione S-transferase in *Synechocystis* PCC 6803. *FEBS Open Bio* **2015**, *5*, 1–7. [CrossRef] [PubMed]
217. Rani, R.; Simarani, K.; Alias, Z. Functional Role of Beta Class Glutathione Transferases and Its Biotechnological Potential (Review). *Biochemistry* **2022**, *49*, S20–S29. [CrossRef]
218. Lederer, B.; Böger, P. A ligand function of glutathione S-transferase. *Z. fur Naturforsch. Sect. C J. Biosci.* **2005**, *60*, 166–171. [CrossRef]
219. Oakley, A. Glutathione transferases: A structural perspective. *Drug Metab. Rev.* **2011**, *43*, 138–151. [CrossRef]

220. Mocchetti, E.; Morette, L.; Mulliert, G.; Mathiot, S.; Guillot, B.; Dehez, F.; Chauvat, F.; Cassier-Chauvat, C.; Brochier-Armanet, C.; Didierjean, C.; et al. Biochemical and Structural Characterization of Chi-Class Glutathione Transferases: A Snapshot on the Glutathione Transferase Encoded by sll0067 Gene in the Cyanobacterium *Synechocystis* sp. Strain PCC 6803. *Biomolecules* **2022**, *12*, 1466. [CrossRef]
221. Bocedi, A.; Fabrini, R.; Farrotti, A.; Stella, L.; Ketterman, A.J.; Pedersen, J.Z.; Allocati, N.; Lau, P.C.K.; Grosse, S.; Eltis, L.D.; et al. The impact of nitric oxide toxicity on the evolution of the glutathione transferase superfamily: A proposal for an evolutionary driving force. *J. Biol. Chem.* **2013**, *288*, 24936–24947. [CrossRef]
222. Lallement, P.A.; Brouwer, B.; Keech, O.; Hecker, A.; Rouhier, N. The still mysterious roles of cysteine-containing glutathione transferases in plants. *Front. Pharmacol.* **2014**, *5*, 192. [CrossRef]
223. Pandey, T.; Singh, S.K.; Chhetri, G.; Tripathi, T.; Singh, A.K. Characterization of a highly pH stable Chi-class glutathione S-transferase from *Synechocystis* PCC 6803. *PLoS ONE* **2015**, *10*, e0126811. [CrossRef] [PubMed]
224. ShylajaNaciyar, M.; Karthick, L.; Prakasam, P.A.; Deviram, G.; Uma, L.; Prabaharan, D.; Saha, S.K. Diversity of glutathione S-transferases (GSTs) in cyanobacteria with reference to their structures, substrate recognition and catalytic functions. *Microorganisms* **2020**, *8*, 712. [CrossRef]
225. Pandey, T.; Shukla, R.; Shukla, H.; Sonkar, A.; Tripathi, T.; Singh, A.K. A combined biochemical and computational studies of the rho-class glutathione s-transferase sll1545 of *Synechocystis* PCC 6803. *Int. J. Biol. Macromol.* **2017**, *94*, 378–385. [CrossRef]
226. Stacpoole, P.W.; Henderson, G.N.; Yan, Z.; James, M.O. Clinical pharmacology and toxicology of dichloroacetate. *Environ. Health Perspect.* **1998**, *106*, 989–994. [CrossRef] [PubMed]
227. Wiktelius, E.; Stenberg, G. Novel class of glutathione transferases from cyanobacteria exhibit high catalytic activities towards naturally occurring isothiocyanates. *Biochem. J.* **2007**, *406*, 115–123. [CrossRef] [PubMed]
228. Kortheerakul, C.; Kageyama, H.; Waditee-Sirisattha, R. Molecular and functional insights into glutathione S-transferase genes associated with salt stress in *Halothece* sp. PCC7418. *Plant. Cell Environ.* **2021**, *44*, 3583–3596. [CrossRef] [PubMed]
229. Galhano, V.; Peixoto, F.; Gomes-Laranjo, J. Bentazon triggers the promotion of oxidative damage in the Portuguese ricefield cyanobacterium *Anabaena cylindrica*: Response of the antioxidant system. *Environ. Toxicol.* **2010**, *25*, 517–526. [CrossRef]
230. Thornalley, P.J. Protein and nucleotide damage by glyoxal and methylglyoxal in physiological systems-role in ageing and disease. *Drug Metab. Drug Interact.* **2008**, *23*, 125–150. [CrossRef]
231. Lee, C.; Park, C. Bacterial responses to glyoxal and methylglyoxal: Reactive electrophilic species. *Int. J. Mol. Sci.* **2017**, *18*, 169. [CrossRef]
232. Mostofa, M.G.; Ghosh, A.; Li, Z.G.; Siddiqui, M.N.; Fujita, M.; Tran, L.S.P. Methylglyoxal—A signaling molecule in plant abiotic stress responses. *Free Radic. Biol. Med.* **2018**, *22*, 96–109. [CrossRef]
233. Schalkwijk, C.G.; Stehouwer, C.D.A. Methylglyoxal, a highly reactive dicarbonyl compound, in diabetes, its vascular complications, and other age-related diseases. *Physiol. Rev.* **2020**, *100*, 407–461. [CrossRef]
234. de Bari, L.; Scirè, A.; Minnelli, C.; Cianfruglia, L.; Kalapos, M.P.; Armeni, T. Interplay among oxidative stress, methylglyoxal pathway and s-glutathionylation. *Antioxidants* **2021**, *10*, 19. [CrossRef] [PubMed]
235. Ramachandra Bhat, L.; Vedantham, S.; Krishnan, U.M.; Rayappan, J.B.B. Methylglyoxal—An emerging biomarker for diabetes mellitus diagnosis and its detection methods. *Biosens. Bioelectron.* **2019**, *133*, 107–124. [CrossRef]
236. Kalapos, M.P. Methylglyoxal in living organisms: Chemistry, biochemistry, toxicology and biological implications. *Toxicol. Lett.* **1999**, *110*, 145–175. [CrossRef] [PubMed]
237. He, Y.; Zhou, C.; Huang, M.; Tang, C.; Liu, X.; Yue, Y.; Diao, Q.; Zheng, Z.; Liu, D. Glyoxalase system: A systematic review of its biological activity, related-diseases, screening methods and small molecule regulators. *Biomed. Pharmacother.* **2020**, *131*, 110663. [CrossRef] [PubMed]
238. Rabbani, N.; Al-Motawa, M.; Thornalley, P.J. Protein glycation in plants—An under-researched field with much still to discover. *Int. J. Mol. Sci.* **2020**, *21*, 3942. [CrossRef] [PubMed]
239. Nemet, I.; Varga-Defterdarovic, L. Methylglyoxal-derived beta-carbolines formed from tryptophan and its derivates in the Maillard reaction. *Amino Acids* **2007**, *32*, 291–293. [CrossRef] [PubMed]
240. Eggen, M.D.; Glomb, M.A. Analysis of Glyoxal- and Methylglyoxal-Derived Advanced Glycation End Products during Grilling of Porcine Meat. *J. Agric. Food Chem.* **2021**, *69*, 15374–15383. [CrossRef]
241. Hopper, D.J.; Cooper, R.A. The regulation of *Escherichia coli* methylglyoxal synthase; A new control of glycolysis? *FEBS Lett.* **1971**, *13*, 213–216. [CrossRef]
242. Kaur, C.; Sharma, S.; Hasan, M.R.; Pareek, A.; Singla-Pareek, S.L.; Sopory, S.K. Characteristic variations and similarities in biochemical, molecular, and functional properties of glyoxalases across prokaryotes and eukaryotes. *Int. J. Mol. Sci.* **2017**, *18*, 250. [CrossRef]
243. Shimakawa, G.; Suzuki, M.; Yamamoto, E.; Nishi, A.; Saito, R.; Sakamoto, K.; Yamamoto, H.; Makino, A.; Miyake, C. Scavenging systems for reactive carbonyls in the cyanobacterium *Synechocystis* sp. PCC 6803. *Biosci. Biotechnol. Biochem.* **2013**, *77*, 2441–2448. [CrossRef] [PubMed]
244. Cassier-Chauvat, C.; Chauvat, F. Cyanobacteria: Wonderful Microorganisms for Basic and Applied Research. *eLS* **2018**, 1–11. [CrossRef]
245. Barseem, N.; Elsamalehy, M. Gene polymorphisms of glutathione S-transferase T1/M1 in egyptian children and adolescents with type 1 diabetes mellitus. *J. Clin. Res. Pediatr. Endocrinol.* **2017**, *9*, 138–143. [CrossRef] [PubMed]

246. Hirokawa, Y.; Goto, R.; Umetani, Y.; Hanai, T. Construction of a novel D-lactate producing pathway from dihydroxyacetone phosphate of the Calvin cycle in cyanobacterium, *Synechococcus elongatus* PCC 7942. *J. Biosci. Bioeng.* **2017**, *124*, 54–61. [CrossRef]
247. Van Loi, V.; Rossius, M.; Antelmann, H. Redox regulation by reversible protein S-thiolation in bacteria. *Front. Microbiol.* **2015**, *6*, 147. [CrossRef]
248. Muller-Schussele, S.J.; Bohle, F.; Rossi, J.; Trost, P.; Meyer, A.; Zaffagnini, M. Plasticity in plastid redox networks: Evolution of glutathione-dependent redox cascades and glutathionylation sites. *BMC Plant Biol.* **2021**, *21*, 322. [CrossRef]
249. Zaffagnini, M.; Marchand, C.H.; Malferrari, M.; Murail, S.; Bonacchi, S.; Genovese, D.; Montalti, M.; Venturoli, G.; Falini, G.; Baaden, M.; et al. Glutathionylation primes soluble glyceraldehyde-3-phosphate dehydrogenase for late collapse into insoluble aggregates. *Proc. Natl. Acad. Sci. USA* **2019**, *116*, 26057–26065. [CrossRef]
250. Ito, H.; Iwabuchi, M.; Ogawa, K. The Sugar-Metabolic Enzymes Aldolase and Triose-Phosphate Isomerase are Targets of Glutathionylation in *Arabidopsis thaliana*: Detection using Biotinylated Glutathione. *Plant Cell Physiol.* **2003**, *44*, 655–660. [CrossRef]
251. Zaffagnini, M.; Michelet, L.; Sciabolini, C.; Di Giacinto, N.; Morisse, S.; Marchand, C.H.; Trost, P.; Fermani, S.; Lemaire, S.D. High-resolution crystal structure and redox properties of chloroplastic triosephosphate isomerase from *Chlamydomonas reinhardtii*. *Mol. Plant* **2014**, *7*, 101–120. [CrossRef]
252. Ye, Z.W.; Zhang, J.; Ancrum, T.; Manevich, Y.; Townsend, D.M.; Tew, K.D. Glutathione S-Transferase P-Mediated Protein S-Glutathionylation of Resident Endoplasmic Reticulum Proteins Influences Sensitivity to Drug-Induced Unfolded Protein Response. *Antioxid. Redox Signal.* **2017**, *26*, 247–261. [CrossRef]
253. Mondal, S.; Kumar, V.; Singh, S.P. Phylogenetic distribution and structural analyses of cyanobacterial glutaredoxins (Grxs). *Comput. Biol. Chem.* **2020**, *84*, 107141. [CrossRef] [PubMed]
254. Holmgren, A. Hydrogen donor system for *Escherichia coli* ribonucleoside diphosphate reductase dependent upon glutathione. *Proc. Natl. Acad. Sci. USA* **1976**, *73*, 2275–2279. [CrossRef] [PubMed]
255. Laurent, T.C.; Moore, E.C.; Reichard, P. Enzymatic Synthesis of Deoxyribonucleotides. Iv. Isolation and and characterization of thioredoxin, the hydrogen donor from *Escherichia coli* B. *J. Biol. Chem.* **1964**, *239*, 3436–3444. [CrossRef] [PubMed]
256. Martin, J.L. Thioredoxin—A fold for all reasons. *Curr. Biol.* **1995**, *3*, 45–250. [CrossRef] [PubMed]
257. Fernandes, A.P.; Holmgren, A. Glutaredoxins: Glutathione-dependent redox enzymes with functions far beyond a simple thioredoxin backup system. *Antioxid. Redox Signal.* **2004**, *6*, 63–74. [CrossRef]
258. Sengupta, R.; Coppo, L.; Mishra, P.; Holmgren, A. Glutathione-glutaredoxin is an efficient electron donor system for mammalian p53R2-R1-dependent ribonucleotide reductase. *J. Biol. Chem.* **2019**, *294*, 12708–12716. [CrossRef]
259. Li, M.; Yang, Q.; Zhang, L.; Li, H.; Cui, Y.; Wu, Q. Identification of novel targets of cyanobacterial glutaredoxin. *Arch. Biochem. Biophys.* **2007**, *458*, 220–228. [CrossRef]
260. Sánchez-Riego, A.M.; López-Maury, L.; Florencio, F.J. Glutaredoxins are essential for stress adaptation in the cyanobacterium *Synechocystis* sp. PCC 6803. *Front. Plant Sci.* **2013**, *4*, 428. [CrossRef]
261. López-Maury, L.; Sánchez-Riego, A.M.; Reyes, J.C.; Florencio, F.J. The glutathione/glutaredoxin system is essential for arsenate reduction in *Synechocystis* sp. strain PCC 6803. *J. Bacteriol.* **2009**, *191*, 3534–3543. [CrossRef]
262. Kim, S.G.; Chung, J.S.; Sutton, R.B.; Lee, J.S.; López-Maury, L.; Lee, S.Y.; Florencio, F.J.; Lin, T.; Zabet-Moghaddam, M.; Wood, M.J.; et al. Redox, mutagenic and structural studies of the glutaredoxin/arsenate reductase couple from the cyanobacterium *Synechocystis* sp. PCC 6803. *Biochim. Biophys. Acta Proteins Proteom.* **2012**, *1824*, 392–403. [CrossRef]
263. Chardonnet, S.; Sakr, S.; Cassier-Chauvat, C.; Le Maréchal, P.; Chauvat, F.; Lemaire, S.D.; Decottignies, P. First proteomic study of S-glutathionylation in cyanobacteria. *J. Proteome Res.* **2015**, *14*, 59–71. [CrossRef] [PubMed]
264. Sakr, S.; Dutheil, J.; Saenkham, P.; Bottin, H.; Leplat, C.; Ortega-Ramos, M.; Aude, J.C.; Chapuis, V.; Guedeney, G.; Decottignies, P.; et al. The activity of the *Synechocystis* PCC6803 AbrB2 regulator of hydrogen production can be post-translationally controlled through glutathionylation. *Int. J. Hydrogen Energy* **2013**, *38*, 13547–13555. [CrossRef]
265. Zaffagnini, M.; Bedhomme, M.; Marchand, C.H.; Morisse, S.; Trost, P.; Lemaire, S.D. Redox regulation in photosynthetic organisms: Focus on glutathionylation. *Antioxid. Redox Signal.* **2012**, *16*, 567–586. [CrossRef] [PubMed]
266. Kumar, C.; Igbaria, A.; D'Autreaux, B.; Planson, A.-G.; Junot, C.; Godat, E.; Bacchawat, A.K.; Delaunay-Moisan, A.; Toledano, M.B. Glutathione revisited: A vital function in iron metabolism and ancillary role in thiol-redox control. *EMBO J.* **2011**, *30*, 2044–2056. [CrossRef] [PubMed]
267. Xu, L.; Liu, Y.; Chen, X.; Zhong, H.; Wang, Y. Ferroptosis in life: To be or not to be. *Biomed. Pharmacol.* **2023**, *159*, 114241. [CrossRef]
268. Aguilera, A.; Berdun, F.; Bartoli, C.; Steelheart, C.; Alegre, M.; Bayir, H.; Tyurina, Y.Y.; Kagan, V.E.; Salerno, G.; Pagnussat, G.; et al. C-ferroptosis is an iron-dependent form of regulated cell death in cyanobacteria. *J. Cell Biol.* **2022**, *221*, e201911005. [CrossRef]
269. Conrad, M.; Kagan, V.E.; Bayir, H.; Pagnussat, G.C.; Head, B.; Traber, M.G.; Stockwell, B.R. Regulation of lipid peroxidation and ferroptosis in diverse species. *Genes Dev.* **2018**, *32*, 602–619. [CrossRef]
270. Lobus, N.V.; Kulikovskly, M.S. The Co-Evolution Aspects of the Biogeochemical Role of Phytoplankton in Aquatic Ecosystems: A Review. *Biology* **2023**, *12*, 92. [CrossRef]
271. Saito, M.A.; Sigman, D.M.; Morel, F.M.M. The bioinorganic chemistry of the ancient ocean: The co-evolution of cyanobacterial metal requirements and biogeochemical cycles at the Archean–Proterozoic boundary? *Inorg. Chim. Acta* **2003**, *356*, 308–318. [CrossRef]

272. Qiu, G.-W.; Koedooder, C.; Qiu, B.-S.; Shaked, Y.; Keren, N. Iron transport in cyanobacteria—From molecules to communities. *Trends Microbiol.* **2022**, *30*, 229–240. [CrossRef]
273. Lill, R. Function and biogenesis of iron–sulphur proteins. *Nature* **2009**, *460*, 831–838. [CrossRef] [PubMed]
274. Przybyla-Toscano, J.; Roland, M.; Gaymard, F.; Couturier, J.; Rouhier, N. Roles and maturation of iron–sulfur proteins in plastids. *J. Biol. Inorg. Chem.* **2018**, *23*, 545–566. [CrossRef] [PubMed]
275. de Bont, L.; Donnay, N.; Couturier, J.; Rouhier, N. Redox regulation of enzymes involved in sulfate assimilation and in the synthesis of sulfur-containing amino acids and glutathione in plants. *Front. Plant Sci.* **2022**, *13*, 958490. [CrossRef] [PubMed]
276. Sipos, K.; Lange, H.; Fekete, Z.; Ullmann, P.; Lill, R.; Kispal, G. Maturation of cytosolic iron-sulfur proteins requires glutathione. *J. Biol. Chem.* **2002**, *277*, 26944–26949. [CrossRef]
277. Srour, N.; Gervason, S.; Hook, M.H.; Monfort, B.; Want, K.; Larkem, D.; Trabelsi, N.; Landrot, G.; Zitolo, A.; Fonda, E.; et al. Iron Insertion at the Assembly Site of the ISCU Scaffold Protein Is a Conserved Process Initiating Fe-S Cluster Biosynthesis. *J. Am. Chem. Soc.* **2022**, *144*, 17496–17515. [CrossRef] [PubMed]
278. Thompson, Z.; Fidai, I.; Wachnowsky, C.; Hendricks, A.L.; Cowan, J.A. Spectroscopic and functional characterization of the [2Fe-2S] scaffold protein Nfu from *Synechocystis* PCC6803. *Biochimie* **2022**, *192*, 51–62. [CrossRef]
279. Lillig, C.H.; Berndt, C.; Vergnolle, O.; Lohn, M.E.; Hudemann, C.; Bill, E.; Holgren, A. Characterization of human glutaredoxin 2 as iron-sulfur protein: A possible role as redox sensor. *Proc. Natl. Acad. Sci. USA* **2005**, *102*, 8168–8173. [CrossRef]
280. Feng, Y.; Zhong, N.; Rouhier, N.; Hase, T.; Kusunoki, M.; Jacquot, J.P.; Jin, C.; Xia, B. Structural insight into poplar glutaredoxin C1 with a bridging iron-sulfur cluster at the active site. *Biochemistry* **2006**, *45*, 7996–8008. [CrossRef]
281. Iwema, T.; Picciocchi, A.; Traore, D.A.K.; Ferrer, J.L.; Chauvat, F.; Jacquamet, L. Structural basis for delivery of the intact [Fe2S2] cluster by monothiol glutaredoxin. *Biochemistry* **2009**, *48*, 6041–6043. [CrossRef]
282. Nishio, K.; Nakai, M. Transfer of iron-sulfur cluster from NifU to apoferredoxin. *J. Biol. Chem.* **2000**, *275*, 22615–22618. [CrossRef]
283. Yeung, N.; Gold, B.; Liu, N.L.; Prathapam, R.; Sterling, H.J.; Willams, E.R.; Butland, G. The *E. coli* monothiol glutaredoxin GrxD forms homodimeric and heterodimeric FeS cluster containing complexes. *Biochemistry* **2011**, *50*, 8957–8969. [CrossRef] [PubMed]
284. Rouhier, N.; Couturier, J.; Johnson, M.K.; Jacquot, J.P. Glutaredoxins: Roles in iron homeostasis. *Trends Biochem. Sci.* **2010**, *35*, 43–52. [CrossRef] [PubMed]
285. Poor, C.B.; Wegner, S.V.; Li, H.; Dlouhy, A.C.; Schuermann, J.P.; Sanisvili, R.; Hinshaw, J.R.; Riggs-Gelasco, P.J.; Outten, C.E.; He, C. Molecular mechanism and structure of the *Saccharomyces cerevisiae* iron regulator Aft2. *Proc. Natl. Acad. Sci. USA* **2014**, *111*, 4043–4048. [CrossRef] [PubMed]
286. Bandyopadhyay, S.; Gama, F.; Molina-navarro, M.M.; Gualberto, J.M.; Claxton, R.; Naik, S.G.; Herrero, E.; Jacquot, J.P.; Johnson, M.K.; Rouhier, N. Chloroplast monothiol glutaredoxins as scaffold proteins for the assembly and delivery of [2Fe-2S] clusters. *EMBO J.* **2008**, *27*, 1122–1133. [CrossRef] [PubMed]
287. Muhlenhoff, U.; Molik, S.; Godoy, J.R.; Uzarska, M.A.; Richter, N.; Seubert, A.; Zhang, Y.; Stubbe, J.; Pierrel, F.; Herrero, E.; et al. Cytosolic Monothiol Glutaredoxins Function in Intracellular Iron Sensing and Trafficking via Their Bound Iron-Sulfur Cluster. *cell Metab.* **2010**, *12*, 373–385. [CrossRef]
288. Banci, L.; Brancaccio, D.; Ciof-Baconi, S.; Del Conte, R.; Gadapelli, R.; Mikolajczyk, M.; Neri, S.; Piccioloi, M.; Winkelman, J. [2Fe-2S] cluster transfer in iron-sulfur protein biogenesis. *Proc. Natl. Acad. Sci. USA* **2014**, *111*, 6203–6208. [CrossRef]
289. Moseler, A.; Aller, I.; Wagner, S.; Nietzel, T.; Przybyta-Toscano, J.; Muhlenhoff, U.; Lill, R.; Berndt, C.; Rouhier, N.; Schwarzlander, M.; et al. The mitochondrial monothiol glutaredoxin S15 is essential for iron-sulfur protein maturation in *Arabidopsis thaliana*. *Proc. Natl. Acad. Sci. USA* **2015**, *112*, 13735–13740. [CrossRef]
290. Banci, L.; Ciofi-Baffoni, S.; Gajda, K.; Muzzioli, R.; Peruzzini, R.; Winkelman, J. N-terminal domains mediate [2Fe-2S] cluster transfer from glutaredoxin-3 to anamorsin. *Nat. Chem. Biol.* **2015**, *11*, 772–778. [CrossRef]

antioxidants

MDPI

Article

Effects of Ultraviolet Radiation on Sediment Burial Parameters and Photo-Oxidative Response of the Intertidal Anemone *Anthopleura hermaphroditica*

Víctor M. Cubillos [1,2,*], Javier A. Álvarez [1,2], Eduardo Ramírez [1,2], Edgardo Cruces [3], Oscar R. Chaparro [1], Jaime Montory [4] and Carlos A. Spano [5]

1 Instituto de Ciencias Marinas y Limnológicas, Facultad de Ciencias, Universidad Austral de Chile, Valdivia 5090000, Chile
2 Laboratorio Costero de Recursos Acuáticos de Calfuco, Facultad de Ciencias, Universidad Austral de Chile, Valdivia 5110566, Chile
3 Centro de Investigaciones Costeras, Universidad de Atacama (CIC-UDA), Avenida Copayapu 485, Copiapó 1530000, Chile
4 Centro i~mar, Universidad de Los Lagos, Casilla 557, Puerto Montt 5480000, Chile
5 Departamento de Oceanografía Biológica, Ecotecnos S.A., Limache 3405, Viña del Mar 2520000, Chile
* Correspondence: victor.cubillos@uach.cl

Citation: Cubillos, V.M.; Álvarez, J.A.; Ramírez, E.; Cruces, E.; Chaparro, O.R.; Montory, J.; Spano, C.A. Effects of Ultraviolet Radiation on Sediment Burial Parameters and Photo-Oxidative Response of the Intertidal Anemone *Anthopleura hermaphroditica*. *Antioxidants* **2022**, *11*, 1725. https://doi.org/10.3390/antiox11091725

Academic Editors: Marcelo Hermes-Lima, Daniel Carneiro Moreira and Tania Zenteno-Savín

Received: 9 August 2022
Accepted: 22 August 2022
Published: 31 August 2022

Abstract: *Anthopleura hermaphroditica* is an intertidal anemone that lives semi-buried in soft sediments of estuaries and releases its brooded embryos directly to the benthos, being exposed to potentially detrimental ultraviolet radiation (UVR) levels. In this study, we investigated how experimental radiation (PAR: photosynthetically active radiation; UVA: ultraviolet A radiation; and UVB: ultraviolet B radiation) influences burrowing (time, depth and speed) in adults and juveniles when they were exposed to PAR (P, 400–700 nm), PAR + UVA (PA, 315–700 nm) and PAR + UVA + UVB (PAB, 280–700 nm) experimental treatments. The role of sediment as a physical shield was also assessed by exposing anemones to these radiation treatments with and without sediment, after which lipid peroxidation, protein carbonyls and total antioxidant capacity were quantified. Our results indicate that PAB can induce a faster burial response compared to those anemones exposed only to P. PAB increased oxidative damage, especially in juveniles where oxidative damage levels were several times higher than in adults. Sediment offers protection to adults against P, PA and PAB, as significant differences in their total antioxidant capacity were observed compared to those anemones without sediment. Conversely, the presence or absence of sediment did not influence total antioxidant capacity in juveniles, which may reflect that those anemones have sufficient antioxidant defenses to minimize photooxidative damage due to their reduced tolerance to experimental radiation. Burrowing behavior is a key survival skill for juveniles after they have been released after brooding.

Keywords: antioxidant metabolism; behavior; oxidative damage; sea anemone; sediment burrowing; UVB radiation

1. Introduction

Many infaunal organisms depend upon burrowing for survival, increasing their levels of food capture [1–3], generating physical protection against predators or minimizing physiological stress, especially in highly unstable environments [4–6]. Estuaries are considered highly stressful areas because tidal cycles generate pronounced environmental changes and high levels of physiological stress in individuals living there [7–9], for example, by exposing sessile or sedentary organisms to the detrimental action of ultraviolet-B radiation (UVB; 280–320 nm), especially when low tide coincides with the daily peak in environmental radiation [10]. The absorption of UVB by different chromophores (i.e., DNA, amino acids, melanin and metabolites) indirectly exacerbates the generation of reactive oxygen species

(ROS) [11–14] that are normally formed as byproducts of oxygen reduction in the cell. Consequently, ROS over-accumulation can induce structural and functional changes in lipids, proteins and DNA [15]. ROS attack polyunsaturated fatty acids; they are incorporated into lipids to form lipid peroxides [16], which can induce the generation of very reactive long-lived byproducts such as aldehydes (malondialdehydes, MDA; 4-hydroxynonenal, HNE and keto fatty acids) [17–19]. Consequently, their generation decreases membrane permeability, inducing electron leakiness in the cell [19]. Proteins with specific amino acids (e.g., proline, arginine, lysine and threonine) can be rapidly affected by ROS, inducing the formation of carbonyl groups [20]. Secondary reactions between specific amino acid side chains and byproducts of lipid peroxidation can induce the formation of carbonyl groups in proteins [20]. Subsequently, protein carbonylation reduces the catalytic activity of enzymes and induces further breakdown of proteins [21]. This results in a crosslinked reaction product, increasing the susceptibility of amino acids and proteins to degradation [18]. Consequently, oxidative damage can trigger physiological [13], anatomical [22], reproductive [23] and ecological [24,25] problems in marine invertebrates, which can cause death in severe cases [26]. However, aerobic organisms can synthesize enzymatic (e.g., superoxide dismutase (SOD), catalase (CAT), glutathione peroxidase (GPx), glutathione reductase (Gr)) and non-enzymatic (glutathione (GSH), oxidized glutathione (GSSG)) antioxidants that minimize cellular damage. Antioxidants can transfer electrons to these oxidizing agents (ROS) by neutralizing them and reducing chain reactions that can cause cell damage [27], reaching cellular homeostasis.

In southern Chile, the sea anemone *Anthopleura hermaphroditica* inhabits estuaries with high levels of ultraviolet B radiation (UVB, 280–315 nm) radiation during summer, which can exceed 50% of values recorded in areas of the Northern Hemisphere for the same latitude and season of the year [28]. This temperate anemone burrows in the sediment, spreading only its tentacles over the surface [29], allowing it to expose its symbiotic dinoflagellate (*Phylozoon anthopleurum*) [30] to environmental radiation. This species incubates its young inside the gastrovascular cavity (GVC) and releases them into the benthos once they reach the tentaculate juvenile stage [31]. Thus, juveniles may be exposed to significantly higher levels of UVB radiation than those experienced inside the incubation cavity, which can generate elevated oxidative damage to them, considering that the initial stages of development are highly affected by UVB radiation [22,32].

In natural conditions, this anemone is found buried in the sediment, projecting only the tentacular crown on the substrate; however, there is no evidence about the reason for this behavior. Although UVB radiation generates cytotoxic effects, we hypothesized that the sediment could provide physical protection against UVB radiation after burial in the sediment, minimizing photo-oxidative damage and the antioxidant response. We also postulated that juvenile anemones are much more susceptible to photo-oxidative damage than adults when exposed to different levels of experimental radiation. This is supported by the fact that under natural conditions, juveniles develop inside the GVC of adults buried in the sediment, always exposed to low radiation conditions. Thus, leaving the GVC and facing an environment with high levels of radiation could generate higher levels of oxidative damage and antioxidant response than in adult individuals. Considering burial in the sediment as a natural process in the life cycle of this cnidarian, it is likely that adult and juvenile anemones exposed to UVB bury themselves in the sediment at higher rates than those exposed without the presence of UVB, as an evasive response to cell damage.

Considering that anemones are conspicuous organisms that play an important role in different aquatic ecosystems supporting different organisms through symbiotic relationships [33–35] and being part of different trophic webs [36–38], it is important to understand the role of sediment in the photooxidative response of *A. hermaphroditica* as an important component of the Quempillén estuary.

In order to test the role of sediment in the photooxidative response of anemones and the elevated susceptibility of juveniles to experimental radiation, adults and juveniles of *A. hermaphroditica* were exposed to a combination of UVB, UVA and PAR radiation with and without sediment and oxidative damage (lipids and proteins), and total antioxidant capacity was quantified. In order to test the effects of experimental radiation on the behavioral response of anemones, juveniles and adults were placed over the substrate under a combination of UVB, UVA and PAR; time, depth and speed were assessed using image analysis.

2. Materials and Methods

2.1. Animal Collection and Maintenance

Approximately 200 individuals of *A. hermaphroditica* and their sediments (60% gravel and 40% sand) were obtained from the Quempillén estuary (41°52′17.36″ S, 73°46′1.02″ W, Figure 1) during the austral summer season, placed in plastic boxes (80 × 45 × 30 cm) and then transferred to the Laboratorio Costero de Recursos Acuáticos de Calfuco (39°46′50″ S, 73°23′34″ W) at the Universidad Austral de Chile (Valdivia, Chile).

Figure 1. Adult of the estuarine anemone *Anthopleura hermaphroditica* (**A**) collected from Quempillén estuary (41°520 S; 73°460 W) of Chiloé Island (**B**), Southern Chile (**C**). Arrow shows location of Chiloé Island in Chile. Map generated using Ocean Data View software (version 5.6.2, Schlitzer, Reiner, Ocean Data View, odv.awi.de, 2021, Germany).

Anemones were acclimated in an aquarium with sediment (collected from the sampling site) where circulating seawater (35 ppt, 13 °C) was provided, with constant aeration for an acclimation period of 3 days and a photoperiod of 12:12 h with photosynthetically available radiation (PAR; 400–700 nm) at an irradiance of 1000 μmol m^{-2} s^{-1}. Prior to the experiments, anemones were separated into two size ranges as follows: juveniles (1–2 cycles of tentacles; <2 mm pedal disc diameter) and adults (>3 cycles of tentacle; >4 mm pedal disc diameter).

2.2. Experimental Treatments

In order to determine the effect of radiation on the substrate burial response, adults and juveniles of *A. hermaphroditica* were exposed to the following experimental radiation conditions: PAR (400–700 nm), PAR + UVA (315–700 nm) and PAR + UVA + UVB (280–700 nm). Three rectangular transparent glass aquaria 30 cm long, 10 cm high and 5 cm wide were set up for the experiment, with a seawater circulation system activated by a submersible pump (220 L h^{-1}) (Figure 2).

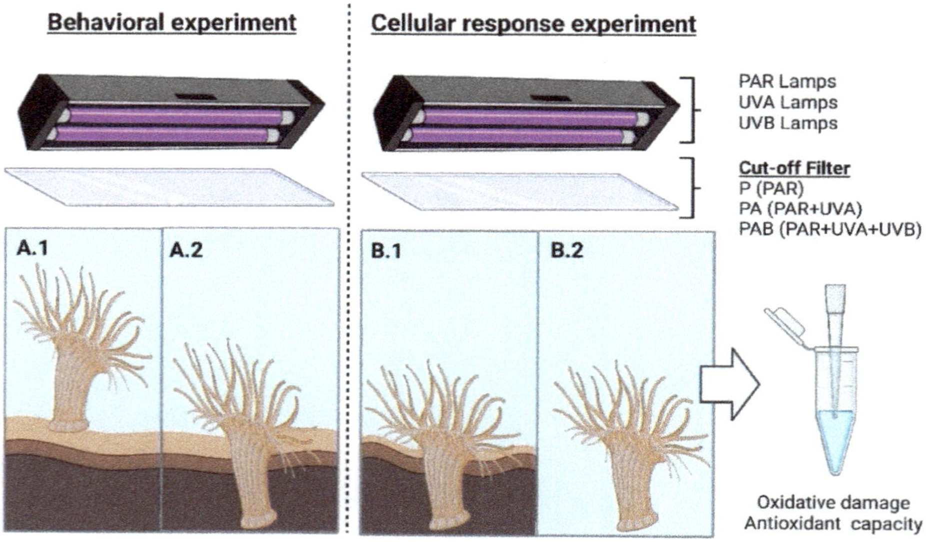

Figure 2. Experimental setup for behavioral ((**A.1**) initial time/(**A.2**) final time) and cellular responses ((**B.1**) with sediment/(**B.2**) without sediment) in adults and juveniles of the sea anemone *A. hermaphroditica* exposed to different radiation conditions generated by a combination of fluorescent lamps (PAR, UVA and UVB) and cut-off filters (P: PAR, PA: PAR + UVA and PAB: PAR + UVA + UVB). Figure created using BioRender.com (accessed on 9 August 2022).

The sediment collected in the field was deposited in each aquarium to a thickness of 4 cm. Anemones were deposited individually on the sediment of the aquaria, which were immediately covered with Clear 226-UV (Chris James Lighting Filters, London, UK), Folanorm SF-AS (0.13 mm; Lüerssen Grafische Vertriebs GmbH, Germany) and cellulose acetate filters (CDA; 0.13 mm; Grafix®, Maple Heights, OH, USA) to generate the radiation treatments P (400–700 nm), PA (315–700 nm) and PAB (280–700 nm), respectively. PAR fluorescent tubes (Philips Day Light, Philips, Eindhoven, The Netherlands), UVA (Philips TL40 W, Philips, Eindhoven, The Netherlands) and UVB (Phillips TL20, Philips, Eindhoven, The Netherlands) were placed 30 cm above the experimental aquaria to simulate the natural irradiance conditions registered in the environment [10].

Anemones were thus exposed to an irradiance of 1000 μmol m^{-2} s^{-1}, 20 W m^{-2} and 2.3 W m^{-2} of PAR, UVA and UVB, respectively, for 3 h. The irradiance of the lamps was quantified using a portable radiometer (PMA-2100, Solar Light, Glenside, PA, USA) fitted with a PAR sensor (PMA-2132 model), UVA sensor (PMA-2110) and UVB sensor (PMA-2106). The experiment was replicated 11 times for each developmental stage. New individuals were used for each experiment.

2.3. Behavioral Responses

In order to quantify sediment burial responses (time, depth and speed) under different experimental radiation treatments, adults and juveniles of *A. hermaphroditica* were recorded using a set of four digital cameras connected to a digital video recorder. The maximum recording time was three hours. The burial parameters were defined as follows:

2.3.1. Burial Time

Burial time was defined as the time elapsed from when the adult and juvenile anemones were placed on the surface of the sediment until the individuals were buried in the sediment in response to the respective radiation treatments.

2.3.2. Burial Depth

Burial depth was measured as the distance between the sediment surface and the distal end of the pedal disc of polyps. Analysis of the images recorded during the burial process allowed calculating the depth to which the body column of adults and juveniles of *A. hermaphroditica* were buried in the sediment. Burial depth was measured using as reference a graduated ruler (mm) previously attached to each experimental aquarium.

2.3.3. Burial Speed

Burial speed was determined in adults and juveniles of *A. hermaphroditica* exposed to different radiation treatments (P, PA and PAB) from the ratio of depth to burial time. The speed data were expressed in mm h^{-1}.

2.4. Cellular Responses

The oxidative damage and the antioxidant capacity of adults and juveniles of *A. hermaphroditica* were quantified using the same aquaria and experimental radiation conditions indicated previously with and without sediment (Figure 2(B.1,B.2)). Anemones were deposited individually in each aquarium and then exposed to P, PA and PAB radiation treatments for 3 h. The experiment was replicated four times for adults and five times for juveniles. After the three experimental hours, individuals from both conditions (with and without sediment) were immediately placed in liquid nitrogen. The tissue was ground using a pestle and mortar to which liquid nitrogen was added. The fine powder obtained during grinding was kept in Eppendorf tubes at −80 °C until oxidative damage (lipid peroxidation and protein carbonyl formation) and total antioxidant capacity were quantified.

2.4.1. Oxidative Damage Assays

Levels of lipid oxidation were determined by malondialdehyde (MDA) quantification using the colorimetric method of thiobarbituric acid reactive substance (TBARS), according to the methodology described by Salama, A. and Pearce, R. [39]. Thirty g (fresh weight, FW) of ground tissue was homogenized with 350 µL of trichloroacetic acid (TCA, 0.1%, Merck KGaA, Darmstadt, Germany) in an Eppendorf tube and centrifuged for 10 min at 13,000 RPM (4 °C) (Labnet, Prism R, Edison, NJ, USA). A volume of 200 µL was extracted from the supernatant and homogenized with 500 µL of TCA (20%) + TBA (thiobarbituric acid, 0.5%) solution in a new Eppendorf tube and then incubated in a ThermoMixer (Eppendorf, Hamburg, Reinbek, Germany) at 80 °C for 30 min. Next, the tubes were incubated on ice for five minutes and centrifuged again for five minutes at 13,000 RPM (4 °C). Finally, malondialdehyde (MDA) levels were quantified by measuring the absorbance at 532 nm of 200 µL of the supernatant in a plate reader (Anthos, Biochrom Ltd., Cambridge, UK).

The concentration of MDA was determined using the molar extinction coefficient, and the results were expressed in nm MDA g^{-1} FW.

Protein carbonyl levels were estimated by the method of dinitrophenylhydrazine (DNPH) used by Levine [40]. An amount of 80 mg of previously ground tissue was homogenized with 2 mL of a 50 mM phosphate buffer solution (0.1 mM EDTA and 1% PVPP polyvinyl polypyroline) at pH 7.0 and centrifuged for 20 min at 13,500 rounds per minute (RPM) (4 °C). Subsequently, 200 μL of the supernatant was homogenized with 200 μL 20% TCA, incubated for 30 min (−20 °C) and centrifuged at 13,500 RPM (4 °C). The resulting supernatant was removed to reserve the pellet. Then 300 μL DNPH solution (100 mM in 2 N HCl) was added to the pellet and incubated in the dark for 1 h. Next, the mixture was precipitated with 500 μL TCA (20%) and incubated again for 15 min at −20 °C. The resulting mixture was centrifuged for 10 min at 13,200 RPM; then, the supernatant was removed carefully to reserve the pellet. The resulting pellet was re-suspended with 500 μL ethanol–ethyl-acetate (1:1) solution and centrifuged for 10 min at 13,200 RPM (4 °C), removing the supernatant again. Next, 2 mL guanidine HCL (6 M) in 20 mM sodium phosphate (pH 6.8) was added to each sample and then centrifuged for 30 min at 9000 RPM (4 °C). Finally, protein carbonyl levels were determined at 380 nm using a Zenyth 100 plate reader (Anthos, Biochrom Ltd., Cambridge, UK), and their concentrations were determined using the molar extinction coefficient.

2.4.2. Total Antioxidant Capacity

The method described by Fukumoto, L. and Mazza, G. [41] was applied for the determination of total antioxidant capacity, which uses 2,2 diphenyl-1-picrylhydrazyl (DPPH). For this quantification, 60 mg of tissue was homogenized in 1 mL of a 70% acetone solution, then the tubes with their content were incubated in an ultrasonic bath for 2 h. Then they were centrifuged for 5 min at 9000 RPM at 4 °C and kept under a fume hood for 90 min, allowing the acetone to evaporate. Then, 200 μL DPPH was added to each tube. Finally, the absorbance was quantified every 10 min for 120 min using a Zenyth 100 plate reader (Anthos, Biochrom Ltd., Cambridge, UK) at 520 nm. That information generated a standard curve of 6-hydroxy-2,5,7,8-tetramethylchromane-2-carboxylic acid (Trolox) for calculating the total antioxidant capacity, expressed as mg Trolox Eq FW g^{-1} of the sample.

2.5. Statistical Analysis

The assumptions of normality and homogeneity of variances of the burial parameters (time, depth and speed), as well as the levels of oxidative damage and total antioxidant capacity for juveniles and adults, were confirmed by Levene's test and the Shapiro–Wilk test, respectively. When necessary, data transformation (Log_{10}) was used to fulfill the assumptions of normality and homogeneity of variances. By using a two-way ANOVA, it was determined whether the development status and/or radiation treatment (P, PA and PAB) influenced burial behavior (i.e., time, depth and speed). Additionally, a three-way ANOVA was used to determine if the levels of oxidative damage (protein carbonyl and lipid peroxidation) in *A. hermaphroditica* were influenced by the developmental state (juveniles and adults), experimental radiation (P, PA and PAB) and level of protection provided by the sediment (presence and absence). Finally, a Tukey *a posteriori* test was used to identify the significant differences. Statistical analyses were carried out with SigmaPlot (Sigma Plot for Windows version 11, Systat Software Inc., Chicago, IL, USA). The significance level was defined at $p < 0.05$.

3. Results

3.1. Behavioral Responses

3.1.1. Burial Time

The two-way ANOVA indicated that radiation treatment and development stage, as well as the interaction of these factors, significantly influenced the burial time of anemones in the sediment (Figure 3A, Table 1). Significant differences in burial time were observed between the developmental stages; juvenile anemones had a shorter burial time than adults (Figure 3A, Table 1). Additionally, experimental radiation treatments that included UV radiation (PA and PAB) had a significant effect in reducing the burial time of adults but not of juveniles (Figure 3A, Table 1).

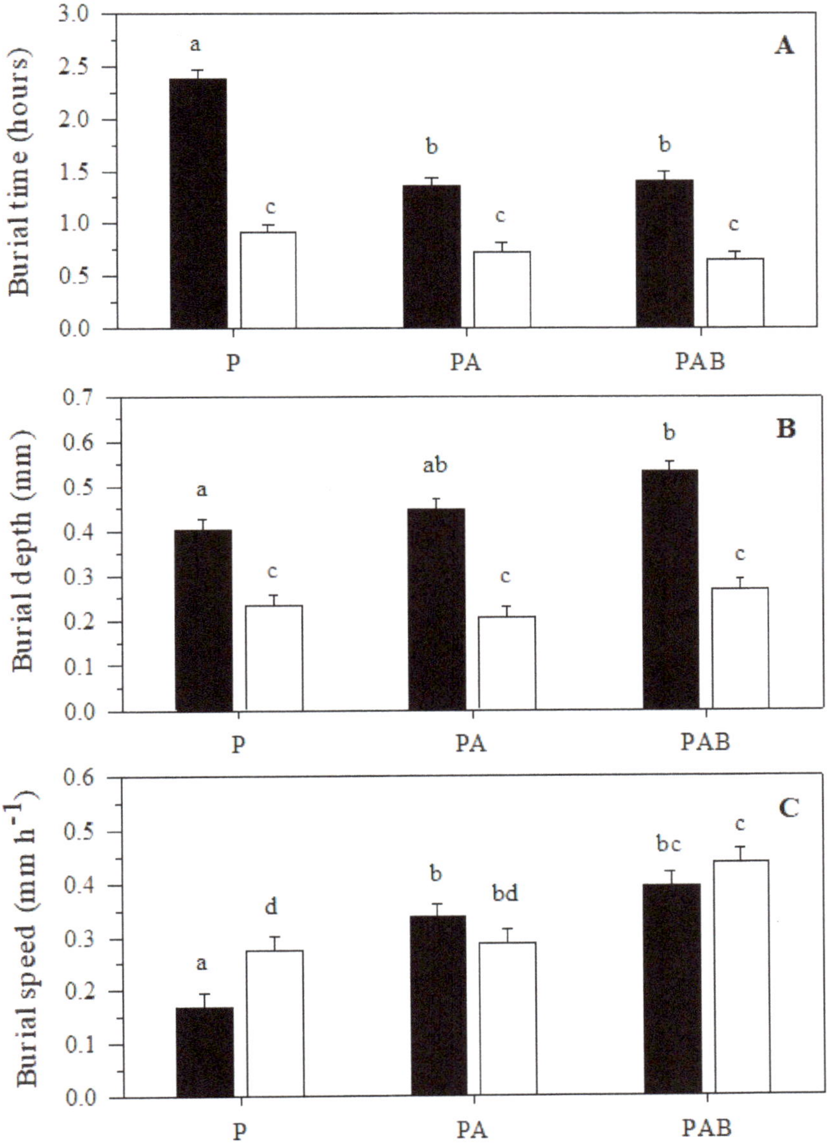

Figure 3. Burial time (**A**), burial depth (**B**) and burial speed (**C**) of adults (black bars) and juveniles (white bars) of the sea anemone *A. hermaphroditica* in the sediment during exposition to P (PAR), PA (PAR + UVA) and PAB (PAR + UVA + UVB) radiation treatment. Bars show mean ± SE. Different letters over bars indicate significant differences ($p < 0.05$).

Table 1. Two-way ANOVA for comparison of burial parameters (time, depth and speed) in different developmental stages (adults and juveniles) of *A. hermaphroditica* exposed to different levels of experimental radiation P (PAR), PA (PAR + UVA) and PAB (PAR + UVA + UVB). Bold numbers indicate significant difference ($p < 0.05$).

Source of Variation	df	ms	f	p
Burial time				
Developmental stage (DS)	1	15.130	212.389	**<0.001**
Radiation treatment (RT)	2	2.783	39.066	**<0.001**
DS × RT	2	1.119	15.71	**<0.001**
Error	60	0.0712		
Burial depth				
Developmental stage (DS)	1	1.370	171.89	**<0.001**
Radiation treatment (RT)	2	0.0663	8.311	**0.001**
DS × RT	2	0.0152	1.913	0.157
Error	60	0.00797		
Burial speed				
Developmental stage (DS)	1	0.0190	2.514	0.118
Radiation treatment (RT)	2	0.206	27.244	**<0.001**
DS × RT	2	0.0338	4.481	**0.015**
Error	60	0.00755		

3.1.2. Burial Depth

The two-way ANOVA indicates that both developmental stages and radiation treatments significantly influenced the depth to which *A. hermaphroditica* was buried in the sediment (Figure 3B, Table 1). Particularly, the PAB radiation treatment caused adult anemones to bury significantly deeper (23%) than those exposed only to P and PA radiation treatments (Figure 3B, Table 1). Although juvenile anemones burrow to shallower depths than adult anemones, there were no significant differences in the depths that they burrowed when they were experimentally exposed to the P, PA and PAB radiation bands (Figure 3B, Table 1).

3.1.3. Burial Speed

Our results indicate that radiation treatment significantly influenced the burrowing speed of *A. hermaphroditica* in the sediment (Figure 3C, Table 1). When adults and juveniles of *A. hermaphroditica* were exposed to the PAB radiation treatment, they increased their mean burial rate by 132% and 36%, respectively, compared to those exposed only to the P-radiation treatment (Figure 3C, Table 1). Radiation treatment PA and PAB showed significant differences in the burial speed of adult sea anemones compared to P (Figure 3C, Table 1). PAB radiation treatment of juveniles induced a significant increase in velocity compared to PA and P. Juvenile sea anemones showed higher burial speed than adults only under P radiation treatment.

3.2. Cellular Responses

3.2.1. Lipid Peroxidation

The three-way ANOVA indicated that type of radiation treatment ($F_{(2,29)} = 25.97$, $p < 0.001$, Table 2), stage of development ($F_{(1,29)} = 158.21$, $p < 0.001$, Table 2) and the sediment condition ($F_{(1,29)} = 12.87$, $p = 0.001$, Table 2) all influenced lipid peroxidation levels in the sea anemone *A. hermaphroditica*. Tukey's *a posteriori* test showed that the presence of UVB through PAB radiation treatment exerted a significant role in the generation of oxidative lipid damage, as opposed to what happens to anemones exposed to P and PA. Juvenile anemones are more susceptible to lipid peroxidation than adult anemones. The presence of sediment in the radiation exposure treatments significantly reduced the level of lipid photo-oxidation in experimental anemones (Table 2).

Table 2. Three-way ANOVA for the estimation of oxidative response (protein carbonyl and lipid peroxidation) and antioxidant capacity in adults and juveniles of the anemone *A. hermaphroditica* exposed to different levels of experimental radiation P (PAR), PA (PAR + UVA) and PAB (PAR + UVA + UVB) with and without sediment. Bold numbers indicate significant difference ($p < 0.05$).

Source of Variation	df	ms	f	p
Lipid peroxidation				
Developmental stage (DS)	1	431.96	158.216	**<0.001**
Sediment condition (SC)	1	35.15	12.875	**0.001**
Radiation treatment (RT)	2	70.90	25.971	**<0.001**
DS × SC	1	0.070	0.0258	0.873
DS × RT	2	30.36	11.121	**<0.001**
SC × RT	2	6.093	2.232	0.125
DS × SC × RT	2	2.257	0.925	0.408
Residual	29	2.730		
Protein carbonyl				
Developmental stage (DS)	1	0.0000765	351.271	**<0.001**
Sediment condition (SC)	1	0.00000193	8.848	**0.006**
Radiation treatment (RT)	2	0.00000260	11.934	**<0.001**
DS × SC	1	0.00000062	3.043	0.091
DS × RT	2	0.00000186	8.564	**<0.001**
SC × RT	2	0.0000000305	0.140	0.870
DS × SC × RT	2	0.000000152	0.700	0.504
Residual	32	0.000000218		
Total antioxidant capacity				
Developmental stage (DS)	1	102.022	4.221	**<0.048**
Sediment condition (SC)	1	463.94	19.191	**<0.001**
Radiation treatment (RT)	2	137.22	5.677	**0.007**
DS × SC	1	674.15	27.891	**<0.001**
DS × RT	2	69.96	2.895	0.069
SC × RT	2	11.307	0.468	0.630
DS × SC × RT	2	0209	0.00864	0.991
Residual	34	25.634		

Only the interaction between the developmental stage and experimental radiation treatment significantly influenced the generation of lipid oxidation ($F_{(2,29)} = 11.12$, $p < 0.001$, Table 2); PAB generated a significant increase in the level of oxidative lipid damage in adult anemones. P, PA and PAB radiation significantly influenced lipid oxidation in juvenile anemones (Figure 4A,B; Table 2). Juveniles exposed with sediment to P, PA and PAB radiation increased their lipid peroxidation levels by four, five and four times, respectively, compared to adult anemones exposed to the same conditions (Figure 4A,B; Table 2). Similarly, juveniles exposed without sediment to P, PA and PAB increased their lipid peroxidation levels by two, three and three times, respectively, compared to adult anemones exposed to the same conditions (Figure 4A,B; Table 2).

3.2.2. Protein Carbonyl

The three-way ANOVA indicated that the generation of protein carbonyl levels in *A. hermaphroditica* in this study were significantly influenced by the radiation type ($F_{(2,32)} = 11.93$, $p < 0.001$, Table 2), developmental stage ($F_{(1,32)} = 351.27$, $p < 0.001$, Table 2) and sediment condition ($F_{(1,32)} = 8.84$, $p = 0.006$, Table 2). PAB generated a significantly higher level of protein carbonyl than those observed under the PA and P radiation treatments (Figure 4C,D; Table 2). Juvenile anemones were significantly more susceptible to protein oxidation than adult anemones, independent of the radiation treatment (Figure 4C,D; Table 2).

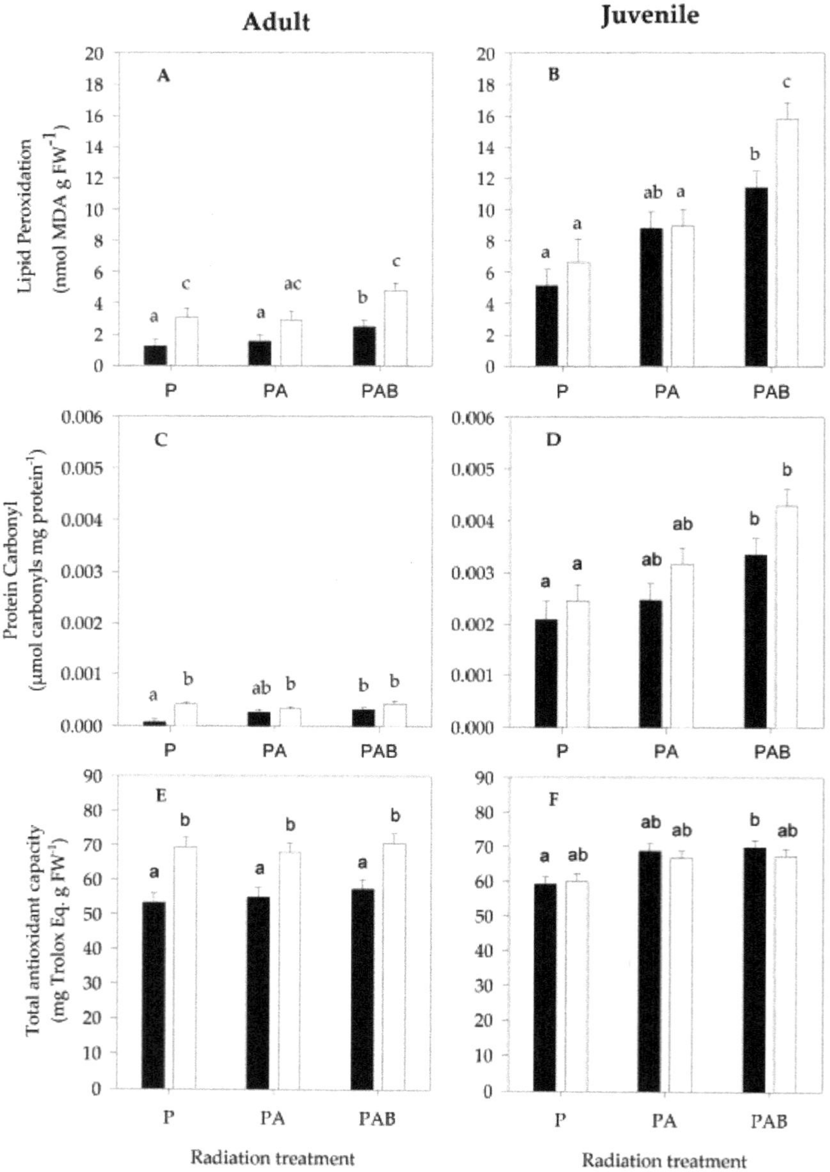

Figure 4. Protein carbonyl (**A**,**B**), lipid peroxidation (**C**,**D**) and total antioxidant capacity (**E**,**F**) in adults and juveniles of the anemone *A. hermaphroditica* exposed to P (PAR), PA (PAR + UVA) and PAB (PAR + UVA + UVB) radiation treatments with sediment (black bars) and without sediment (white bars). Error bars indicate ±SE. Different letters over bars indicate significant differences ($p < 0.05$).

Particularly, the presence or absence of sediment had a significant effect on the level of protein oxidation in adults and juveniles of *A. hermaphroditica* exposed to experimental radiation. The presence of sediment in the experimental aquaria protected anemones against protein oxidation, especially when they were exposed to PAB radiation treatment. The absence of sediment in the experimental aquaria produced significant differences in the level of protein carbonyls in adults and juveniles. Juveniles exposed to experimental radiation without sediment significantly increased the level of oxidative damage to proteins more than adults.

The interaction of the developmental stage and radiation treatment generated a significant impact on protein oxidation levels (Figure 4C,D; Table 2). Significant differences in carbonyl levels were observed when juveniles and adults were exposed to either PA or PAB. Adults, but not juveniles, showed significant differences in protein carbonyl levels when anemones with and without sediment were exposed to P radiation. A significant difference in protein carbonyl level of juvenile anemones was observed between P and PAB radiation treatment. Juveniles in sediment directly exposed to P, PA and PAB generated carbonyl levels 28, 9 and 10 times greater, respectively, than adults exposed under the same conditions (Figure 4C,D; Table 2). When juveniles without sediment were directly exposed to P, PA and PAB, they generated carbonyl levels 6, 10 and 10 times higher, respectively, than adults exposed to the same experimental radiations (Figure 4C,D; Table 2).

3.2.3. Total Antioxidant Capacity

Three-way ANOVA indicated that the experimental radiation type ($F_{(2,34)} = 5.67$, $p = 0.007$, Table 2), developmental stage ($F_{(1,34)} = 4.22$, $p = 0.048$, Table 2) and sediment condition ($F_{(1,34)} = 19.19$, $p < 0.001$, Table 2) significantly influenced total antioxidant capacity in *A. hermaphroditica* (Figure 4E,F). Juveniles exhibited greater total antioxidant capacity than adults. PAB radiation treatment significantly increased the total antioxidant capacity response in *A. hermaphroditica* compared to individuals exposed to P. Adults exposed to experimental PAB radiation treatment generated significant differences in total antioxidant capacity only when compared to P-exposed anemones. Juvenile sea anemones showed higher antioxidant capacity levels when they were exposed to PA and PAB compared to P-exposed anemones.

The presence or absence of sediment significantly influenced the total antioxidant response in anemones. A significant difference in total antioxidant capacity was observed between adults and juveniles exposed to experimental radiation with sediment. The interaction of developmental stage and sediment condition significantly influenced the total antioxidant capacity response in *A. hermaphroditica* ($F_{(1,34)} = 27.89$, $p < 0.001$, Table 2).

4. Discussion

4.1. Burrowing Behavior against Experimental Radiation

UVB radiation is an abiotic factor known to control the distribution and abundance of different marine organisms in a bathymetric profile due to its negative effects on cells, with physiological and behavioral consequences [42]. Evasive responses to UVR include habitat selection processes [43,44], vertical/horizontal migration in the water column [45], burial in the sediment [46] and body covering with physical structures that minimize damage induced by direct or indirect exposure [47,48]. UVR evasive responses described previously in sea anemones include covering the body column with either gravel or debris and contraction of its body column [49].

In the Quempillén estuary, the sea anemone *A. hermaphroditica* inhabits the sediment, semi-buried, protecting its complete body column from direct UVR radiation, projecting only its tentacles over the surface. Our results demonstrate for the first-time burial in the sediment of the intertidal anemone *A. hermaphroditica* as an evasive response to UVB radiation. Similar avoidance responses were found in the mantis shrimp *Haptosquilla trispinosa*, which possesses negative phototaxis when exposed to UVB and UVA radiation, sheltering in its burrow constructed in the sediment [46]. Exposure of *A. hermaphroditica* adults to PAR + UVA and PAR + UVA + UVB results in a reduction in burial time. Our results indicate that juvenile anemones have increased susceptibility to radiation; thus, they keep their burial times almost constant under different combinations of PAR, PAR + UVA and PAR + UVA + UVB experimental radiation. This indicates that there is a need for juveniles to bury as quickly as possible to reduce radiation exposure times and thus minimize photo-oxidative damage. Previous observations in the intertidal anemone *Actinia tenebrosa* indicate that exposure times are determinant in the levels of DNA damage through the generation of CPDs, which can increase by approximately 200% and 300% in anemones exposed to PAR + UVA and PAR + UVA + UVB compared to those exposed only to PAR (control) for a period of 96 h [50]. Thus, adults and juveniles of *A. hermaphroditica* should increase their burial speeds, minimizing exposure time, especially when exposed to UVB radiation. Rapid evasion to UVB exposure was identified in early developmental stages (zoea I and zoea II) of the Patagonian crustacean *Cyrtograpsus altimanus*, which increased evasion speed by 50% compared to those exposed only to PAR radiation [51].

The ability of anemones to bury themselves in sediment appears to be inversely related to the development of body outgrowths. *A. hermaphroditica* tends to adhere to gravel or pebbles, preventing overexposure to the environment (i.e., avoiding desiccation; [52]). This adherence process is much more common in the rocky intertidal zone; therefore, the possibility of burrowing seems to be an adaptive response to the environment where the sediments are too fine to be attached to the body column. In adults, burrowing has not only been shown to minimize photo-oxidation during incubation, but also it seems to be a more successful strategy than sticking stones to the body wall, at least when comparing population densities reached in hard and soft-bottom habitats [29,31].

4.2. Cellular Response in A. hermaphroditica

The levels of lipid peroxidation and protein carbonyls observed in this study are lower than those previously reported for *A. hermaphroditica* in the Quempillén river estuary [10]. Our study demonstrates that exposure to PAR + UVA and PAR + UVA + UVB radiation generated a significant increase in the levels of oxidative damage to lipids and proteins in *A. hermaphroditica*. Similar results were observed in the intertidal anemone *Actinia tenebrosa*, where exposure to P, PA and PAB conditions generated lipid peroxidation levels of 3.8, 6.0 and 10 nmol mg FW^{-1}, respectively [50]. In the same species, the levels of protein carbonyls were approximately 1.8, 5.0 and 6.0 for anemones exposed to P, PA and PAB, respectively, after 96 h of exposure. It was well established that the presence of photosensitizing molecules generates greater absorption of UVB radiation in the cell, producing photo-oxidation of different macromolecules [53]. In this study, higher levels of photo-oxidative damage were found in juvenile anemones than in adult anemones exposed to P, PA and PAB treatments, confirming their high susceptibility to photo-oxidation. The above is consistent with previous studies, showing high susceptibility of the initial developmental stages of aquatic organisms to UVB, which includes pluteus larvae of urchins [54], zoea I larvae of crustaceans [55–57], fish larvae [58,59] and amphibian larvae [60]. Meta-analysis studies support that early stages of development are much more prone to damage than adult organisms [61–63], mainly due to their limited energy reserve [64].

Prolonged exposure of larvae of the Antarctic urchin *Sterechinus neumayerii* to PAB radiation generates increases in the levels of lipid peroxidation and protein carbonyls by approximately 3-fold compared to larvae exposed to P radiation [65]. However, when larvae of this echinoderm were cultured under the Antarctic ice cover, the levels of abnormalities in PAB were like those generated under P radiation. Thus, the Antarctic ice cover may act as a physical barrier, minimizing the levels of lipid peroxidation and protein carbonyls in the early stages of development. The above indicates the importance for some marine invertebrates, which lack external physical protection such as shells or body plates, to seek shelter that minimizes UVB-induced photo-oxidative damage. In the present study, the sediment constituted a refuge and played an important role in the physical protection of both adults and juveniles of *A. hermaphroditica* against radiation, minimizing photo-oxidative damage.

A previous study in infaunal organisms (the amphipod *Chaetocorophium lucasi* and juvenile stages of the bivalves *Macomona liliana* and *Austrovenus stutchburyi*) showed that exposure to UVR in the absence of sediment generates between 5 and 10 times more phototoxicity (photoactivation of polycyclic aromatic hydrocarbons (PAH)) than in control organisms (without UVR). This indicates the high susceptibility of this type of organism to UVR radiation, and it is the sediment that acts as a physical barrier to radiation, which increases the survival of this type of animal by more than 50% [66].

In our study, the increased susceptibility to photo-oxidation in juveniles of *A. hermaphroditica* may be associated with acclimation to low levels of environmental radiation offered by the GVC during embryonic development, coupled with the presence of zooxanthellae that are acquired during the hatching process [67]. Internal incubation is part of the embryonic development strategy, which implies that the planulae larvae develop into tentaculate juveniles inside the GVC of an adult (Cubillos pers. obs.). This process is conceived as a strategy that allows embryonic care under high levels of environmental stress [68], which is consistent with the protection identified in this study when juveniles were confronted with UVR exposure.

A previous study carried out with the symbiont anemone *Anthopleura elegantissima* showed that the presence of zooxanthellae generates a negative phototactic response of the host anemone when exposed to high levels of radiation [69], a response that coincides with that of the symbiont anemone *A. hermaphroditica*. In photoautotrophic organisms, the photosynthetic apparatus represents the most delicate and fragile cellular structure that can be affected by UVB and UVA [70]. For example, chlorophyll levels in zooxanthellae of the symbiotic anemone *Aiptasia pallida* were significantly reduced due to photo-oxidation because of increased UVB levels [71], resulting in high levels of oxidative damage in the dinoflagellate *Symbiodinium bermudense*, reducing its photosynthetic capacity and impacting RUBISCO activity [72]. Previous studies indicated that overexposure of primary producers to UVB generates structural damage to the D1 protein of PSII, which is the molecule responsible for maintaining the photoprotection process in primary producers [73–75]. Although the spectral intensity of UVA is 10 times higher than that of UVB, it is also responsible for inducing photo-oxidative damage to PSII and PSI in photosynthetic organisms [76,77].

Therefore, when juvenile anemones of *A. hermaphroditica* leave the GVC where they are brooded, they must abruptly face a new environment with high levels of UVB and UVA, which could induce the high levels of photo-oxidative damage observed in this study. A similar situation was identified in the anemone *Actinia tenebrosa*, where the rapid transition from an environment with low levels of UVB to one with increased levels triggered the production of higher amounts of lipid peroxidation and protein carbonyl, increasing the basal levels by five and six times, respectively [50]. *A. hermaphroditica* adults, but mainly juveniles, increase the burial rate in the sediment, minimizing photo-oxidative damage when the substrate acts as a protective barrier for their tissues.

Levels of total antioxidant capacity observed in *A. hermaphroditica* during experimental exposure to UVB with sediment (68 mg Trolox Eq g^{-1} FW) are consistent with those observed (35 to 70 mg Trolox Eq g^{-1} FW) in the field for the same species in the Quempillén estuary [10]. Significant increases in total antioxidant capacity at PAB for adults and PA and PAB for juveniles reinforce the fact that juveniles of *A. hermaphroditica* show greater susceptibility to experimental radiation. Similar results were observed in the intertidal anemone *A. tenebrosa*, where controlled exposures to PA and PAB generated significant increases in GPox and GSH levels compared to those exposed only to P radiation [50]. Similarly, increases in SOD and CAT enzyme activity levels of 61% and 42% were recorded when pluteus larvae of *Sterechinus neumayerii* were exposed to P and PAB radiation for a period of 96 h [65]. Our results indicate that the level of antioxidant response in *A. hermaphroditica* was mainly associated with the developmental stage and the physical sediment protection with which they were provided. The sediment acted as a physical barrier for adults exposed to UVR by reducing total antioxidant defenses by 20–27% compared to those anemones exposed without sediment.

Although burial is a strategy to minimize UVB absorption damage, the use of photoprotective compounds such as mycosporine-like amino acids (MAAs) can minimize the effects of UVR exposure thanks to their high photo-stability since they can absorb wavelengths between 295 and 365 nm and dissipate the energy without the generation of

ROS [78–80]. These secondary metabolites were widely used by sea anemones [81,82] as a photo-protective mechanism to minimize the generation of cyclobutane pyrimidine dimmers (CPDs) in the DNA [83], mutations that affect the replication and transcription of the genetic material [84]. The presence of MAAs in sea anemones was particularly associated with trophic transference and the presence of symbiotic zooxanthellae that are involved in the de novo synthesis of these compounds [82].

Although the photodynamic response of MAAs was not analyzed in this study, it is necessary to indicate that in *A. hermaphroditica*, the presence of Mycosporine 2-glyicine (Myc 2-gly) was described, which can absorb mostly λ_{max} = 334 nm, a compound that would allow minimizing the damage due to UVR absorption [10]. Previous observations carried out in sea anemones of the *Anthopleura* genus indicate that the presence of Myc 2-gly is common among different species, and their presence is associated with those that exhibit symbiotic relations with zooxanthellae [82]. Considering that *A. hermaphroditica* is a symbiont anemone [30], the transfer of zooxanthellae to hatched juveniles inside the GVC (Cubillos pers. obs.) could provide some degree of protection when facing a new environment characterized by high levels of UVR radiation. However, a previous study indicates that the density of zooxanthellae in the larvae of the sea anemone *A. elegantissima* is directly related to the radiation levels to which they were experimentally exposed [85]. Thus, low levels of radiation within the GVC of adults of *A. hermaphroditica* semi-buried in the sediment would generate reduced levels of zooxanthellae infestation in newly hatched juveniles. Consequently, this may induce a low level of MAAs in juvenile tissues, increasing the levels of molecular damage when exposed to increased levels of UVR radiation.

Consequently, higher levels of oxidative damage in newly hatched juveniles compared to adults of *A. hermaphroditica* exposed to UVB would have serious implications for the energy budget since the de novo synthesis of enzymes and antioxidant compounds implies a high demand for ATP [86,87]. The above involves an imbalance in the energy budget, affecting the cellular processes destined to cell repair activities, as well as the degradation and synthesis of new cell components damaged by the photo-oxidative effect [88–90]. Additional studies are needed to determine whether the body walls of adult *A. hermaphroditica* can protect incubated embryos against UVB when they are removed from the sediment by bioturbation processes.

5. Conclusions

The present study showed that the burrowing activity of the intertidal anemone *A. hermaphroditica* is a behavioral response to UVR radiation. Independent of the developmental stage, PAR + UVA + UVB radiation caused adults and juveniles to conceal their bodies by burrowing faster into the sediment compared to those exposed only to PAR radiation, providing evidence that these anemones are evading UVB, reducing oxidative damage. A general pattern of adult and juvenile anemones exposed to PAB is to suffer higher levels of oxidative damage than those generated by P radiation treatment alone, indicating its noxious effect. Although the presence or absence of sediment generated a clear photo-oxidative response in adults under each radiation treatment, the total antioxidant capacity was significantly higher when anemones were exposed to different radiations without sediment. The presence of sediment seems to be important to juvenile anemones, especially when they are exposed to UVB radiation, possibly due to their elevated photosensitivity, considering that naturally, they develop inside of the GVC of burrowed adults, a microenvironment characterized by reduced or null radiation conditions. Further studies should focus on understanding the effect of brooding in the GVC and its role in protecting embryos against UVB radiation.

Author Contributions: Conceptualization, V.M.C. and J.A.Á.; methodology, V.M.C., J.A.Á. and E.R.; software, J.A.Á. and E.R.; validation, V.M.C., E.R. and E.C.; formal analysis, V.M.C.; investigation, V.M.C.; resources, V.M.C. and O.R.C.; data curation, V.M.C. and E.C.; writing—original draft preparation, V.M.C. and O.R.C.; writing—review and editing, E.C., O.R.C., J.M. and C.A.S.; visualization, V.M.C.; supervision, V.M.C.; project administration, V.M.C.; funding acquisition, V.M.C. All authors have read and agreed to the published version of the manuscript.

Funding: This research was funded by FONDECYT (ANID-Chile), grant number 1190875 to V.M.C. and FONDECYT (ANID-Chile), grant number 1180643 to O.Ch.

Institutional Review Board Statement: The study was conducted in accordance with the Ethics Committee of UNIVERSIDAD AUSTRAL DE CHILE. The animal study protocol was approved by the Ethics Committee of VICERRECTORÍA DE INVESTIGACION Y CREACION ARTÍSTICA (VIDCA) (protocol code 356/2019 of 24/5/2019) for studies involving animals.

Informed Consent Statement: Not applicable.

Data Availability Statement: The data presented in this study are available in a data repository.

Acknowledgments: We acknowledge the logistic support given by the Laboratorio Costero de Recursos Acuáticos de Calfuco (LCRAC)—Universidad Austral de Chile.

Conflicts of Interest: The authors declare no conflict of interest.

References

1. Dorgan, K.; D'Amelio, C.; Lindsay, S. Strategies of burrowing in soft muddy sediments by diverse polychaetes. *Invertebr. Biol.* **2016**, *135*, 287–301. [CrossRef]
2. Ansell, A.; Peck, L. Burrowing in the Antarctic anemone, *Halcampoides* sp., from Signy Island, Antarctica. *J. Exp. Mar. Biol. Ecol.* **2000**, *252*, 45–55. [CrossRef]
3. Sassa, S.; Watabe, Y.; Yang, S.; Kuwae, T. Burrowing criteria and burrowing mode adjustment in bivalves to varying geoenvironmental conditions in intertidal flats and beaches. *PLoS Biol.* **2011**, *6*, e25041. [CrossRef] [PubMed]
4. Hörtnagl, P.; Sommaruga, R. Photo-oxidative stress in symbiotic and aposymbiotic strains of the cyliate *Paramecium bursaria*. *Photochem. Photobiol. Sci.* **2007**, *6*, 842–847. [CrossRef]
5. Kinzie, R. Effects of ambient levels of solar ultraviolet radiation on zooxanthellae and photosynthesis of the reef coral *Montipora verrucosa*. *Mar. Biol.* **1993**, *116*, 319–327. [CrossRef]
6. Pardo, L.; González, K.; Fuentes, J.; Paschke, K.; Chaparro, O. Survival and behavioral responses of juvenile crabs of *Cancer edwardsii* to severe hyposalinity events triggered by increased runoff at an estuarine nursery ground. *J. Exp. Mar. Biol. Ecol.* **2011**, *404*, 33–39. [CrossRef]
7. Montory, J.; Chaparro, O.; Navarro, J.; Pechenik, J.; Cubillos, V. Post-metamorphic impact of brief hyposaline stress on recently hatched veligers of the gastropod *Crepipatella peruviana* (Calyptraeidae). *Mar. Biol.* **2016**, *163*, 6–16. [CrossRef]
8. Chaparro, O.; Segura, C.; Montory, J.; Navarro, J.; Pechenik, J. Brood chamber isolation during salinity stress in two estuarine mollusk species: From a protective nursery to a dangerous prison. *Mar. Ecol. Prog. Ser.* **2009**, *374*, 145–155. [CrossRef]
9. Segura, C.; Pechenik, J.; Montory, J.; Navarro, J.; Paschke, K.; Cubillos, V.; Chaparro, O. The cost of brooding in an estuary: Implications of declining salinity for gastropod females and their brooded embryos. *Mar. Ecol. Prog. Ser.* **2016**, *543*, 187–199. [CrossRef]
10. Cubillos, V.; Ramírez, E.; Cruces, E.; Montory, J.; Segura, C.; Mardones, D. Temporal changes in environmental conditions of a mid-latitude estuary (southern Chile) and its influences in the cellular response of the euryhaline anemone *Anthopleura hermaphroditica*. *Ecol. Indic.* **2018**, *88*, 169–180. [CrossRef]
11. Young, A. Chromophores in human skin. *Phys. Med. Biol.* **1997**, *42*, 789–802. [CrossRef]
12. Abele, D.; Vázquez-Medina, J.; Zenteno-Savín, T. *Oxidative Stress in Aquatic Ecosystems*; Wiley-Blackwell: Oxford, UK, 2012; p. 548.
13. Lesser, M. Oxidative stress in marine environments: Biochemistry and physiological ecology. *Annu. Rev. Physiol.* **2006**, *68*, 253–278. [CrossRef]
14. Madkour, L. *The Roles and Mechanisms of Ros, Oxidative Stress, and Oxidative Damage*; Springer: Cham, Switzerland, 2020; pp. 139–191.
15. Sies, H. Role of reactive oxygen species in biological processes. *J. Mol. Med.* **1991**, *69*, 965–968. [CrossRef]
16. Halliwell, B.; Chirico, S. Lipid peroxidation: Its mechanism, measurement and significance. *Am. J. Clin. Nutr.* **1993**, *57*, 715S–724S. [CrossRef]
17. Ahmad, S. Antioxidant Mechanisms of Enzymes and Proteins. In *Oxidative Stress and Antioxidant Defenses in Biology*; Amhad, S., Ed.; Chapman & Hall: New York, NY, USA, 1995; pp. 238–272.
18. Cabiscol, E.; Tamarit, J.; Ros, J. Oxidative stress in bacteria and protein damage by reactive oxygen species. *Int. Microbiol.* **2000**, *3*, 3–8.
19. Møller, I.; Jensen, P.; Hansson, A. Oxidative modifications to cellular components in plants. *Annu. Rev. Plant Biol.* **2007**, *58*, 459–481. [CrossRef]
20. Dalle-Donne, I.; Rossi, R.; Giustarini, D.; Milzani, A.; Colombo, R. Protein carbonyl groups as biomarkers of oxidative stress. *Clin. Chim. Acta* **2003**, *329*, 23–38. [CrossRef]
21. Almroth, B.; Sturve, J.; Berglund, A.; Forlin, L. Oxidative damage in eelpout (*Zoarces viviparus*), measured as protein carbonyls and TBARS, as biomarkers. *Aquat. Toxicol.* **2005**, *73*, 171–180. [CrossRef]

22. Lister, K.; Lamare, M.; Burritt, D. Oxidative damage in response to natural levels of UV–B radiation in larvae of the tropical sea urchin *Tripneustes gratilla*. *Photochem. Photobiol.* **2010**, *86*, 1091–1098. [CrossRef]
23. Victor, S.; Richmond, R. Effect of copper on fertilization success in the reef coral *Acropora surculosa*. *Mar. Pollut. Bull.* **2005**, *50*, 1448–1451. [CrossRef]
24. Negri, A.; Heyward, A. Inhibition of coral fertilisation and larval metamorphosis by tributyltin and copper. *Mar. Environ. Res.* **2001**, *51*, 17–27. [CrossRef]
25. Ross, C.; Ritson-Williams, R.; Olsen, K.; Paul, V.J. Short-term and latent post-settlement effects associated with elevated temperature and oxidative stress on larvae from the coral *Porites astreoides*. *Coral Reefs* **2013**, *32*, 71–79. [CrossRef]
26. Yakovleva, I.; Baird, A.; Yamamoto, H.; Bhagooli, R.; Nonaka, M.; Hidaka, M. Algal symbionts increase oxidative damage and death in coral larvae at high temperatures. *Mar. Ecol. Prog. Ser.* **2009**, *378*, 105–112. [CrossRef]
27. Rasheed, A.; Azeez, R. A Review on Natural Antioxidants. In *Traditional and Complementary Medicine*; Mordeniz, C., Ed.; IntechOpen: London, UK, 2019; pp. 1–24.
28. Seckmeyer, G.; Glandorf, M.; Wichers, C.; McKenzie, R.; Henriques, D.; Carvalho, F.; Webb, A.; Siani, A.; Bais, A.; Kjeldstad, B.; et al. Europe's darker atmosphere in the UV-B. *Photochem. Photobiol. Sci.* **2008**, *7*, 925–930. [CrossRef] [PubMed]
29. Schories, D.; Reise, K.; Sanamyan, K.; Sanamyan, N.; Clasing, E.; Reise, A. Actinian dominated intertidal mudflats: A new case of an extraordinary rare phenomenon from Southern Chile. *J. Sea Res.* **2011**, *65*, 304–314. [CrossRef]
30. LaJeunesse, T.C.; Wiedenmann, J.; Casado-Amezúa, P.; D'Ambra, I.; Turnham, K.E.; Nitschke, M.R.; Oakley, C.A.; Goffredo, S.; Spano, C.A.; Cubillos, V.M.; et al. Revival of Philozoon Geddes for host-specialized dinoflagellates, 'zooxanthellae', in animals from coastal temperate zones of northern and southern hemispheres. *Eur. J. Phycol.* **2022**, *57*, 166–180. [CrossRef]
31. Spano, C.; Rozbaczylo, N.; Häussermann, V.; Bravo, R. Redescription of the sea anemones *Anthopleura hermaphroditica* and *Bunodactis hermafroditica* (Cnidaria: Anthozoa: Actiniaria) from Chile. *Rev. Biol. Mar. Oceanogr.* **2013**, *48*, 521–534. [CrossRef]
32. Steeger, H.; Freitag, J.; Michl, S.; Wiemer, M.; Paul, R. Effects of UV-B radiation on embryonic, larval and juvenile stages of North Sea plaice (*Pleuronectes platessa*) under simulated ozone-hole conditions. *Helgol. Mar. Res.* **2001**, *55*, 56–66. [CrossRef]
33. Shick, M. *A Functional Biology of Sea Anemones*; Functional Biology Series Springer-Science + Business Media, B.V.: Hong Kong, China, 1991; p. 395.
34. Huebner, L.; Shea, C.; Schueller, P.; Terrell, A.D.; Ratchford, S.; Chadwick, N. Crustacean symbiosis with Caribbean sea anemones *Bartholomea annulata*: Occupancy modeling, habitat partitioning, and persistence. *Mar. Ecol. Prog. Ser.* **2019**, *631*, 99–116. [CrossRef]
35. Émie, A.-G.; François-Étienne, S.; Sidki, B.; Nicolas, D. Microbiomes of clownfish and their symbiotic host anemone converge before their first physical contact. *Microbiome* **2021**, *9*, 109. [CrossRef]
36. Augustine, L.; Muller-Parker, G. Selective predation by the mosshead sculpin *Clinocottus globiceps* on the sea anemone *Anthopleura elegantissima* and its two algal symbionts. *Limnol. Oceanogr.* **1998**, *43*, 711–715. [CrossRef]
37. Bachman, S.; Muller-Parker, G. Viable algae released by the seastar *Dermasterias imbricata* feeding on the symbiotic sea anemone *Anthopleura elegantissima*. *Mar. Biol.* **2007**, *150*, 369–375. [CrossRef]
38. Ottaway, J. Predators of sea anemones. *Tautara* **1977**, *22*, 213–220.
39. Salama, A.; Pearce, R. Ageing of cucumber and onion seeds: Phospholipase d, lipoxygenase activity and changes in phospholipid content. *J. Exp. Bot.* **1993**, *44*, 1253–1265. [CrossRef]
40. Levine, R.L.; Wehr, N.; Williams, J.A.; Stadman, E.R.; Shacter, E. Determination of Carbonyl Groups in Oxidized Proteins. In *Stress Response: Methods and Protocols*; Keyse, S.M., Ed.; Humana Press: Totowa, NJ, USA, 2000.
41. Fukumoto, L.; Mazza, G. Assessing antioxidant and prooxidant activities of phenolic compounds. *J. Agric. Food Chem.* **2000**, *48*, 3597–3604. [CrossRef]
42. Huovinen, P.; Gómez, I. Photosynthetic characteristics and UV stress tolerance of Antarctic seaweeds along the depth gradient. *Polar Biol.* **2013**, *36*, 1319–1332. [CrossRef]
43. Gleason, D.; Edmunds, P.; Gates, R. Ultraviolet radiation effects on the behavior and recruitment of larvae from the reef coral *Porites astreoides*. *Mar. Biol.* **2005**, *148*, 503–512. [CrossRef]
44. Brown, W. The effects of uv radiation on adult and larval behavior and chromatophore size in *Octopus rubescens*. *Explor. Undergrad. Res. J.* **2015**, *17*, 1–13.
45. Leech, D.; Johnsen, S. Behavioral Responses—UVR Avoidance and Vision. In *UV Effects in Aquatic Organisms and Ecosystems*; Helbling, W., Zagarese, H., Eds.; Royal Society of Chemistry: Cambridge, UK, 2003; pp. 455–481.
46. Bok, M.J.; Roberts, N.W.; Cronin, T.W. Behavioural evidence for polychromatic ultraviolet sensitivity in mantis shrimp. *Proc. R. Soc. B Biol. Sci.* **2018**, *285*, 20181384. [CrossRef]
47. Adams, N. UV radiation evokes negative phototaxis and covering behavior in the sea urchin *Strongylocentrotus droebachiensis*. *Mar. Ecol. Prog. Ser.* **2001**, *213*, 87–95. [CrossRef]
48. Dumont, P.; Drolet, D.; Deschênes, I.; Himmelman, J. Multiple factors explain the covering behaviour in the green sea urchin, *Strongylocentrotus droebachiensis*. *Anim. Behav.* **2007**, *73*, 979–986. [CrossRef]
49. Dykens, J.; Shick, J. Photobiology of the symbiotic sea anemone, *Anthopleura elegantissima*: Defenses against photodynamic effects, and seasonal photoacclimatization. *Biol. Bull.* **1984**, *167*, 683–697. [CrossRef]
50. Cubillos, V.; Lamare, M.; Peake, B.; Burritt, D. Cellular changes associated with the acclimation of the intertidal sea anemone *Actinia tenebrosa* to ultraviolet radiation. *Photochem. Photobiol.* **2014**, *90*, 1314–1323. [CrossRef] [PubMed]

51. Hernández-Moresino, R.; Gonçalves, R.; Helbling, E. Sublethal effects of ultraviolet radiation on crab larvae of *Cyrtograpsus altimanus*. *J. Exp. Mar. Biol. Ecol.* **2011**, *407*, 363–369. [CrossRef]
52. Hart, C.; Crowe, J. The effect of attached gravel on survival of intertidal anemones. *Trans. Am. Microsc Soc.* **1977**, *96*, 28–41. [CrossRef]
53. de Mora, S.; Demers, S.; Vernet, M. *The Effects of UV Radiation in the Marine Environment*; Cambridge Environmental Chemistry Series; Cambridge University Press: Cambridge, UK, 2004.
54. Lamare, M.; Barker, M.; Lesser, M.; Marshall, C. DNA photorepair in echinoid embryos: Effects of temperature on repair rate in Antarctic and non—Antarctic species. *J. Exp. Biol.* **2006**, *209*, 5017–5028. [CrossRef] [PubMed]
55. Hernández Moresino, R.; Helbling, E. Combined effects of UVR and temperature on the survival of crab larvae (Zoea I) from Patagonia: The role of UV-absorbing compounds. *Mar. Drugs* **2010**, *8*, 1681–1698. [CrossRef] [PubMed]
56. Carreja, B.; Fernandez, M.; Agustí, S. Joint additive effects of temperature and UVB radiation on zoeae of the crab *Taliepus dentatus*. *Mar. Ecol. Prog. Ser.* **2016**, *550*, 135–145. [CrossRef]
57. Wubben, D. UV-induced mortality of zoea I larvae of brown shrimp *Crangon crangon* (Linnaeus, 1758). *J. Plankton Res.* **2000**, *22*, 2095–2104. [CrossRef]
58. Fukunishi, Y.; Browman, H.; Durif, C.; Bjelland, R.; Skiftesvik, A. Effect of sub–lethal exposure to ultraviolet radiation on the escape performance of Atlantic cod larvae (*Gadus morhua*). *PLoS Biol.* **2012**, *7*, e35554. [CrossRef]
59. Alves, R.; Agustí, S. Effect of ultraviolet radiation (UVR) on the life stages of fish. *Rev. Fish Biol. Fish.* **2020**, *30*, 335–372. [CrossRef]
60. Lundsgaard, N.; Cramp, R.; Franklin, C. Effects of ultraviolet–B radiation on physiology, immune function and survival is dependent on temperature: Implications for amphibian declines. *Conserv. Physiol.* **2020**, *8*, coaa002. [CrossRef]
61. Darling, E.S.; Côté, I.M. Quantifying the evidence for ecological synergies. *Ecol. Lett.* **2008**, *11*, 1278–1286. [CrossRef]
62. Harvey, B.P.; Gwynn-Jones, D.; Moore, P.J. Meta-analysis reveals complex marine biological responses to the interactive effects of ocean acidification and warming. *Ecol. Evol.* **2013**, *3*, 1016–1030. [CrossRef]
63. Kroeker, K.J.; Kordas, R.L.; Crim, R.; Hendriks, I.E.; Ramajo, L.; Singh, G.S.; Duarte, C.M.; Gattuso, J.P. Impacts of ocean acidification on marine organisms: Quantifying sensitivities and interaction with warming. *Glob. Chang. Biol.* **2013**, *19*, 1884–1896. [CrossRef]
64. Hunter, E.; Okano, K.; Tomono, Y.; Fusetani, N. Functional partitioning of energy reserves by larvae of the marine bryozoan *Bugula neritina* (L.). *J. Exp. Biol.* **1998**, *201*, 2857–2865. [CrossRef]
65. Lister, K.; Lamare, M.; Burritt, D. Sea ice protects the embryos of the Antarctic sea urchin *Sterechinus neumayeri* from oxidative damage due to naturally enhanced levels of UV-B radiation. *J. Exp. Biol.* **2010**, *213*, 1967–1975. [CrossRef]
66. Ahrens, M.J.; Hickey, C.W. UV-Photoactivation of Polycyclic Aromatic Hydrocarbons and the Sensitivity of Sediment-Dwelling Estuarine Organisms. In *Transcript of the National Institute of Water and Atmospheric Research Workshop on UV-Radiation and Its Effects: An Update*; Antarctic Center: Christchurch, New Zealand, 2002.
67. Achituv, Y.; Benayahu, Y.; Hanania, J. Planulae brooding and acquisition of zooxanthellae in *Xenia macrospiculata* (Cnidaria: Octocorallia). *Helgoländer Meeresunters.* **1992**, *46*, 301–310. [CrossRef]
68. Larson, P. Brooding sea anemones (Cnidaria: Anthozoa: Actiniaria): Paragons of diversity in mode, morphology, and maternity. *Invertebr. Biol.* **2017**, *136*, 92–112. [CrossRef]
69. Pearse, V. Modification of sea anemone behavior by symbiotic zooxanthellae: Phototaxis. *Biol. Bull.* **1974**, *147*, 630–640. [CrossRef]
70. Hilal, M.; Rodríguez-Montelongo, L.; Rosa, M.; Gallardo, M.; González, J.; Interdonato, R.; Rapisarda, V.; Prado, F. Solar and supplemental UV-B radiation effects in lemon peel UV-B-absorbing compound content—Seasonal variations. *Photochem. Photobiol.* **2008**, *84*, 1480–1486. [CrossRef] [PubMed]
71. Lesser, M. Photobiology of natural populations of zooxanthellae from the sea anemone *Aiptasia pallida*: Assessment of the host's role in protection against ultraviolet radiation. *Cytometry* **1989**, *10*, 653–658. [CrossRef] [PubMed]
72. Lesser, M. Elevated temperatures and ultraviolet radiation cause oxidative stress and inhibit photosynthesis in symbiotic dinoflagellates. *Limnol. Oceanogr.* **1996**, *41*, 271–283. [CrossRef]
73. Lupínková, L.; Komenda, J. Oxidative modifications of the photosystem II D1 protein by reactive oxygen species: From isolated protein to cyanobacterial cells. *Photochem. Photobiol.* **2004**, *79*, 152–162. [CrossRef]
74. Bouchard, J.; Roy, S.; Campbell, D. UVB effects on the photosystem II–D1 protein of phytoplankton and natural phytoplankton communities. *Photochem. Photobiol.* **2006**, *82*, 936–951. [CrossRef]
75. Aro, E.; McCaffery, S.; Anderson, J. Photoinhibition and D1 protein degradation in peas acclimated to different growth irradiances. *Plant Physiol.* **1993**, *103*, 835–843. [CrossRef]
76. Sicora, C.; Szilárd, A.; Sass, L.; Turcsányi, E.; Máté, Z.; Vass, I. UV-B and UV-A Radiation Effects on Photosynthesis at the Molecular level. In *Environmental UV Radiation: Impact on Ecosystems and Human Health and Predictive Models*; Springer: Berlin/Heidelberg, Germany, 2006; pp. 121–135.
77. Turcsányi, E.; Vass, I. Inhibition of Photosynthetic Electron Transport by UV—A Radiation Targets the Photosystem II Complex. *Photochem. Photobiol.* **2000**, *72*, 513–520. [CrossRef]
78. Conde, F.; Churio, M.; Previtali, C. Experimental study of the excited-state properties and photostability of the mycosporine-like amino acid palythine in aqueous solution. *Photochem. Photobiol. Sci.* **2007**, *6*, 669–674. [CrossRef]
79. Conde, F.; Churio, M.; Previtali, C. The deactivation pathways of the excited-states of the mycosporine-like amino acids shinorine and porphyra-334 in aqueous solution. *Photochem. Photobiol. Sci.* **2004**, *3*, 960–967. [CrossRef]

80. Conde, F.; Churio, M.; Previtali, C. The photoprotector mechanism of mycosporine-like amino acids. Excited-state properties and photostability of porphyra-334 in aqueous solution. *J. Photochem. Photobiol. B Biol.* **2000**, *56*, 139–144. [CrossRef]
81. Arbeloa, E.; Carignan, M.; Acuña, F.; Churio, M.; Carreto, J. Mycosporine-like amino acid content in the sea anemones *Aulactinia marplatensis*, *Oulactis muscosa* and *Anthothoe chilensis*. *Comp. Biochem. Physiol. B Biochem. Mol. Biol.* **2010**, *156*, 216–221. [CrossRef]
82. Shick, M.; Dunlap, W.; Pearse, J.; Pearse, V. Mycosporine-like amino acid content in four species of sea anemones in the genus *Anthopleura* reflects phylogenetic but not environmental or symbiotic relationships. *Biol. Bull.* **2002**, *203*, 315–330. [CrossRef]
83. Cubillos, V.; Burritt, D.; Lamare, M.; Peake, B. The relationship between UV-irradiance, photoprotective compounds and DNA damage in two intertidal invertebrates with contrasting mobility characteristics. *J. Photochem. Photobiol. B Biol.* **2015**, *149*, 280–288. [CrossRef]
84. Setlow, R. Cyclobutane-type pyrimidine dimers in polynucleotides. *Science* **1966**, *153*, 379–386. [CrossRef]
85. Weis, V.; Verde, E.; Pribyl, A.; Schwarz, J. Aspects of the larval biology of the sea anemones *Anthopleura elegantissima* and *A. artemisia*. *Invertebr. Biol.* **2002**, *121*, 190–201. [CrossRef]
86. Pamplona, R.; Costantini, D. Molecular and structural antioxidant defenses against oxidative stress in animals. *Am. J. Physiol. Regul. Integr. Comp. Physiol.* **2011**, *301*, R843–R863. [CrossRef]
87. Cubillos, V. Temporal Variation of UV–B: Photoprotection and Photodamage in Intertidal Organisms from the New Zealand Coast. Ph.D. Thesis, University of Otago, Dunedin, New Zealand, 2013.
88. Selman, C.; McLaren, J.; Collins, A.; Duthie, G.; Speakman, J. The Impact of Experimentally Elevated Energy Expenditure on Oxidative Stress and Lifespan in the Short-Tailed Field Vole *Microtus agrestis*. *Proc. Biol. Sci.* **2008**, *275*, 1907–1916.
89. Selman, C.; Blount, J.; Nussey, D.; Speakman, J. Oxidative damage, ageing, and life-history evolution: Where now? *Trends Ecol. Evol.* **2012**, *27*, 570–577. [CrossRef]
90. Hochachka, P.; Somero, G. *Bio-Chemical Adaptation: Mechanism and Process in Physiological Evolution*; Oxford University Press: Oxford, UK, 2002; p. 466.

antioxidants

Article

Glutathione Depletion Disrupts Redox Homeostasis in an Anoxia-Tolerant Invertebrate

Marlize Ferreira-Cravo [1,2,†], Daniel C. Moreira [1,3,†] and Marcelo Hermes-Lima [1,*]

1 Department of Cell Biology, University of Brasilia, Brasilia 70910-900, Brazil
2 Federal University of Mato Grosso do Sul, Campo Grande 79070-900, Brazil
3 Research Center in Morphology and Applied Immunology, Faculty of Medicine, University of Brasilia, Brasilia 70910-900, Brazil
* Correspondence: hermes@unb.br
† These authors contributed equally to this work.

Abstract: The upregulation of endogenous antioxidants is a widespread phenomenon in animals that tolerate hypoxia/anoxia for extended periods. The identity of the mobilized antioxidant is often context-dependent and differs among species, tissues, and stresses. Thus, the contribution of individual antioxidants to the adaptation to oxygen deprivation remains elusive. This study investigated the role of glutathione (GSH) in the control of redox homeostasis under the stress of anoxia and reoxygenation in *Helix aspersa*, an animal model of anoxia tolerance. To do so, the total GSH (tGSH) pool was depleted with L-buthionine-(S, R)-sulfoximine (BSO) before exposing snails to anoxia for 6 h. Then, the concentration of GSH, glutathione disulfide (GSSG), and oxidative stress markers (TBARS and protein carbonyl) and the activity of antioxidant enzymes (catalase, glutathione peroxidase, glutathione transferase, glutathione reductase, and glucose 6-phosphate dehydrogenase) were measured in foot muscle and hepatopancreas. BSO alone induced tGSH depletion by 59–75%, but no other changes happened in other variables, except for foot GSSG. Anoxia elicited a 110–114% increase in glutathione peroxidase in the foot; no other changes occurred during anoxia. However, GSH depletion before anoxia increased the GSSG/tGSH ratio by 84–90% in both tissues, which returned to baseline levels during reoxygenation. Our findings indicate that glutathione is required to withstand the oxidative challenge induced by hypoxia and reoxygenation in land snails.

Keywords: hypoxia; ischemia; preparation for oxidative stress; reoxygenation; reperfusion

Citation: Ferreira-Cravo, M.; Moreira, D.C.; Hermes-Lima, M. Glutathione Depletion Disrupts Redox Homeostasis in an Anoxia-Tolerant Invertebrate. *Antioxidants* **2023**, *12*, 1197. https://doi.org/10.3390/antiox12061197

Academic Editor: Stanley Omaye

Received: 21 April 2023
Revised: 27 May 2023
Accepted: 29 May 2023
Published: 31 May 2023

1. Introduction

Many animal species tolerate drastic changes in the natural environment, which may become unsuitable for development, growth, and reproduction within varying ranges of time, from hours to weeks. Such harsh conditions include wide fluctuations in temperature, salinity, air humidity, oxygen availability, food and/or water supply, and UV incidence [1–3]. The biochemical adaptation program activated under such conditions includes the activation of endogenous antioxidants to cope with large fluctuations in the generation of reactive oxygen and nitrogen species (RONS) that are expected to accompany these natural, but potentially stressful, events. The upregulation of antioxidant systems under environmental stress was coined "Preparation for Oxidative Stress" (POS) [2,4,5], which has been described to occur in over 80 animal species from 8 phyla, under laboratory simulations of harsh environmental conditions [6,7]. More recently, the POS phenotype has also been observed to happen naturally in the field for a few animal taxa, including mollusks [8–11].

Since its conception, the mechanisms leading to POS remained elusive for a long time [12]. Only a few year ago, we proposed that the POS phenotype (i.e., the activation of endogenous antioxidants) is the result of RONS-induced activation of redox-sensitive transcriptional factors, prompting the increase in the expression of enzymes involved in the management of oxidative stress [13]. Experimental evidence suggests that RONS-mediated activation of kinases may also activate the POS phenotype through a hormetic process [14]. Together, these processes ultimately lead to the increase in the expression and/or activity of enzymes involved in the management of reactive species and oxidative stress [13]. Rather than a global upregulation of all endogenous antioxidants, often the identity of the upregulated antioxidant is context-dependent, depending on tissue, species, and stress [2,15]. For example, glutathione peroxidase and catalase are the most commonly upregulated antioxidants in animals exposed to anoxia [2], highlighting the relevance of catalase- and glutathione-dependent systems in the POS response and adaptation to anoxia.

One challenge in the field lies in demonstrating the relevance of individual antioxidants for the functionality of the POS mechanism and overall fitness of animals, as most studies typically measure the levels of only a few selected antioxidants rather than multiple components of antioxidant systems simultaneously. A few years ago, this issue was addressed by studying the effects of injecting 3-amino-1,2,4-triazole (ATZ), a catalase inhibitor, on redox balance. When land snails (*Helix aspersa*) or Nile tilapia (*Oreochromis niloticus*) were exposed to low or very low oxygen tensions in the presence of pharmacologically inhibited catalase, significant changes in markers of redox metabolism were observed [16,17]. Specifically, animals injected with ATZ showed signs of oxidative stress compared to animals injected with saline. These studies highlighted the role of catalase and H_2O_2 in the biochemical mechanism of POS in both species. Here, we aimed to investigate the role of glutathione, another antioxidant, in the biochemical response of *H. aspersa* to anoxia. To achieve this, we depleted glutathione levels by injecting buthionine sulfoximine (BSO), an inhibitor of γ-glutamylcysteine synthetase. The animals were then exposed to anoxia and reoxygenation. Subsequently, the activity of antioxidant enzymes and levels of oxidative stress markers were measured at various time points in the hepatopancreas and skeletal muscle. Animals that evolved under the pressure of periodic events of oxygen deprivation have been widely used as models of hypoxia tolerance, an ability that most mammal cells lack [18–22]. Land snails *Helix* spp. are representative of such species that withstand long periods of complete lack of oxygen [17,23–25], being experimental models to study biochemical adaptations associated with anoxia tolerance.

2. Materials and Methods

2.1. Animals

The study utilized *Helix aspersa* maxima (Müller, 1774) (currently renamed *Cornu aspersum*) as a resilient experimental model for hypoxia. This species can tolerate up to 15 h of anoxia [26]. Snails weighing approximately 15 g were obtained from a commercial supplier located in the state of Sao Paulo, Brazil (Helix Escargots). Upon arrival at the laboratory, the snails were placed in transparent plastic boxes and provided with a balanced diet (prepared as described in Ref. [17]) and unlimited access to water. The animal boxes were replaced daily with clean ones. The laboratory conditions included a photoperiod of 12 h of light and 12 h of darkness, with the temperature maintained at 23 ± 1 °C. The snails were acclimatized to these conditions for at least three weeks before any experimentation took place. The experiments were conducted during March and April.

2.2. Glutathione Depletion

The levels of glutathione in *H. aspersa* tissues were pharmacologically depleted by injecting L-buthionine-(S,R)-sulfoximine (BSO), a specific inhibitor of γ-glutamate-cysteine ligase (GCL) activity. This enzyme catalyzes the rate-limiting step of glutathione synthesis [27,28]. Based on previous studies involving invertebrates and fish [29–31], snails were injected with a dose of 1 g BSO per 1 kg of body weight. Pilot experiments indicated that this dosage was sufficient to reduce total glutathione levels by 60–70% in the foot muscle and hepatopancreas of snails 72–96 h after injection. A stock solution of BSO was prepared in physiological saline and injected at a volume of 12.5 µL per gram of body weight, excluding the shell mass (considering that the shell accounts for 23% of the total weight, as per [17]).

One group of animals received the BSO injection at the aforementioned dosage (BSO group), while another group received a parallel injection of saline using the same volume-to-weight ratio (saline group). Control animals did not receive any injections (control group). The injection was conducted using a Hamilton syringe through the foot muscle, which allowed the drug to enter the celomic cavity of the snail. The injection was performed using a Hamilton syringe through the foot muscle, entering the celomic cavity of the snail, simulating what would be an intraperitoneal injection in mammals. This method has been previously effective in delivering drugs to the hepatopancreas and skeletal muscle of snails [17].

2.3. Anoxia and Reoxygenation

After the injection of either saline or BSO, the snails were maintained under normoxia for 88.5 h before being exposed to anoxia. This synchronization was necessary to align the peak of glutathione depletion with the onset of oxygen stress. The snails were placed in glass containers measuring 20 cm × 19.5 cm × 12.5 cm, with one opening connected to a nitrogen gas cylinder (in) and another opening for gas flow (out). Anoxia was achieved by flushing nitrogen gas for 30 min, after which the containers were sealed and maintained under anoxia for 5.5 h. At the end of the anoxia period, a subset of animals was immediately euthanized, while another subset was allowed to reoxygenate for up to 2 h.

Within each glass container, a group of 5 animals was placed, representing the anoxia, reoxygenation for 15 min, reoxygenation for 30 min, reoxygenation for 1 h, and reoxygenation for 2 h groups, respectively, for both the BSO and saline injection. The experiment included normoxia groups as well, with animals (control, BSO, and saline) being kept under normoxia for a total of 96 h. The anoxia groups (BSO and saline) consisted of animals exposed to 5.5 h of anoxia. The reoxygenation groups (BSO and saline) comprised animals exposed to 5.5 h of anoxia followed by reoxygenation for 15–120 min.

At their respective endpoints, the snails were euthanized by decapitation, and their hepatopancreas and foot muscle were dissected, washed in saline, blotted dry, frozen in liquid nitrogen, and stored at −80 °C.

2.4. Antioxidant Enzymes

The activities of catalase, glutathione transferase (GST), glutathione peroxidase (GPX), glutathione reductase (GR), and glucose 6-phosphate dehydrogenase (G6PD) were measured using spectrophotometric kinetic assays as previously described [17]. Briefly, tissue samples were homogenized in ice-cold 50 mM potassium phosphate (KPi) buffer (pH 7.2) containing 0.5 mM EDTA and 0.1 µmol/g of tissue PMSF using an ULTRA-TURRAX tissue homogenizer (IKA, Staufen, Germany). The homogenates were then centrifuged at 10,000× *g* for 15 min at 4 °C, and the resulting supernatant was collected and immediately utilized for enzymatic assays.

Catalase activity was determined by monitoring the consumption of hydrogen peroxide at 240 nm in a reaction medium consisting of 50 mM KPi, 0.5 mM EDTA, and 10 mM H_2O_2 [32]. Glutathione S-transferase (GST) activity was measured by monitoring the production of a glutathione (GSH) conjugate at 340 nm in a reaction medium comprising 50 mM KPi, 0.5 mM EDTA, 1 mM GSH, and 1 mM 1-chloro-2,4-dinitrobenzene [33]. Glutathione reductase (GR) activity was assessed by monitoring the consumption of NADPH at 340 nm in a reaction medium containing 50 mM KPi, 0.5 mM EDTA, 1 mM oxidized glutathione (GSSG), and 0.1 mM NADPH [34]. Glucose-6-phosphate dehydrogenase (G6PD) activity was determined by monitoring the production of NADPH at 340 nm in a reaction medium consisting of 50 mM KPi, 0.5 mM EDTA, 5 mM $MgSO_4$, 1 mM glucose 6-phosphate, and 0.2 mM $NADP^+$ [35]. Glutathione peroxidase (GPX) activity was measured by monitoring the consumption of NADPH in a coupled reaction system containing 50 mM KPi, 0.5 mM EDTA, 4 mM NaN_3, 5 mM GSH, 0.1 U/mL GR (Baker's yeast), 0.2 mM NADPH, and 0.073 mM H_2O_2 [36].

For all enzymes, one unit of enzymatic activity (U) was defined as the amount of enzyme that consumes or produces substrate or product at a rate of 1 μmol/min. Enzymatic activity was normalized to the total protein content, which was measured using the Bradford protein assay with bovine serum albumin as the reference protein for constructing the standard curves [37].

2.5. Oxidative Stress Markers and Glutathione

Colorimetric assays were used to quantify the levels of thiobarbituric acid reactive substances (TBARS) [38], protein carbonyl [39], and glutathione [40,41], as previously described [17]. Glutathione was measured as total glutathione (tGSH) and disulfide glutathione (GSSG), from which reduced glutathione (GSH) and GSSG/tGSH levels were derived.

Briefly, tissue samples were homogenized in 10% (*w*/*v*) trichloroacetic acid (TCA) using a Ten Broeck tissue homogenizer (PYREX®, Corning, NY, USA) on ice. Aliquots of the crude homogenate were used for thiobarbituric acid reactive substances (TBARS) quantification, while other aliquots were centrifuged at 10,000× *g* for 6 min at 4 °C. The resulting supernatant was collected and immediately utilized for glutathione measurement and kept on ice. The pellet was frozen and stored at −80 °C for protein carbonyl analysis.

For TBARS quantification, the crude homogenate was mixed with thiobarbituric acid (TBA) and HCl to achieve final concentrations of 0.25% (*w*/*v*) TBA, 0.17 M HCl, and 10% (*w*/*v*) TCA [38]. The reaction mixture was then incubated at 96 °C for 15 min, followed by centrifugation at 10,000× *g* for 6 min. The absorbance of the supernatant was measured at 532 nm and 600 nm and corrected for the "blank" tubes containing no TBA (only sample + HCl + TCA). An absorptivity coefficient of 156 mM^{-1} cm^{-1} was used to calculate TBARS levels.

Protein carbonyl levels were measured by reacting protein pellets with 10 mM 2,4-dinitrophenylhydrazine (DNPH) in 0.5 M HCl for 1 h at room temperature with vigorous vortex agitation every 15 min. The protein samples were then washed three times with 1:1 (*v*/*v*) ethanol:ethyl acetate and resolubilized with 6 M guanidine chloride in 20 mM KPi (pH 2.3). After centrifugation at 10,000× *g* for 6 min at 4 °C to remove any insoluble material, the solubilized protein samples were read at 370 nm (protein carbonyl) and 280 nm (total protein content). An absorptivity coefficient of 22,000 M^{-1} cm^{-1} was used to calculate protein carbonyl levels [39].

Glutathione levels were measured using the enzymatic recycling method. For total glutathione (tGSH), a standard curve ranging from 0.375 μM to 1.5 μM GSH was constructed using a reaction medium containing 100 mM KPi (pH 7.0), 1 mM EDTA, 0.1 mM NADPH, 0.1 mM DTNB, and 0.05 U/mL GR. For the disulfide form (GSSG), a standard curve ranging from 0.025 μM to 0.2 μM GSH was constructed using a reaction medium containing 100 mM KPi (pH 7.0), 1 mM EDTA, 0.1 mM NADPH, 0.1 mM DTNB, and 0.3 U/mL GR. The rate of increase in absorbance at 412 nm was plotted against the concentration of GSH to construct the calibration curves. For each sample, one aliquot of the supernatant was directly used to measure tGSH, while another aliquot was incubated with 2-vinylpyridine before the measurement of GSSG [42].

2.6. Statistics

Data normality was assessed using the Shapiro-Wilk test. The effect of BSO injection was evaluated using Kruskal-Wallis tests, comparing the following groups: control, normoxia-saline, normoxia-BSO, anoxia-saline, and anoxia-BSO (presented as column graphs). Subsequently, two-way ANOVA tests were conducted, considering oxygen and BSO as sources of variation, followed by Tukey's multiple comparisons test (represented as point graphs). Statistical analyses and figure production were performed using GraphPad Prism 9 version 9.5.1 (GraphPad Software, San Diego, CA, USA).

3. Results

BSO injection significantly reduced total glutathione (tGSH) levels in both the hepatopancreas and foot muscle (Figure 1). In the foot muscle, tGSH content decreased by 75% in BSO-injected snails under normoxia and by 71% in BSO-injected snails exposed to anoxia (Figure 1A). Similarly, in the hepatopancreas, tGSH content decreased by 66% in BSO-injected snails under normoxia and by 59% in BSO-injected snails exposed to anoxia compared to the control group (Figure 1C). In both tissues, tGSH levels remained lower in BSO-injected snails than in saline-injected snails (Figure 1B,D). Similar trends were observed for reduced glutathione (GSH). GSH levels in the foot muscle decreased by 76% in BSO-injected snails under normoxia and by 75% in BSO-injected snails exposed to anoxia compared to the control group (Figure 1E). In the hepatopancreas, GSH concentration decreased by 70% in BSO-injected snails under normoxia and by 64% in BSO-injected snails exposed to anoxia compared to the control group (Figure 1G). Reduced glutathione levels in BSO-treated snails remained lower than those in saline-treated snails during reoxygenation in both the foot muscle and hepatopancreas (Figure 1F,H). These results demonstrate the effective depletion of glutathione by BSO treatment and indicate that anoxia alone did not affect tGSH or GSH levels.

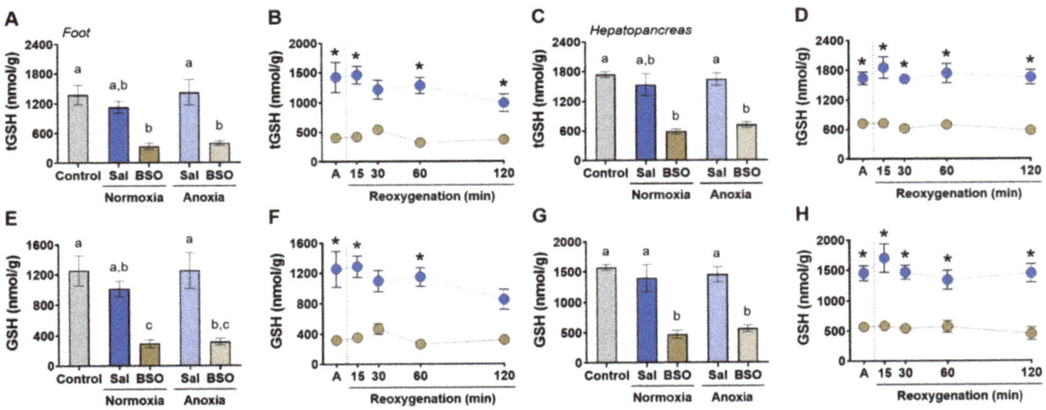

Figure 1. Glutathione levels in foot muscle and hepatopancreas of snails *Helix aspersa*, injected with saline or buthionine sulfoximine and maintained in normoxia or exposed to anoxia and reoxygenation. Control animals were maintained in normoxia and not injected with any substance. (**A**,**B**) Total glutathione (tGSH = GSH + 2GSSG) concentration in foot muscle (n = 4–7). (**C**,**D**) Total glutathione (tGSH = GSH + 2GSSG) concentration in hepatopancreas (n = 4–8). (**E**,**F**) Reduced glutathione (GSH) concentration in foot muscle (n = 4–7). (**G**,**H**) Reduced glutathione (GSH) concentration in hepatopancreas (n = 4–8). Groups that do not share letters are significantly different from each other ($p < 0.05$). The asterisk (*) denotes significant differences in relation to the respective BSO group ($p < 0.05$).

In the foot muscle, the levels of disulfide glutathione (GSSG) were lower in BSO-treated snails compared to saline-treated snails in both the normoxia and anoxia groups (Figure 2A). During reoxygenation, GSSG levels in the foot muscle of BSO-injected snails were lower than those in saline-injected snails (Figure 2B). In the hepatopancreas, however, there were no significant differences between BSO-injected snails and saline-injected snails in GSSG levels, although there was a trend of BSO decreasing hepatic GSSG levels in snails maintained in normoxia (Figure 2C). Anoxia and subsequent reoxygenation did not affect hepatic GSSG levels (Figure 2D). These results indicate that the levels of GSSG generally followed the pattern observed for tGSH and GSH.

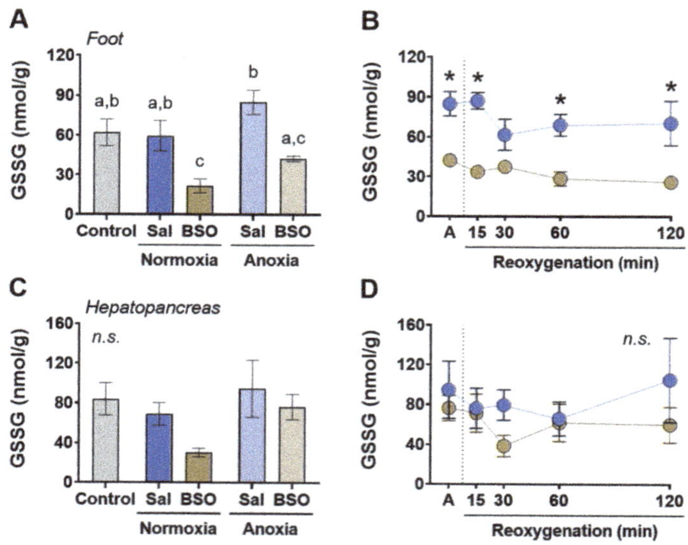

Figure 2. Disulfide glutathione (GSSG) levels in foot muscle and hepatopancreas of snails *Helix aspersa*, injected with saline or buthionine sulfoximine and maintained in normoxia or exposed to anoxia and reoxygenation. Control animals were maintained in normoxia and not injected with any substance. (**A**,**B**) Disulfide glutathione (GSSG) levels in foot muscle (n = 4–7). (**C**,**D**) Disulfide glutathione (GSSG) levels in hepatopancreas (n = 4–8). Groups that do not share letters are significantly different from each other ($p < 0.05$). The asterisk (*) denotes significant differences in relation to the respective BSO group ($p < 0.05$).

Redox balance, as indicated by GSSG/tGSH levels, remained unaffected by anoxia alone in both hepatopancreas and foot muscle. There were no significant differences observed between the control, saline-normoxia, and saline-anoxia groups (Figure 3A,C). Similarly, glutathione depletion caused by BSO alone did not alter GSSG/tGSH levels in either tissue (Figure 3A,C). However, when BSO-treated snails were exposed to anoxia, redox imbalance occurred in both tissues. GSSG/tGSH levels nearly doubled, increasing by 84% in foot muscle (Figure 3A) and by 90% in hepatopancreas (Figure 3C) compared to saline-injected snails under anoxia. During reoxygenation, GSSG/tGSH levels returned to baseline in both tissues (Figure 3B,D). Furthermore, GSSG/tGSH levels in the BSO-anoxia groups were higher than those in the control group. These findings suggest that neither glutathione depletion nor anoxia alone induces significant redox imbalance. It is only when both factors occur simultaneously that GSSG/tGSH shifts towards a more oxidized state.

There were no significant changes in the levels of protein carbonyl and thiobarbituric acid reactive substances (TBARS) during anoxia or reoxygenation, regardless of saline or BSO treatment (Table 1). These results suggest that the disturbance in redox balance did not reach a level of severity that would cause excessive oxidative damage to lipids and proteins.

Regarding antioxidant enzyme activities, catalase activity slightly decreased in the foot muscle of saline-injected snails under normoxia (Figure 4A) but not significantly in hepatopancreas (Figure 4B). Anoxia exposure led to an increase in glutathione peroxidase (GPX) activity in the foot muscle of snails, regardless of saline or BSO injection (Figure 4C). The GPX activity in both anoxia groups (saline- and BSO-injected) was elevated by approximately 110% and 114% compared to the control group (Figure 4C). In hepatopancreas, GPX activity remained unchanged (Figure 4D).

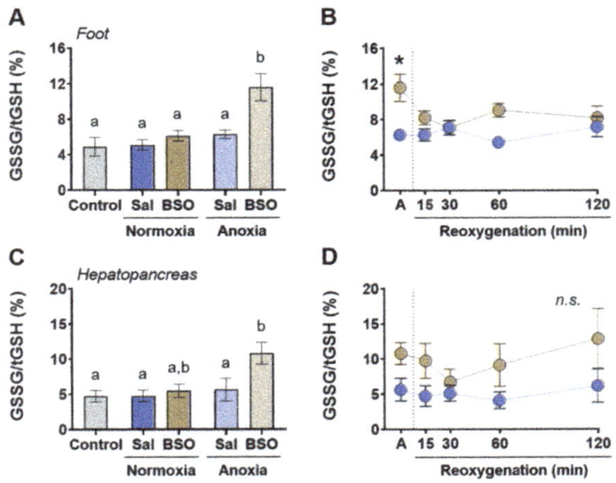

Figure 3. Redox balance in foot muscle and hepatopancreas of snails *Helix aspersa*, injected with saline or buthionine sulfoximine and maintained in normoxia or exposed to anoxia and reoxygenation. Control animals were maintained in normoxia and not injected with any substance. (**A**,**B**) Ratio between disulfide glutathione (GSSG) and total glutathione (tGSH = GSH + 2GSSG) in foot muscle (n = 4–7). (**C**,**D**) Ratio between disulfide glutathione (GSSG) and total glutathione (tGSH = GSH + 2GSSG) in hepatopancreas (n = 4–8). Groups that do not share letters are significantly different from each other ($p < 0.05$). The asterisk (*) denotes significant differences in relation to the respective BSO group ($p < 0.05$).

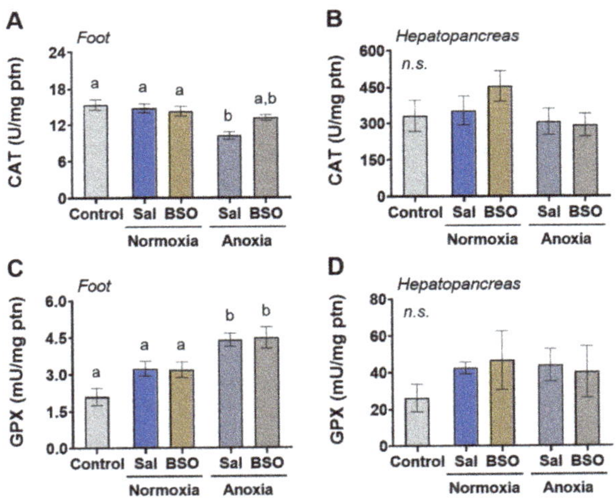

Figure 4. Activity of peroxide-detoxifying enzymes in foot muscle and hepatopancreas of snails *Helix aspersa*, injected with saline or buthionine sulfoximine and maintained in normoxia or exposed to anoxia. Control animals were maintained in normoxia and not injected with any substance. (**A**) Catalase activity in foot muscle (n = 5–7). (**B**) Catalase activity in hepatopancreas (n = 6–8). (**C**) Glutathione peroxidase (GPX) activity in foot muscle (n = 5–7). (**D**) Glutathione peroxidase (GPX) activity in hepatopancreas (n = 5–8). Groups that do not share letters are significantly different from each other ($p < 0.05$).

Table 1. Oxidative stress markers (TBARS and protein carbonyl) in foot muscle and hepatopancreas of snails *Helix aspersa*, previously injected with saline or buthionine sulfoximine, exposed to anoxia and reoxygenation. Control animals were maintained in normoxia and not injected with any substance.

Group	Foot Muscle				Hepatopancreas			
	TBARS (nmol/gww)		Carbonyl (nmol/mg prot.)		TBARS (nmol/gww)		Carbonyl (nmol/mg prot.)	
Control	14.85 ± 2.73	5	10.41 ± 1.51	5	23.27 ± 9.41	8	11.98 ± 5.04	8
Normoxia								
Saline	16.27 ± 2.98	6	12.24 ± 2.53	5	20.13 ± 5.09	6	15.40 ± 8.64	6
BSO	15.06 ± 3.64	6	11.80 ± 1.41	6	19.87 ± 7.05	6	11.72 ± 4.07	6
Anoxia								
Saline	17.82 ± 2.94	6	9.65 ± 5.76	6	20.58 ± 6.92	8	10.78 ± 1.47	8
BSO	22.09 ± 12.37	7	10.09 ± 2.77	7	24.23 ± 6.00	7	12.97 ± 8.62	7
Reoxygenation (15 min)								
Saline	18.57 ± 3.50	7	9.82 ± 2.45	7	20.77 ± 6.45	7	10.19 ± 4.66	7
BSO	18.24 ± 5.67	7	11.16 ± 2.47	7	25.08 ± 3.71	5	8.32 ± 3.33	5
Reoxygenation (30 min)								
Saline	17.18 ± 2.91	6	10.86 ± 1.59	6	20.64 ± 11.21	6	9.87 ± 3.44	7
BSO	16.35 ± 2.19	4	8.35 ± 2.44	4	24.00 ± 10.23	5	9.26 ± 2.64	5
Reoxygenation (60 min)								
Saline	17.80 ± 4.40	7	12.08 ± 5.31	6	20.69 ± 7.17	5	8.23 ± 2.88	6
BSO	16.43 ± 2.80	7	11.58 ± 3.20	7	20.29 ± 10.69	4	9.29 ± 1.69	4
Reoxygenation (120 min)								
Saline	16.00 ± 2.74	5	11.48 ± 3.44	5	25.77 ± 6.48	5	8.24 ± 1.16	5
BSO	17.58 ± 3.66	7	10.35 ± 1.82	7	26.10 ± 12.39	7	10.33 ± 3.23	7

Data are shown as mean ± standard deviation and *n*.

The activities of glutathione S-transferase (GST) remained stable in both tissues in response to anoxia and BSO treatment (Figure S1). Glucose-6-phosphate dehydrogenase (G6PD) also remained stable in hepatopancreas (Figure S1). In the foot muscle, G6PD activity was 31–33% lower in both saline- and BSO-injected snails exposed to anoxia compared to the other groups (Figure S1). Glutathione reductase (GR) activity was 26–29% lower in the foot muscle of saline- and BSO-injected snails under anoxia compared to the control group (Figure S1). In the hepatopancreas, BSO-injected snails under normoxia exhibited higher GR activity (83% increase) compared to the control group (Figure S1). These oscillations in enzyme activities suggest that the foot muscle is more sensitive and responsive to stimuli compared to the hepatopancreas.

4. Discussion

Glutathione has a significant role in maintaining redox homeostasis in animals experiencing hypoxia/anoxia, aerial exposure (in aquatic species), and estivation and during their subsequent recovery. Indeed, glutathione levels have been shown to be upregulated in diverse animal species, from tardigrades [43] to lizards [44], exposed to environmental stresses [2]. For example, *Littorina littorea* exhibited elevated GSH levels in the muscle and hepatopancreas during anoxia exposure [45]. Similarly, mussels *Brachidontes solisianus* showed a 60% increase in whole-body GSH levels after 4 h of aerial exposure during low tide [46]. These findings indicate an enhanced antioxidant defense system to counteract redox imbalances during anoxia (in the case of *L. littorea*) or physiological hypoxia (in the case of *B. solisianus*). These biochemical responses serve as characteristic indicators of the presence of preparation for oxidative stress (POS), wherein other antioxidant defenses, including enzymes, may or may not be involved.

In our previous studies on *Helix aspersa*, we observed an increase in GSH levels during estivation, which would help regulate redox imbalance during the cycles of estivation and arousal [47,48]. The POS phenotype during estivation also involves the participation of antioxidant enzymes such as GPX and SOD. Our previous work also revealed an increase in muscle GPX levels during anoxia exposure in saline-injected *H. aspersa* [17], and this finding was replicated in the current study (Figure 4C). Most other antioxidant enzymes remained largely unchanged during anoxia in saline-injected snails. Although GSH levels did not show significant alterations during anoxia, it still plays a crucial role in the overall POS phenotype. This is because the observed increase in muscle GPX activity is effective in vivo only if GSH levels, which serve as a GPX substrate, are maintained at adequate levels. Therefore, we further investigated how the pharmacological depletion of GSH could impact the redox metabolism in snails during anoxia and reoxygenation.

Despite the significant depletion of glutathione (GSH) in BSO-injected animals, only minor changes were observed in antioxidant enzymes compared to controls or saline-injected snails. The increase in muscle GPX activity observed in saline-injected animals was also present in BSO-injected snails. Additionally, markers of oxidative stress such as TBARS and carbonyl protein remained unchanged during anoxia and reoxygenation in both saline- and BSO-injected snails. Although lipid and protein oxidation mediated by reactive oxygen and nitrogen species (RONS) appeared unaltered in GSH-depleted snails, the ratio of oxidized GSSG to total GSH (GSSG/tGSH) significantly increased during anoxia in BSO-injected snails in both tissues. However, this increased GSSG/tGSH ratio was reversed within 15–30 min of reoxygenation. These observations indicate a severe redox imbalance during anoxia in GSH-depleted animals. The decrease in GSH levels caused by BSO during anoxia led to an increased conversion of GSH into GSSG, indicating a more oxidized cellular environment proportional to the initial amount of GSH. Interestingly, GSSG levels decreased in BSO-injected normoxic or anoxic snails (reaching significance only in muscle) due to the overall depletion of glutathione induced by the drug. However, the percentage of GSSG relative to total GSH nearly doubled under anoxia in glutathione-depleted animals, providing indirect evidence of increased RONS levels and redox imbalance. It is important to note that when animals are exposed to an anoxic environment, their tissues gradually become hypoxic rather than instantly anoxic, as there is still residual oxygen present in the tissues [1]. This residual oxygen is gradually consumed over time, although not completely, and the unconsumed oxygen, being hydrophobic, is expected to be preferentially located in membranes [49,50]. The increase in the GSSG/tGSH ratio has long been recognized as a marker of disruption in redox homeostasis [51].

BSO-induced GSH depletion is associated with higher ROS levels [52–55] and activation of redox-related transcription factors [54,56]. Still, we did not observe major alterations in antioxidant systems and oxidative stress markers elicited by glutathione depletion alone in *H. aspersa*. Such a lack of antioxidant response has also been observed in other experimental systems [57,58]. The depletion of the total glutathione pool has been associated with increased tissue sensitivity to additional stresses [59–62], but not a toxic condition per se [63,64]. Studies have demonstrated that the pharmacological depletion of glutathione (GSH) exacerbates hypoxic/ischemic stress [65,66]. One notable study reported a significant increase in the GSSG/tGSH ratio in the muscle of mice exposed to 24 h of hypobaric hypoxia, which is equivalent to an altitude of 7000 m [67]. This increase in the GSSG/tGSH ratio was twice as large in glutathione-depleted mice caused by BSO injection compared to saline-injected mice. While hypoxia alone induced an increase in muscle GSSG levels, the levels of GSSG were lower in hypoxic BSO-injected mice than in saline-injected animals. These results indicate that hypoxia alone induces redox imbalance and oxidative stress (evidenced by increased TBARS levels) and that the combination of hypoxia and GSH depletion imposes even greater stress on the muscle, as indicated by the higher GSSG/tGSH ratio [67]. This redox imbalance was attributed to a higher rate of GSH oxidation to GSSG due to oxidative stress, as well as altered glutathione exchange between the muscle and circulation in response to hypoxic stress.

In snails *H. aspersa*, the increase in muscle GPX activity may have prevented oxidative stress (indicated by TBARS and carbonyl levels) in that organ in both saline- and BSO-injected anoxic animals. In the case of the hepatopancreas, the high constitutive levels of catalase may have provided protection. This is supported by the observation that catalase-depleted snails injected with ATZ exhibited increased carbonyl levels during reoxygenation [17]. However, these mechanisms were unable to prevent the redox imbalance in GSH-depleted anoxic snails. It is worth noting that redox imbalance occurs before the onset of uncontrolled oxidative stress and oxidative damage to cellular components [68,69]. It is possible that if GSH levels had been further diminished (more than 70–75% depletion) by employing higher levels of BSO, the redox scenario could have shifted towards actual oxidative stress. This hypothesis should be explored in future experiments.

At first glance, a shift towards a more oxidized state in a condition of low oxygen availability might seem counterintuitive. Initially, it was believed that exposure to hypoxia would lead to a decrease in the formation of reactive oxygen and nitrogen species (RONS) due to limited oxygen availability [7,70]. However, accumulating evidence from various animal species and cell biology experiments has shown an increase in the GSSG/tGSH ratio and oxidative damage to lipids, proteins, and DNA under hypoxic and/or anoxic conditions [13]. These findings, along with measurements of RONS using fluorescent probes, suggest increased RONS formation under hypoxia, with the mitochondrial respiratory chain being a major source of reactive species [71–74]. These pieces of evidence support the molecular basis of the POS mechanism, which involves increased RONS formation under anoxia/hypoxia or hypoxic-like conditions, followed by the activation of redox-sensitive transcription factors and subsequent increase in antioxidant defenses [7,13]. The results from anoxia-exposed snails in the present study suggest that BSO-treated animals have higher intracellular levels of RONS due to the diminished GSH pool. However, the steady-state levels of reactive species are not sufficiently high to cause oxidative damage to lipids and proteins during oxygen depletion and reoxygenation in snails.

Although the present study investigated several classic antioxidant systems, other antioxidants that were not analyzed could play a role in *H. aspersa* resistance to oxidative stress. Examples of such antioxidants include uric acid, ascorbic acid, ovothiols, and other thiol-containing molecules [75,76]. The presence of significant amounts of these molecules could potentially explain how *H. aspersa* maintains redox homeostasis without experiencing oxidative damage to proteins and lipids under anoxia/reoxygenation stress, even when GSH is depleted by 60–70%. These non-enzymatic antioxidants could compensate for such pro-oxidizing conditions. Uric acid has been associated with the POS strategy in tissues of snails *Pomacea canaliculata* during estivation [77,78] and hibernation [78,79]. Additionally, levels of ascorbic acid increase in *Pila globosa* snails during estivation [80]. These findings suggest a potential role of ascorbate and uric acid in anoxia/hypoxia tolerance in gastropods. Moreover, although ovothiols have not yet been determined in land snails, they are present in high concentrations in the organs of mussels [81] and other aquatic mollusks [82], indicating their potential relevance for redox control under anoxia/hypoxia stress. Therefore, evidence from other experimental systems suggests that these antioxidant molecules and their potential role in the POS mechanism should be investigated in *H. aspersa* and other animal models in further research.

5. Conclusions

In conclusion, our study demonstrates that the pharmacological inhibition of glutathione synthesis does not lead to redox imbalance or oxidative stress in land snails *H. aspersa*. Anoxia exposure and reoxygenation did not induce oxidative stress, but anoxia did trigger an antioxidant response in the muscle of these snails, characterized by increased GPX activity. However, glutathione depletion rendered the snails more susceptible to the oxidative challenge posed by oxygen depletion, resulting in a disturbance of redox homeostasis, as indicated by an elevated GSSG/tGSH ratio in both tissues of snails exposed to anoxia after BSO treatment. Importantly, our findings highlight the robust antioxidant systems present in snails *H. aspersa*, enabling them to withstand low oxygen levels and reoxygenation, as well as glutathione depletion, without experiencing significant oxidative stress. Even when exposed to both anoxia and GSH depletion, there is no evidence of oxidative damage to lipids and proteins; instead, only a shift in glutathione redox balance towards a more oxidized cellular state is observed. This remarkable ability to maintain redox homeostasis under potentially pro-oxidant conditions likely contributes to the anoxia/hypoxia tolerance observed in *H. aspersa*. These observations provide further insights into the mechanism of the POS phenomenon in invertebrates experiencing low oxygenation.

Supplementary Materials: The following supporting information can be downloaded at https://www.mdpi.com/article/10.3390/antiox12061197/s1: Figure S1: Enzymatic activity in foot muscle and hepatopancreas of *Helix aspersa* snails, previously injected with saline or buthionine sulfoximine, exposed to anoxia and reoxygenation.

Author Contributions: Conceptualization, M.H.-L. and M.F.-C.; methodology, M.F.-C.; formal analysis, M.F.-C. and D.C.M.; investigation, M.F.-C. and D.C.M.; resources, M.H.-L.; data curation, M.F.-C. and D.C.M.; writing—original draft preparation, M.H.-L. and D.C.M.; writing—review and editing, M.H.-L. and D.C.M.; visualization, M.H.-L. and D.C.M.; supervision, M.H.-L.; project administration, M.H.-L.; funding acquisition, M.H.-L. All authors have read and agreed to the published version of the manuscript.

Funding: This research was funded by Fundação de Apoio à Pesquisa do Distrito Federal (FAPDF, Brazil), grant number 193.00000219/2019–71, and Conselho Nacional de Desenvolvimento Científico e Tecnológico (CNPq, Brazil), grant 421384/2018–2.

Institutional Review Board Statement: Ethical review and approval were waived for this study due to the use of a non-cephalopod invertebrate species as animal model.

Informed Consent Statement: Not applicable.

Data Availability Statement: Data are available within the article or its Supplementary Materials.

Acknowledgments: The authors thank Carlos Alberto F. Funcia (Helix Escargots, São Paulo, Brazil), Francisco Erivan Alves (University of Brasília), and Maximiliano Giraud-Billoud (Universidad Nacional de Cuyo, Mendoza, Argentina) for their contributions to this study. This article is dedicated to Cláudio Mário Guimarães, a retired biology teacher now in his 90s, who imparted the first notions of science to M.H-L. and Marianita Ferreira-Cravo, teacher and an inspiration for her family (M.F-C.) and friends. This paper is also dedicated to the memory of José G. Dórea (1944–2023), good friend, great toxicology scientist and enthusiastic teacher at the university.

Conflicts of Interest: The authors declare no conflict of interest.

References

1. Welker, A.F.; Moreira, D.C.; Campos, É.G.; Hermes-Lima, M. Role of Redox Metabolism for Adaptation of Aquatic Animals to Drastic Changes in Oxygen Availability. *Comp. Biochem. Physiol. A* **2013**, *165*, 384–404. [CrossRef]
2. Moreira, D.C.; Venancio, L.P.R.; Sabino, M.A.C.T.; Hermes-Lima, M. How Widespread Is Preparation for Oxidative Stress in the Animal Kingdom? *Comp. Biochem. Physiol. A* **2016**, *200*, 64–78. [CrossRef] [PubMed]
3. Somero, G.N.; Lockwood, B.L.; Tomanek, L. *Biochemical Adaptation: Response to Environmental Challenges, from Life's Origins to the Anthropocene*; Sinauer Associates, Inc. Publishers: Sunderland, MA, USA, 2017; ISBN 978-1-60535-564-1.
4. Hermes-Lima, M.; Zenteno-Savín, T. Animal Response to Drastic Changes in Oxygen Availability and Physiological Oxidative Stress. *Comp. Biochem. Physiol. C* **2002**, *133*, 537–556. [CrossRef] [PubMed]
5. Moreira, D.C.; Campos, É.G.; Giraud-Billoud, M.; Storey, K.B.; Hermes-Lima, M. Commentary: On the Merit of an Early Contributor of the "Preparation for Oxidative Stress" (POS) Theory. *Comp. Biochem. Physiol. A* **2023**, *276*, 111341. [CrossRef] [PubMed]
6. Moreira, D.C.; Oliveira, M.F.; Liz-Guimarães, L.; Diniz-Rojas, N.; Campos, É.G.; Hermes-Lima, M. Current Trends and Research Challenges Regarding "Preparation for Oxidative Stress". *Front. Physiol.* **2017**, *8*, 702. [CrossRef] [PubMed]

7. Giraud-Billoud, M.; Rivera-Ingraham, G.A.; Moreira, D.C.; Burmester, T.; Castro-Vazquez, A.; Carvajalino-Fernández, J.M.; Dafre, A.; Niu, C.; Tremblay, N.; Paital, B.; et al. Twenty Years of the 'Preparation for Oxidative Stress' (POS) Theory: Ecophysiological Advantages and Molecular Strategies. *Comp. Biochem. Physiol. A* **2019**, *234*, 36–49. [CrossRef] [PubMed]
8. Istomina, A.; Belcheva, N.; Chelomin, V. Antioxidant System of the Intertidal Mollusk *Littorina kurila* in Its Natural Habitat. *J. Environ. Health Sci. Eng. A* **2013**, *2*, 713–718.
9. Moreira, D.C.; Carvajalino-Fernández, J.M.; Navas, C.A.; de Carvalho, J.E.; Hermes-Lima, M. Metabolic and Redox Biomarkers in Skeletal Muscle Underlie Physiological Adaptations of Two Estivating Anuran Species in a South American Semi-Arid Environment. *Front. Physiol.* **2021**, *12*, 769833. [CrossRef]
10. Patnaik, P.; Sahoo, D.D. Variations in Oxidative Stress and Antioxidant Defense Level during Different Phases of Hibernation in Common Asian Toad, *Duttaphrynus melanostictus*. *Biol. Open.* **2021**, *10*, bio058567. [CrossRef]
11. Moreira, D.C.; Aurélio Da Costa Tavares Sabino, M.; Minari, M.; Torres Brasil Kuzniewski, F.; Angelini, R.; Hermes-Lima, M. The Role of Solar Radiation and Tidal Emersion on Oxidative Stress and Glutathione Synthesis in Mussels Exposed to Air. *PeerJ* **2023**, *11*, e15345. [CrossRef]
12. Hermes-Lima, M.; Storey, J.M.; Storey, K.B. Antioxidant Defenses and Metabolic Depression. The Hypothesis of Preparation for Oxidative Stress in Land Snails. *Comp. Biochem. Physiol. B* **1998**, *120*, 437–448. [CrossRef]
13. Hermes-Lima, M.; Moreira, D.C.; Rivera-Ingraham, G.A.; Giraud-Billoud, M.; Genaro-Mattos, T.C.; Campos, É.G. Preparation for Oxidative Stress under Hypoxia and Metabolic Depression: Revisiting the Proposal Two Decades Later. *Free Radic. Biol. Med.* **2015**, *89*, 1122–1143. [CrossRef] [PubMed]
14. Oliveira, M.F.; Geihs, M.A.; França, T.F.A.; Moreira, D.C.; Hermes-Lima, M. Is "Preparation for Oxidative Stress" a Case of Physiological Conditioning Hormesis? *Front. Physiol.* **2018**, *9*, 945. [CrossRef] [PubMed]
15. Rivera-Ingraham, G.A.; Lignot, J.-H. Osmoregulation, Bioenergetics and Oxidative Stress in Coastal Marine Invertebrates: Raising the Questions for Future Research. *J. Exp. Biol.* **2017**, *220*, 1749–1760. [CrossRef] [PubMed]
16. Welker, A.F.; Campos, É.G.; Cardoso, L.A.; Hermes-Lima, M. Role of Catalase on the Hypoxia/Reoxygenation Stress in the Hypoxia-Tolerant Nile Tilapia. *Am. J. Physiol.-Regul. Integr. Comp. Physiol.* **2012**, *302*, R1111–R1118. [CrossRef]
17. Welker, A.F.; Moreira, D.C.; Hermes-Lima, M. Roles of Catalase and Glutathione Peroxidase in the Tolerance of a Pulmonate Gastropod to Anoxia and Reoxygenation. *J. Comp. Physiol. B* **2016**, *186*, 553–568. [CrossRef] [PubMed]
18. Hochachka, P.W.; Buck, L.T.; Doll, C.J.; Land, S.C. Unifying Theory of Hypoxia Tolerance: Molecular/Metabolic Defense and Rescue Mechanisms for Surviving Oxygen Lack. *Proc. Natl. Acad. Sci. USA* **1996**, *93*, 9493–9498. [CrossRef]
19. Hochachka, P.W.; Lutz, P.L. Mechanism, Origin, and Evolution of Anoxia Tolerance in Animals. *Comp. Biochem. Physiol. Part B Biochem. Mol. Biol.* **2001**, *130*, 435–459. [CrossRef]
20. Larson, J.; Drew, K.L.; Folkow, L.P.; Milton, S.L.; Park, T.J. No Oxygen? No Problem! Intrinsic Brain Tolerance to Hypoxia in Vertebrates. *J. Exp. Biol.* **2014**, *217*, 1024–1039. [CrossRef]
21. Country, M.W.; Jonz, M.G. Goldfish and Crucian Carp Are Natural Models of Anoxia Tolerance in the Retina. *Comp. Biochem. Physiol. Part A Mol. Integr. Physiol.* **2022**, *270*, 111244. [CrossRef] [PubMed]
22. Bundgaard, A.; Ruhr, I.M.; Fago, A.; Galli, G.L.J. Metabolic Adaptations to Anoxia and Reoxygenation: New Lessons from Freshwater Turtles and Crucian Carp. *Curr. Opin. Endocr. Metab. Res.* **2020**, *11*, 55–64. [CrossRef]
23. Kluytmans, J.H.; Zandee, D.I. Comparative Study of the Formation and Excretion of Anaerobic Fermentation Products in Bivalves and Gastropods. *Comp. Biochem. Physiol. Part B Comp. Biochem.* **1983**, *75*, 729–732. [CrossRef]
24. Pedrini-Martha, V.; Niederwanger, M.; Kopp, R.; Schnegg, R.; Dallinger, R. Physiological, Diurnal and Stress-Related Variability of Cadmium-Metallothionein Gene Expression in Land Snails. *PLoS ONE* **2016**, *11*, e0150442. [CrossRef] [PubMed]
25. Vorhaben, J.E.; Klotz, A.V.; Campbell, J.W. Activity and Oxidative Metabolism of the Land Snail Helix Aspersa. *Physiol. Zool.* **1984**, *57*, 357–365. [CrossRef]
26. Ramos-Vasconcelos, G.R. Metabolismo de Radicais Livres Em Gastrópodes Terrestres. Ph.D. Thesis, University of Brasilia, Brasilia, Brazil, 2005.
27. Griffith, O.W. Biologic and Pharmacologic Regulation of Mammalian Glutathione Synthesis. *Free Radic. Biol. Med.* **1999**, *27*, 922–935. [CrossRef]
28. Sies, H. Glutathione and Its Role in Cellular Functions. *Free Radic. Biol. Med.* **1999**, *27*, 916–921. [CrossRef]
29. Canesi, L.; Ciacci, C.; Betti, M.; Gallo, G. Growth Factor-Mediated Signal Transduction and Redox Balance in Isolated Digestive Gland Cells from Mytilus Galloprovincialis Lam. *Comp. Biochem. Physiol. C* **2000**, *125*, 355–363. [CrossRef]
30. Mitchelmore, C.L.; Ringwood, A.H.; Weis, V.M. Differential Accumulation of Cadmium and Changes in Glutathione Levels as a Function of Symbiotic State in the Sea Anemone *Anthopleura elegantissima*. *J. Exp. Mar. Biol. Ecol.* **2003**, *284*, 71–85. [CrossRef]
31. Lushchak, V.I.; Hermes-Lima, M. The Effect of Buthionine Sulfoximine on the Glutathione Level in Goldfish Tissues. *Ukr. Biokhim. Zh.* **2005**, *77*, 35–38.
32. Aebi, H. Catalase in vitro. *Methods Enzymol.* **1984**, *105*, 121–126. [CrossRef]
33. Habig, W.H.; Pabst, M.J.; Jakoby, W.B. Glutathione S-Transferases. The First Enzymatic Step in Mercapturic Acid Formation. *J. Biol. Chem.* **1974**, *249*, 7130–7139. [CrossRef] [PubMed]
34. Racker, E. Glutathione Reductase from Bakers' Yeast and Beef Liver. *J. Biol. Chem.* **1955**, *217*, 855–865. [CrossRef]
35. Glock, G.E.; McLean, P. Further Studies on the Properties and Assay of Glucose 6-Phosphate Dehydrogenase and 6-Phosphogluconate Dehydrogenase of Rat Liver. *Biochem. J.* **1953**, *55*, 400–408. [CrossRef]

36. Paglia, D.E.; Valentine, W.N. Studies on the Quantitative and Qualitative Characterization of Erythrocyte Glutathione Peroxidase. *J. Lab. Clin. Med.* **1967**, *70*, 158–169.
37. Bradford, M.M. A Rapid and Sensitive Method for the Quantitation of Microgram Quantities of Protein Utilizing the Principle of Protein-Dye Binding. *Anal. Biochem.* **1976**, *72*, 248–254. [CrossRef]
38. Buege, J.A.; Aust, S.D. Microsomal Lipid Peroxidation. *Methods Enzymol.* **1978**, *52*, 302–310. [CrossRef] [PubMed]
39. Levine, R.L.; Garland, D.; Oliver, C.N.; Amici, A.; Climent, I.; Lenz, A.-G.; Ahn, B.-W.; Shaltiel, S.; Stadtman, E.R. Determination of Carbonyl Content in Oxidatively Modified Proteins. *Methods Enzymol.* **1990**, *186*, 464–478. [CrossRef]
40. Tietze, F. Enzymic Method for Quantitative Determination of Nanogram Amounts of Total and Oxidized Glutathione: Applications to Mammalian Blood and Other Tissues. *Anal. Biochem.* **1969**, *27*, 502–522. [CrossRef] [PubMed]
41. Rahman, I.; Kode, A.; Biswas, S.K. Assay for Quantitative Determination of Glutathione and Glutathione Disulfide Levels Using Enzymatic Recycling Method. *Nat. Protoc.* **2006**, *1*, 3159–3165. [CrossRef]
42. Griffith, O.W. Determination of Glutathione and Glutathione Disulfide Using Glutathione Reductase and 2-Vinylpyridine. *Anal. Biochem.* **1980**, *106*, 207–212. [CrossRef]
43. Rizzo, A.M.; Negroni, M.; Altiero, T.; Montorfano, G.; Corsetto, P.; Berselli, P.; Berra, B.; Guidetti, R.; Rebecchi, L. Antioxidant Defences in Hydrated and Desiccated States of the Tardigrade *Paramacrobiotus richtersi*. *Comp. Biochem. Physiol. B* **2010**, *156*, 115–121. [CrossRef] [PubMed]
44. Zhang, Y.; Liang, S.; He, J.; Bai, Y.; Niu, Y.; Tang, X.; Li, D.; Chen, Q. Oxidative Stress and Antioxidant Status in a Lizard Phrynocephalus Vlangalii at Different Altitudes or Acclimated to Hypoxia. *Comp. Biochem. Physiol. A* **2015**, *190*, 9–14. [CrossRef] [PubMed]
45. Pannunzio, T.M.; Storey, K.B. Antioxidant Defenses and Lipid Peroxidation during Anoxia Stress and Aerobic Recovery in the Marine Gastropod *Littorina littorea*. *J. Exp. Mar. Biol. Ecol.* **1998**, *221*, 277–292. [CrossRef]
46. Moreira, D.C.; Sabino, M.A.C.T.; Kuzniewski, F.T.B.; Furtado-Filho, O.V.; Carvajalino-Fernández, J.M.; Angelini, R.; Freire, C.A.; Hermes-Lima, M. Redox Metabolism in Mussels (*Brachidontes solisianus*) under the Influence of Tides in a Rocky Beach in Southern Brazil. *Estuar. Coast. Shelf Sci.* **2021**, *258*, 107424. [CrossRef]
47. Ramos-Vasconcelos, G.R.; Hermes-Lima, M. Hypometabolism, Antioxidant Defenses and Free Radical Metabolism in the Pulmonate Land Snail *Helix aspersa*. *J. Exp. Biol.* **2003**, *206*, 675–685. [CrossRef]
48. Ramos-Vasconcelos, G.R.; Cardoso, L.A.; Hermes-Lima, M. Seasonal Modulation of Free Radical Metabolism in Estivating Land Snails Helix Aspersa. *Comp. Biochem. Physiol. C* **2005**, *140*, 165–174. [CrossRef]
49. Windrem, D.A.; Plachy, W.Z. The Diffusion-Solubility of Oxygen in Lipid Bilayers. *Biochim. Biophys. Acta* **1980**, *600*, 655–665. [CrossRef]
50. Dzikovski, B.G.; Livshits, V.A.; Marsh, D. Oxygen Permeation Profile in Lipid Membranes: Comparison with Transmembrane Polarity Profile. *Biophys. J.* **2003**, *85*, 1005–1012. [CrossRef]
51. Schafer, F.Q.; Buettner, G.R. Redox Environment of the Cell as Viewed through the Redox State of the Glutathione Disulfide/Glutathione Couple. *Free Radic. Biol. Med.* **2001**, *30*, 1191–1212. [CrossRef]
52. Yoshida, M.; Saegusa, Y.; Fukuda, A.; Akama, Y.; Owada, S. Measurement of Radical-Scavenging Ability in Hepatic Metallothionein of Rat Using in vivo Electron Spin Resonance Spectroscopy. *Toxicology* **2005**, *213*, 74–80. [CrossRef]
53. Denniss, S.G.; Levy, A.S.; Rush, J.W.E. Effects of Glutathione-Depleting Drug Buthionine Sulfoximine and Aging on Activity of Endothelium-Derived Relaxing and Contracting Factors in Carotid Artery of Sprague–Dawley Rats. *J. Cardiovasc. Pharmacol.* **2011**, *58*, 272–283. [CrossRef]
54. Abalenikhina, Y.V.; Myl'nikov, P.Y.; Shchul'kin, A.V.; Chernykh, I.V.; Yakusheva, E.N. Regulation and Role of Hypoxia-Induced Factor 1α (HIF-1α) under Conditions of Endogenous Oxidative Stress in vitro. *Bull. Exp. Biol. Med.* **2022**, *173*, 312–316. [CrossRef] [PubMed]
55. Merad-Boudia, M.; Nicole, A.; Santiard-Baron, D.; Saille, C.; Ceballos-Picot, I. Mitochondrial Impairment as an Early Event in the Process of Apoptosis Induced by Glutathione Depletion in Neuronal Cells: Relevance to Parkinson's Disease. *Biochem. Pharmacol.* **1998**, *56*, 645–655. [CrossRef]
56. Abdelhamid, G.; El-Kadi, A.O.S. Buthionine Sulfoximine, an Inhibitor of Glutathione Biosynthesis, Induces Expression of Soluble Epoxide Hydrolase and Markers of Cellular Hypertrophy in a Rat Cardiomyoblast Cell Line: Roles of the NF-KB and MAPK Signaling Pathways. *Free Radic. Biol. Med.* **2015**, *82*, 1–12. [CrossRef]
57. Morales, N.P.; Yamaguchi, Y.; Murakami, K.; Kosem, N.; Utsumi, H. Hepatic Reduction of Carbamoyl-PROXYL in Ferric Nitrilotriacetate Induced Iron Overloaded Mice: An in vivo ESR Study. *Biol. Pharm. Bull.* **2012**, *35*, 1035–1040. [CrossRef]
58. Li, Y.; Liang, K.; Yuan, L.; Gao, J.; Wei, L.; Zhao, L. The Role of Thioredoxin and Glutathione Systems in Arsenic-Induced Liver Injury in Rats under Glutathione Depletion. *Int. J. Environ. Health Res.* **2022**, 1–17. [CrossRef]
59. Ono, K.; Komuro, C.; Nishidai, T.; Shibamoto, Y.; Tsutsui, K.; Takahashi, M.; Abe, M. Radiosensitizing Effect of Misonidazole in Combination with an Inhibitor of Glutathione Synthesis in Murine Tumors. *Int. J. Radiat. Oncol. Biol. Phys.* **1986**, *12*, 1661–1666. [CrossRef]
60. Nakagawa, I.; Suzuki, M.; Imura, N.; Naganuma, A. Enhancement of Paraquat Toxicity by Glutathione Depletion in Mice in vivo and in vitro. *J. Toxicol. Sci.* **1995**, *20*, 557–564. [CrossRef] [PubMed]

61. Li, S.; Chen, J.; Xie, P.; Guo, X.; Fan, H.; Yu, D.; Zeng, C.; Chen, L. The Role of Glutathione Detoxification Pathway in MCLR-Induced Hepatotoxicity in SD Rats: Glutathione Detoxification Pathway of Microcystin-LR. *Environ. Toxicol.* **2015**, *30*, 1470–1480. [CrossRef] [PubMed]
62. Mello, D.F.; Trevisan, R.; Danielli, N.M.; Dafre, A.L. Vulnerability of Glutathione-Depleted *Crassostrea gigas* Oysters to Vibrio Species. *Mar. Environ. Res.* **2020**, *154*, 104870. [CrossRef]
63. Standeven, A.M.; Wetterhahn, K.E. Tissue-Specific Changes in Glutathione and Cysteine after Buthionine Sulfoximine Treatment of Rats and the Potential for Artifacts in Thiol Levels Resulting from Tissue Preparation. *Toxicol. Appl. Pharmacol.* **1991**, *107*, 269–284. [CrossRef] [PubMed]
64. Watanabe, T.; Sagisaka, H.; Arakawa, S.; Shibaya, Y.; Watanabe, M.; Igarashi, I.; Tanaka, K.; Totsuka, S.; Takasaki, W.; Manabe, S. A Novel Model of Continuous Depletion of Glutathione in Mice Treated with L-Buthionine(S,R)-Sulfoximine. *J. Toxicol. Sci.* **2003**, *28*, 455–469. [CrossRef] [PubMed]
65. Mizui, T.; Kinouchi, H.; Chan, P.H. Depletion of Brain Glutathione by Buthionine Sulfoximine Enhances Cerebral Ischemic Injury in Rats. *Am. J. Physiol.-Heart Circ. Physiol.* **1992**, *262*, H313–H317. [CrossRef] [PubMed]
66. Lluis, J.M.; Morales, A.; Blasco, C.; Colell, A.; Mari, M.; Garcia-Ruiz, C.; Fernandez-Checa, J.C. Critical Role of Mitochondrial Glutathione in the Survival of Hepatocytes during Hypoxia. *J. Biol. Chem.* **2005**, *280*, 3224–3232. [CrossRef] [PubMed]
67. Magalhães, J.; Ascensão, A.; Soares, J.M.C.; Neuparth, M.J.; Ferreira, R.; Oliveira, J.; Amado, F.; Duarte, J.A. Acute and Severe Hypobaric Hypoxia-Induced Muscle Oxidative Stress in Mice: The Role of Glutathione against Oxidative Damage. *Eur. J. Appl. Physiol.* **2004**, *91*, 185–191. [CrossRef] [PubMed]
68. Lushchak, V.I. Free Radicals, Reactive Oxygen Species, Oxidative Stress and Its Classification. *Chem. Biol. Interact.* **2014**, *224*, 164–175. [CrossRef]
69. Lushchak, V.I. Time-Course and Intensity-Based Classifications of Oxidative Stresses and Their Potential Application in Biomedical, Comparative and Environmental Research. *Redox Rep.* **2016**, *21*, 262–270. [CrossRef]
70. Hermes-Lima, M.; Storey, K.B. Antioxidant Defenses in the Tolerance of Freezing and Anoxia by Garter Snakes. *Am. J. Physiol. Integr. Comp. Physiol.* **1993**, *265*, 646–652. [CrossRef]
71. Chandel, N.S.; Maltepe, E.; Goldwasser, E.; Mathieu, C.E.; Simon, M.C.; Schumacker, P.T. Mitochondrial Reactive Oxygen Species Trigger Hypoxia-Induced Transcription. *Proc. Natl. Acad. Sci. USA* **1998**, *95*, 11715–11720. [CrossRef]
72. Chandel, N.S.; McClintock, D.S.; Feliciano, C.E.; Wood, T.M.; Melendez, J.A.; Rodriguez, A.M.; Schumacker, P.T. Reactive Oxygen Species Generated at Mitochondrial Complex III Stabilize Hypoxia-Inducible Factor-1α during Hypoxia. *J. Biol. Chem.* **2000**, *275*, 25130–25138. [CrossRef]
73. Clanton, T. Yet Another Oxygen Paradox. *J. Appl. Physiol.* **2005**, *99*, 1245–1246. [CrossRef] [PubMed]
74. Hernansanz-Agustín, P.; Izquierdo-Álvarez, A.; Sánchez-Gómez, F.J.; Ramos, E.; Villa-Piña, T.; Lamas, S.; Bogdanova, A.; Martínez-Ruiz, A. Acute Hypoxia Produces a Superoxide Burst in Cells. *Free Radic. Biol. Med.* **2014**, *71*, 146–156. [CrossRef] [PubMed]
75. Gerdol, M.; Sollitto, M.; Pallavicini, A.; Castellano, I. The Complex Evolutionary History of Sulfoxide Synthase in Ovothiol Biosynthesis. *Proc. R. Soc. B* **2019**, *286*, 20191812. [CrossRef]
76. Castellano, I.; Seebeck, F.P. On Ovothiol Biosynthesis and Biological Roles: From Life in the Ocean to Therapeutic Potential. *Nat. Prod. Rep.* **2018**, *35*, 1241–1250. [CrossRef]
77. Giraud-Billoud, M.; Vega, I.A.; Rinaldi Tosi, M.E.; Abud, M.A.; Calderón, M.L.; Castro-Vazquez, A. Antioxidant and Molecular Chaperone Defenses during Estivation and Arousal in the South American Apple-Snail *Pomacea canaliculata*. *J. Exp. Biol.* **2013**, *216*, 614–622. [CrossRef]
78. Rodriguez, C.; Campoy-Diaz, A.D.; Giraud-Billoud, M. Short-Term Estivation and Hibernation Induce Changes in the Blood and Circulating Hemocytes of the Apple Snail Pomacea Canaliculata. *Metabolites* **2023**, *13*, 289. [CrossRef] [PubMed]
79. Giraud-Billoud, M.; Castro-Vazquez, A.; Campoy-Diaz, A.D.; Giuffrida, P.M.; Vega, I.A. Tolerance to Hypometabolism and Arousal Induced by Hibernation in the Apple Snail Pomacea Canaliculata (Caenogastropoda, Ampullariidae). *Comp. Biochem. Physiol. Part B Biochem. Mol. Biol.* **2018**, *224*, 129–137. [CrossRef]
80. Krishnamoorthy, R.V. Hepatopancreatic Unsaturated Fatty Acids during Aestivation of the Snail, Pila Globosa. *Comp. Biochem. Physiol.* **1968**, *24*, 279–282. [CrossRef]
81. Murano, C.; Zuccarotto, A.; Leone, S.; Sollitto, M.; Gerdol, M.; Castellano, I.; Palumbo, A. A Survey on the Distribution of Ovothiol and OvoA Gene Expression in Different Tissues and Cells: A Comparative Analysis in Sea Urchins and Mussels. *Mar. Drugs* **2022**, *20*, 268. [CrossRef] [PubMed]
82. Castellano, I.; Migliaccio, O.; D'Aniello, S.; Merlino, A.; Napolitano, A.; Palumbo, A. Shedding Light on Ovothiol Biosynthesis in Marine Metazoans. *Sci. Rep.* **2016**, *6*, 21506. [CrossRef]

antioxidants

MDPI

Article

Tissue Oxidative Ecology along an Aridity Gradient in a Mammalian Subterranean Species

Paul J. Jacobs [1,*], Daniel W. Hart [1], Hana N. Merchant [2], Andries K. Janse van Vuuren [1], Chris G. Faulkes [3], Steven J. Portugal [2], Barry Van Jaarsveld [4] and Nigel C. Bennett [1]

[1] Department of Zoology and Entomology, Mammal Research Institute, University of Pretoria, Pretoria 0002, South Africa
[2] Department of Biological Sciences, School of Life and Environmental Sciences, Royal Holloway University of London, Egham, Surrey TW20 0EX, UK
[3] School of Biological and Behavioural Sciences, Queen Mary University of London, Mile End Road, London E1 4NS, UK
[4] Department of Physical Geography, Faculty of Geosciences, Utrecht University, Princetonlaan 8a, 3584 CB Utrecht, The Netherlands
* Correspondence: u10533207@tuks.co.za

Abstract: Climate change has caused aridification which can alter habitat vegetation, soil and precipitation profiles potentially affecting resident species. Vegetation and soil profiles are important for subterranean mole-rats as increasing aridity causes soils to become harder and geophytes less evenly distributed, and the inter-geophyte distance increases. Mole-rats obtain all water and dietary requirements from geophytes, and thus digging in harder soils may amplify stressors (hyperthermia, dehydration- or exercise-induced damage). This study assessed the oxidative status of the wild common mole-rat along an aridity gradient (arid, semi-arid and mesic). Kidney and liver oxidative markers, including total oxidant status (TOS), total antioxidant capacity (TAC), oxidative stress index (OSI), malondialdehyde (MDA) and superoxide dismutase (SOD) were measured. Liver oxidative status did not demonstrate any significance with the degree of the aridity gradient. Aridity affected the TAC and OSI of the kidney, with individuals in the most arid habitats possessing the highest TAC. The evolution of increased group size to promote survival in African mole-rats in arid habitats may have resulted in the additional benefit of reduced oxidative stress in the kidneys. The SOD activity of the kidneys was higher than that of the liver with lower oxidative damage, suggesting this species pre-emptively protects its kidneys as these are important for water balance and retention.

Keywords: oxidative stress; redox balance; oxidative status; life history; survival; precipitation; sociality; aridity; water balance; mole-rat

Citation: Jacobs, P.J.; Hart, D.W.; Merchant, H.N.; Janse van Vuuren, A.K.; Faulkes, C.G.; Portugal, S.J.; Van Jaarsveld, B.; Bennett, N.C. Tissue Oxidative Ecology along an Aridity Gradient in a Mammalian Subterranean Species. *Antioxidants* **2022**, *11*, 2290. https://doi.org/10.3390/antiox11112290

Academic Editors: Marcelo Hermes-Lima, Daniel Carneiro Moreira and Tania Zenteno-Savín

Received: 23 September 2022
Accepted: 15 November 2022
Published: 18 November 2022

1. Introduction

Animals have evolved specific behavioural, morphological, physiological and molecular adaptations to aid survival in their respective habitats [1–4]. Of late, organism adaptations to aridity (a region with low and unpredictable precipitation often accompanied by extreme temperatures) have become a focal point of research [5]. The aridity index (AI) indicates how wet or dry a region is; the lower the AI, the drier or less wet the region [5,6]. Aridification is the process of the AI declining; for example, a mesic region (a region with a moderate or well-balanced supply of water and a moderate temperature range) can become semi-arid through the process of aridification, changing temperature and precipitation profiles [5,6]. Consequences of aridification, as the result of climate change, may include changes to a habitat's soil drought profile, overall water availability [6–10], vegetation types and floristic biodiversity [11–14]. Understanding adaptations of organisms in different habitat aridities could be pivotal to our understanding of the possible changes required by organisms towards future aridification and climate extremes.

Animals that may be sensitive to changes in aridity are subterranean rodents, where a significant contributor to their increased sensitivity is the energetic costs associated with digging and the corresponding heat generation [15–17]. Excavation of tunnel systems is energetically more expensive (up to 3.6 times) than aboveground exploration [15,16]. Aridification may force subterranean rodents to dig in harder soils more frequently, which can result in a detrimental elevation of core body temperatures [17,18], with the possibility of exercise-induced hyperthermia [19,20]. Furthermore, aridification may affect the foraging of subterranean rodents; food sources would become more clumped and dispersed due to changes in flora, while foraging would become difficult due to increased soil hardness [21].

Oxidative stress has previously been used as a measure of the consequences of heat stress and/or dehydration [22–26] and exercise-induced damage [19,20,27–31]. Oxidative stress occurs when the oxidative balance is compromised in favour of reactive oxygen species (ROS) at the expense of antioxidant activity, which cannot prevent the overproduction of free radicals [32–36]. Additionally, oxidative stress can bring about physiological changes to lipids, DNA and proteins, which can compromise or disrupt cellular function and signalling [34,35,37,38]. Although physiological parameters have been investigated along an aridity gradient in mammals [39–44], to date, no study has investigated the oxidative status parameters of a mammal across an aridity gradient. The family Bathyergidae occupies a wide range of habitats from mesic to hyper-arid [45–47], as well as displaying a unique cline of sociality ranging from solitary, social and some species considered eusocial [48–50]. As such, this study set out to determine the variation in oxidative status of a single subterranean mammalian species, the common mole-rat (*Cryptomys hottentotus hottentotus*), across an aridity gradient, where this species occupies arid, semi-arid and mesic habitats [21]. A single species was chosen to avoid species-specific variation in oxidative status.

The ecological constraints and foraging risks associated with digging in harder soils in arid environments were proposed as conditions for mole-rat social behaviour termed the arid food distribution hypothesis [21,51]. Social behaviour would mitigate the energy expenditure required for digging in harder soils and allow groups to be more efficient in obtaining all their nutrient and water requirements from the storage organs of underground geophytes [21,51]. Thus, group living would alleviate the effects of aridification, where a group of mole-rats would be better equipped to deal with harsher environments [21,51]. Despite the benefit of group living, the subterranean niche has disadvantages, including hypoxic and hypercapnic conditions with poor ventilation and high humidity, which is exacerbated by increased group sizes [52–56]. The thermal environment of the burrow is fairly uniform, and temperatures are muted depending on soil depth, resulting in a limited preference for a thermal environment to potentially dissipate heat [57]. This suggests heat generation during digging is unlikely to be offloaded readily, resulting in possible hyperthermic changes in a subterranean animal when digging [16].

African mole-rats can utilise behavioural and physiological means of mitigating the harsh effects of an arid environment [48,49,57–60]. Firstly, a regular food resource is particularly important, as African mole-rats do not drink free-standing water but obtain all of their water requirements from the underground geophytes upon which they feed [61–63]. It was proposed by Hart et al. [60] that group living in mole-rats allows for increased efficiency in foraging for food, which culminates in year-round water availability, effectively resulting in behavioural osmoregulation. Mole-rats can readily use this excess water to urinate on themselves to increase evaporative cooling [59]. This is supported by a very high-water economy index (units of mL water used per kJ energy metabolized) in Damaraland mole-rats (*Fukomys damarensis*) despite inhabiting an arid environment [60]. African mole-rats can also selectively forage during post-rainfall periods, as moist soil is preferable to work with [64,65]. This allows digging in cooler soils to result in heat transfer via conduction, possibly allowing the mole-rat to dig more intensely and for longer periods [66]. Additionally, mole-rats can limit activity times during cooler periods regardless of rainfall, where heat stress may be less impactful [59,67,68]. Physiological adjustments include a higher conductance than rodents of comparable body sizes, which can assist in additional energy and water saving, as well as not requiring investment in energetically costly cooling mechanisms [48,57,59,69,70]. Additionally, mole-rats do not typically lower their metabolic rate due to living in an arid environment as surface-dwelling rodents do, as no correlation between aridity and metabolic rate occurs in mole-rats [61,71].

The current study investigated the oxidative status of the liver and kidney of the common mole-rat along an aridity gradient. Although the skeletal muscle and the heart [29] are vulnerable to exercise-induced oxidative stress [72–75], the liver [20,29,76,77] and the kidney [20,78–80] have also demonstrated vulnerability to exercise-induced oxidative stress; importantly, however, the kidney [80–83] and liver [22,84,85] are also susceptible to hyperthermia [86] and/or dehydration stress [22], making these tissues more likely to be stressed during hyperthermia-induced oxidative stress [87]. Since the kidneys are the most important organ for water retention and osmoregulation [88–90] and the liver susceptibility to hyperthermia [87], and a source of glutathione, an essential thiol for free radical scavenging during exercise [91], these two tissues were chosen for the current study. Differences in oxidative status and antioxidant activity of the liver and the kidney are provided as supplementary material (Supplementary S1). Markers used to investigate oxidative stress include antioxidant activity through non-enzymatic total antioxidant activity (TAC) and enzymatic antioxidant activity through the superoxide dismutase (SOD) enzyme. We measured total oxidant status (TOS) and malondialdehyde (MDA) as measures of oxidative damage potential and lipid oxidative damage, respectively. Lastly, we used the TOS:TAC measurement as an indicator of the oxidative stress index (OSI).

In accordance with behavioural osmoregulation and the aridity food-distribution hypothesis, possible oxidative stress caused by exercise-induced hyperthermia, temperature and/or dehydration brought about by digging in arid conditions would be negated as individuals that live in a larger group size would share digging duties with their colony mates and have a constant supply of food and water. Therefore, we predict that although habitats differ in aridity, the effect of differing aridity on oxidative damage (TOS and MDA), antioxidant activity (TAC and SOD) and oxidative stress index (OSI) for each tissue (kidney and liver) will have similar concentrations. We intend to discuss the results of our findings in light of aridification post hoc.

2. Materials and Methods

2.1. Study Area, Sampling Sites and Sampling Sites' Climate Data and Analysis

The arid site (Steinkopf) is defined as the Richtersveld bioregion. The Richtersveld is a desert landscape characterised by rugged kloofs and high mountains, situated in the north-western corner of South Africa's Northern Cape province. The Richtersveld plant life grows in rocky surfaces, gravel slopes, sandy sheet washplains and riverbeds in nutrient-poor soils (similar to Fynbos regions). The Richtersveld plant life is diverse and mainly endemic to South Africa comprising mainly succulents and aloe species, interspersed with succulent geophytes. This area receives a winter rainfall.

The semi-arid site (Darling) is defined as the west coast Renosterveld bioregion. Renosterveld plants grow in rich soil (in comparison to Fynbos regions). Typically, Renosterveld is largely confined to fine-grained soils—mainly clays and silts. The vegetation type is dominated by a species of grey-coloured plant called the *renosterbos*. However, the typical vegetation of the Fynbos tends to occur in very low abundance in Renosterveld. The types of plants are grasses, shrubs and small trees and perennials with numerous geophytes. This area also receives winter rainfall.

The mesic site (Somerset West) is defined as the Southwest Fynbos bioregion. This area is predominantly coastal and mountainous, with a Mediterranean climate and rainy winters. Fynbos plant life grows in nutrient-poor soils. The Fynbos plant life is one of the most diverse in the world and mainly endemic to South Africa, and comprises mainly succulent and aloe species with an array of geophytes. The most conspicuous components of the flora are evergreen sclerophyllous plants, many with ericoid leaves and gracile habit, as opposed to timber forest. Several plant families are conspicuous in fynbos; the Proteaceae are prominent, with genera such as Protea, Leucospermum (the "pincushions") and Leucadendron (the silver tree and "cone bushes").

According to Colantoni et al. [92], Somerset West (AI: 0.95 ± 0.03) is hyper-mesic, while Darling (AI: 0.51 ± 0.02) is semi-arid and Steinkopf (AI: 0.08 ± 0.01) is arid. Furthermore, the consequences of anthropogenic climate change within the last six years (2016-current) have exacerbated aridification, resulting in decreased AI in both the Western and Northern Cape (Figure 1). As such, Somerset West is currently classified as mesic, while Darling (AI: 0.51 ± 0.02) remains semi-arid, and Steinkopf (AI: 0.08 ± 0.01) is now hyper-arid (Figure 1).

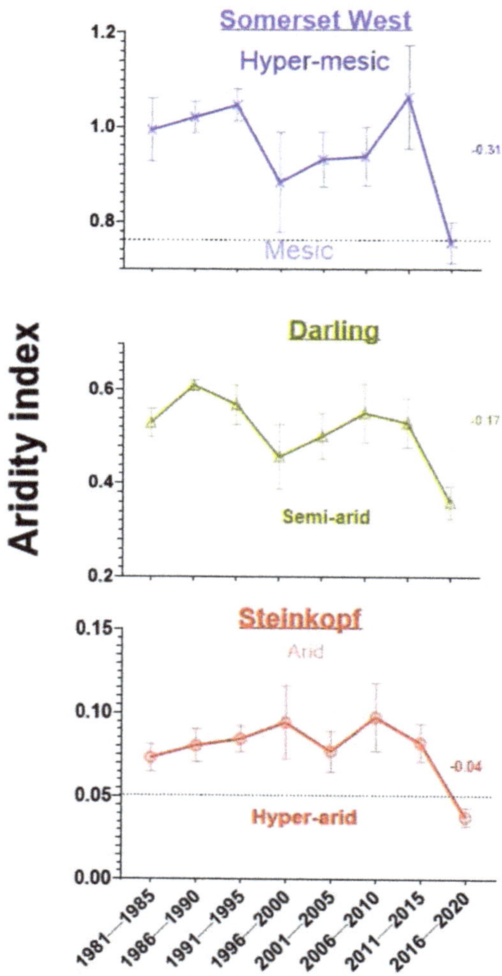

Figure 1. The aridity index for the last 40 years split into 5-year periods for three regions in South Africa, namely Somerset West (34.0757° S, 18.8433° E), Darling (33.3756° S, 18.3861° E) in the Western Cape and Steinkopf (29.2602° S, 17.7340° E) in the Northern Cape. Climate classification is according to Colantoni et al. [92]. Mean ± s.e.m.

Climate data for each site were retrieved from ERA5-Land of the European Centre for Medium-Range Weather Forecasts—the latest generation created by the Copernicus Climate Change Service [93]. The spatial resolution is 0.1 degree by 0.1 degree. Monthly averaged temperature (T_{air}; °C), total precipitation (tp; m), and dew point temperature (d2m; °C) from 1981 to 2020 were used. These data were used to calculate the annual aridity index (AI) (Equation (1)), where tp was directly obtained from ERA5-Land and potential evapotranspiration (PET) was calculated from the well-known Romanenko estimation (Equation (2)) [94]. For Equation (2), relative humidity (RH) was calculated from ERA5-Land d2m (Equation (3)).

$$AI = \frac{tp}{PET} \quad (1)$$

$$PET = 0.00006 \times (100 - RH) \times (25 + Tair)^2 \quad (2)$$

$$RH = 100 \times 10^{7.591386\left(\frac{d2m}{d2m+240.763} - \frac{Tair}{Tair+240.7263}\right)} \quad (3)$$

2.2. Animal Capture

The common mole-rats were captured from three localities in South Africa, namely Somerset West (34.0757° S, 18.8433° E; Western Cape), Darling (33.3756° S, 18.3861° E; Western Cape) and Steinkopf (29.2602° S, 17.7340° E; Northern Cape) between November 2021 and March 2022. Animals were captured using Hickman live traps [95] baited with sweet potatoes. Social African mole-rats show an extreme reproductive division of labour, whereby only one female and one to three males breed. At the same time, the remaining colony members are reproductively suppressed and would not have achieved sexual maturity or participated in breeding whilst in the confines of their natal colony [49]. All animals used in this study were considered adults and reproductively inactive (non-breeders) as a consequence of reproductive suppression [96,97]. Therefore, only male and female non-breeders were used in this study (see Hart, et al. [98] on how reproductive status was determined). The use of non-breeders only circumvents complications of oxidative stress associated with reproduction [99,100].

2.3. Animal Housing

Animals were transported back to the laboratory at the Department of Zoology and Entomology at the University of Pretoria (25.7545° S, 28.2314° E), South Africa, where the mole-rats were then housed. At the laboratory, the mole-rats were placed in a climate-controlled room, with a temperature of 25 °C and a humidity between 40–60%. All individuals were housed separately in large polyurethane containers (70 cm × 34 cm × 54 cm) containing wood shavings and paper towelling for nesting material. The animals were fed daily with sweet potatoes and apple ad libitum.

2.4. Age

Relative age was determined using tooth eruption and wear as outlined by Bennett, et al. [101], where animals were placed in relative age classes. Age has been observed to affect oxidative stress [102–105]; therefore, relative age was included in all analyses.

2.5. Reagents

Unless otherwise stated, all chemicals, tools and reagents in this study were obtained from Merck (Pty) Ltd. (Gauteng, South Africa).

2.6. Euthanasia and Tissue Excision

Samples were collected from 10 individuals (all non-breeders) per site (Somerset West: 6 males and 4 females; Darling: 5 males and 5 females; Steinkopf: 4 males and 6 females). All mole-rats came from at least five or more colonies from each site. All individuals were in captivity for no longer than a month (22–28 days) prior to being euthanised with an overdose of isoflurane. All animals were weighed immediately after death (Somerset West: 57.5 ± 3.33 g; Darling: 58.3 ± 4.31 g; Steinkopf: 74.7 ± 6.58 g). All liver and kidney samples were collected at the same time of the day (10:00–13:00) to prevent daily rhythm effects. Furthermore, tissues were collected in the same order (liver then kidney) within

10 min post-mortem, with an approximate 1-min interval between tissues. This was done to prevent and/or minimise proteins and metabolites from denaturing following dissection. The liver and kidney were then washed in 20 mM phosphate-buffered saline (PBS) to remove blood and flash-frozen in liquid nitrogen, and subsequently stored at −80 °C until analysis.

2.7. Tissue Homogenization Procedure

Liver and kidney were homogenised on ice by 10% weight per volume of 20 mM ice-cold PBS buffer in an 8 mL Potter–Elvehjem tissue grinder. Homogenates were split into separate tubes and centrifuged according to kit specifications. The supernatant was removed and stored in a −80 °C freezer until the time of analysis.

2.8. TOS

Tissue supernatant TOS levels were measured through Erel's method [106]. Briefly, this method is based on the oxidation of ferrous ion to ferric ion in the presence of various oxidative species. The oxidation reaction is enhanced by glycerol molecules, which are abundantly present in the reaction medium. The ferric ion makes a coloured complex with xylenol orange in an acidic medium. The colour intensity, measured spectrophotometrically, is related to the total amount of oxidant molecules that are present in the sample. The results are expressed in terms of micromole hydrogen peroxide equivalent per g tissue. Samples were run in duplicate and only once per plate with a repeatability of $r = 0.89$. Intra-assay variability was 4.18%.

2.9. MDA

The concentration of MDA was measured in all liver and kidney samples collected and was quantified using a commercially available kit (Sigma-Aldrich, cat. No. MAK085, A6283, 258105, and 360465), following standard procedures [107]. Polyunsaturated fatty acids (lipids) are susceptible to oxidative attack through ROS, resulting in MDA. The kit determines MDA content by reacting with thiobarbituric acid (TBA) to form a colorimetric complex at 532 nm. Absorbance was read using a Spectramax M2 plate reader (Molecular Devices Corp., Sunnyvale, CA, USA) and compared to a 2 mM MDA standard (2–10 nmol/mL). Results are expressed as nmol/g tissue. Samples were run in duplicate with repeatability of $r = 1.0$ and the intra-assay variability was 1.37%.

2.10. TAC

TAC in homogenates of liver and kidney were quantified using a commercially available kit (Antioxidant Assay Kit, Cayman Chemical Co., Ann Arbor, MI, USA) which measures the oxidation of ABTS (2.29-Azino-di- 3-ethybenzthiazoline sulphonate]) by metmyoglobin, which is inhibited by non-enzymatic antioxidants contained in the sample. Oxidized ABTS is measured by spectrophotometry at a wavelength of 750 nm. The capacity of antioxidants in the sample to inhibit oxidation of ABTS is compared with the capacity of known concentrations of Trolox. The results are expressed as µmol of Trolox equivalents per g tissue. Absorbance was read using a Spectramax M2 plate reader (Molecular Devices Corp., Sunnyvale, CA, USA). Samples were run in duplicate and only once per plate with a repeatability of $r = 0.99$. Intra-assay variability was 2.65%.

2.11. SOD

We measured the total SOD activity of liver and kidney samples, where SOD is an enzymatic antioxidant that catalyses the dismutation of superoxide anions to oxygen and hydrogen peroxide [25]. We measured SOD content with a commercially available kit (Superoxide Dismutase (SOD) Colorimetric Activity Kit, Arbor Assay, Arbor Assays, Ann Arbor, MI, USA) that measures the percentage of superoxide radicals that undergo dismutation in a given sample. Absorbances were read at 450 nm using a Spectramax M2 plate reader (Molecular Devices Corp., Sunnyvale, CA, USA). The results are expressed as

SOD activity per milligram tissue (SOD activity/mg tissue). Samples were run in duplicate and only once per plate with a repeatability of $r = 0.99$. Intra-assay variability was 5.07%.

2.12. OSI

The oxidative stress index (OSI) value was calculated according to the following formula: OSI (arbitrary unit) = [(TOS, µmol H_2O_2 equivalent/g tissue)/(TAC, µmol Trolox equivalent/g tissue)] $*$ 100, where the mmol value in the TAC test unit was converted to µmol units as in the TOS test [108,109].

2.13. Statistical Analyses

All statistical analyses were performed in R 4.2.1 [110]. Tissues were analysed separately and not compared due to the discrepancy of inherent tissue differences such as cellular turnover and accumulation or repair of damage [34]. As such, tissues were not compared statistically, and only patterns between the liver and kidney were considered. The normality of the response variables TOS, TAC, OSI, MDA and SOD was determined using Shapiro–Wilk tests. The homogeneity of all dependent variables was confirmed with Levene's test. Log transformation was used to normalise all non-normal data. Data were analysed using a linear model using the *lme4* package [111]. All initial models contained aridity; sex and aridity*sex interaction were run as predictors, with body mass and age as covariates (Table S2). Backwards elimination of linear models were performed using the *step* function of the lmerTest package in order to determine the best model for each response variable determined through the AIC criterion [112]. The best models of the backward elimination of all response variables are presented in Tables 1 and 2, while full backward elimination of models can be viewed in the Supplementary Material (Table S2). Significant variables in the regression models were followed up with post hoc comparisons, conducted using Tukey's HSD pairwise comparisons using the *emmeans* package [113]. Furthermore, Pearson correlations were conducted between SOD and MDA for kidney and liver from each population differing in aridity. Data are presented as mean ± standard error (s.e.m), and a p-value of ≤ 0.05 was defined as significant. Raw data can be observed in the Supplementary Material (Table S3).

Table 1. The linear best model output for kidney oxidative markers, namely total oxidant status (TOS), non-enzymatic total antioxidant activity (TAC), oxidative stress index (OSI), malondialdehyde (MDA) and superoxide dismutase (SOD), for the common mole-rat (*Cryptomys hottentotus hottentotus*) in response to body mass, age, aridity, sex and the interaction between sex and aridity determined through backwards elimination based on the Akaike information criterion.

Kidney	Variables Kept in Best Model	R^2	Adjusted R^2	F (df Regression, df Residual)	p (<0.05)
TOS	Aridity + Sex + Aridity*Sex	0.22	0.06	1.34 (5, 24)	0.28
TAC	Aridity + Sex	0.41	0.35	6.13 (3, 26)	0.003 *
OSI	Aridity + Sex	0.32	0.24	4.02 (3, 26)	0.02 *
MDA	No variable	0.13	−0.15	0.46 (7, 22)	0.86
SOD	Age	0.11	0.08	3.53 (1, 28)	0.07

Note: Full linear model before backwards elimination includes body mass, age, aridity, sex and sex*aridity as variables. * indicates significance ($p \leq 0.05$).

Table 2. The linear best model output for liver oxidative markers, namely total oxidant status (TOS), non-enzymatic total antioxidant activity (TAC), oxidative stress index (OSI), malondialdehyde (MDA) and superoxide dismutase (SOD), for the common mole-rat (*Cryptomys hottentotus hottentotus*) in response to body mass, age, aridity, sex and the interaction between sex and aridity determined through backwards elimination based on the Akaike information criterion.

Liver	Variables Kept in Best Model	R^2	Adjusted R^2	F (df Regression, df Residual)	p (<0.05)
TOS	No variable	0.15	−0.13	0.53 (7, 22)	0.80
TAC	No variable	0.09	−0.20	0.31 (7, 22)	0.94
OSI	No variable	0.09	−0.20	0.32 (7, 22)	0.94
MDA	No variable	0.21	−0.04	0.84 (7, 22)	0.56
SOD	Aridity + Sex + Aridity*Sex	0.29	0.14	1.98 (5, 24)	0.12

Note: Full linear model before backwards elimination includes body mass, age, aridity, sex and sex*aridity as variables. * indicates significance ($p \leq 0.05$).

3. Results

3.1. Oxidative Damage

3.1.1. Total Oxidant Status (TOS)

Age and body mass were not included in the kidney TOS best model (Table 1). At the same time, the best model contained aridity, sex and sex*aridity; none of these significantly affected kidney TOS levels (Table 1; Figure 2A).

The overall regression was not statistically significant, where no variable explained variation in liver TOS levels (Table 2; Figure 2A).

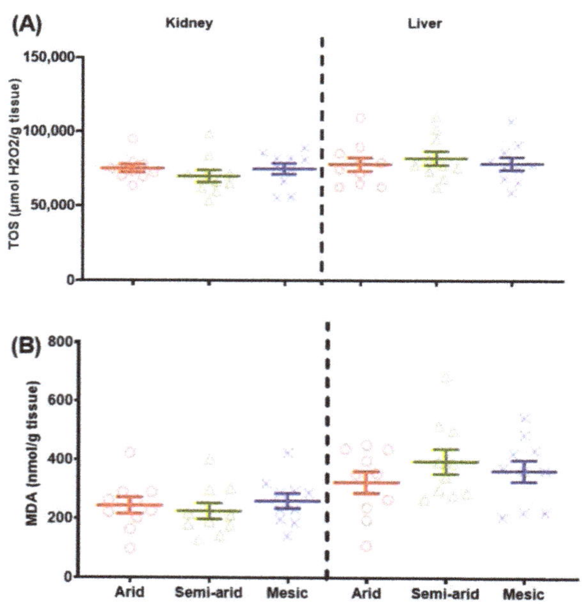

Figure 2. Kidney and liver (**A**) total oxidant status (TOS—µmol H_2O_2/g tissue) and (**B**) malondialdehyde (MDA—nmol/g tissue) of the common mole-rat (*Cryptomys hottentotus hottentotus*) for each population along an aridity gradient (arid—red (circles), semi-arid—moss (triangle), mesic—blue (x)). Mean ± s.e.m. An asterisk (*) indicates significance ($p \leq 0.05$).

3.1.2. Malondialdehyde (MDA)

The final regression models for kidney MDA (Table 1) and liver (Table 2) were not significant for any variable explaining variation in MDA levels (Figure 2B).

3.2. Antioxidant Defense

3.2.1. Total Antioxidant Capacity (TAC)

The overall regression was statistically significant for kidney TAC, where aridity and sex were kept in the model, with aridity being a significant contributor to the variation observed ($F_{(2,26)} = 7.5692$, $p = 0.003$) (Table 1). Post hoc analyses for kidney TAC show that the most arid individuals had the highest TAC as compared to semi-arid (t = −2.104, $p = 0.1086$) individuals and significantly higher compared to mesic individuals, which possessed the lowest TAC (t = 4.133, $p = 0.0009$; Figure 3A). Mesic and semi-arid populations did not differ significantly from each other (t = 2.071, $p = 0.1157$) (Figure 3A).

In the final regression model, no variables explained any variation observed in liver TAC levels (Table 2; Figure 3A).

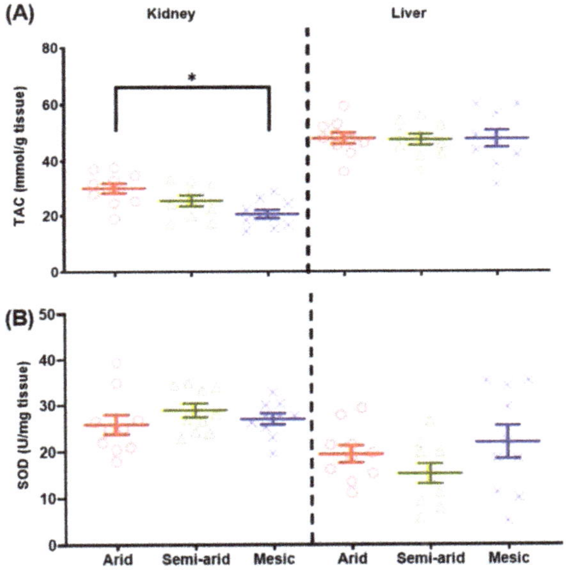

Figure 3. Kidney and liver (**A**) non-enzymatic total antioxidant activity (TAC—mmol/g tissue) and (**B**) enzymatic antioxidant activity through superoxide dismutase (SOD—U/mg tissue) of the common mole-rat (*Cryptomys hottentotus hottentotus*) for each population along an aridity gradient (arid—red (circles), semi-arid—moss (triangle), mesic—blue (x)). Mean ± s.e.m. An asterisk (*) indicates significance ($p \leq 0.05$).

3.2.2. Superoxide Dismutase (SOD)

Age was the only variable to explain any variation in kidney SOD levels and was negatively correlated with age ($r = -0.3$, $p = 0.07$), but this variation was not significant (Table 1; Figure 3B).

The best model for liver SOD levels included aridity, sex and aridity*sex, where aridity explained significant variation in the main model (t = 2.886, $p = 0.008$) (Table 2; Figure 3B). Post hoc analyses of aridity showed that no population was significantly affected by aridity (arid–mesic (t = 1.043, $p = 0.56$), semi-arid–mesic (t = 2.267, $p = 0.08$) and arid–semi-arid (t = 1.213, $p = 0.46$)) (Figure 3B).

3.3. Oxidative Stress: OSI

The kidney OSI regression model retained body mass, aridity and sex as variables contributing to the variation in kidney OSI, where only aridity was found to be a significant contributor to kidney OSI ($F(_{(2,25)} = 3.9429$, $p = 0.03$) (Table 1; Figure 4). Post hoc comparisons show that mesic individuals had the highest OSI compared to semi-arid individuals (t = −2.928, p = 0.0977) and significantly higher compared to arid individuals (t = −2.928, p = 0.019) (Figure 4). Conversely, arid individuals had the lowest OSI compared to semi-arid individuals, but this was not significant (t = 1.033, p = 0.5633) (Figure 4).

No variable explained any variation in liver OSI, with the regression model being not significant (Table 2; Figure 4).

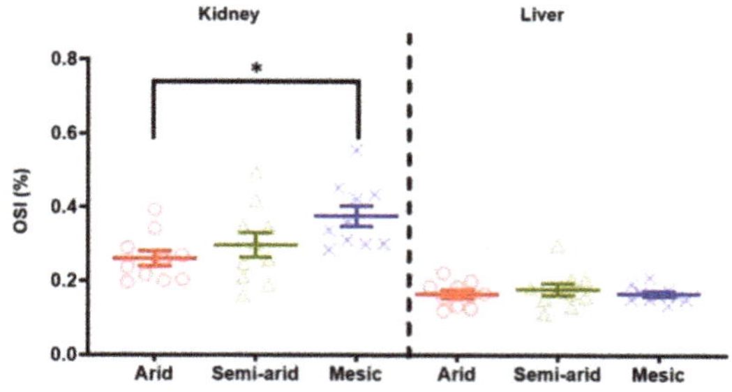

Figure 4. Kidney and liver oxidative stress index (OSI—%) of the common mole-rat (*Cryptomys hottentotus hottentotus*) for each population along an aridity gradient (arid—red (circles), semi-arid—moss (triangle), mesic—blue (x)). The OSI is the ratio of total oxidant status (TOS) to non-enzymatic total antioxidant activity (TAC). Mean ± s.e.m. An asterisk (*) indicates significance ($p \leq 0.05$).

3.4. SOD-MDA

A negative correlation was observed between kidney SOD and MDA for all three aridity index regions, which was significant only for populations from the arid (r = −0.71, p = 0.02) and semi-arid (r = −0.69, p = 0.03) regions, but not from the mesic region (r = −0.42, p = 0.22) (Figure 5).

No significant correlation was observed between liver SOD and MDA for arid, semi or mesic regions ($r \leq 0.08$, $p \geq 0.71$, for all three, Figure 5).

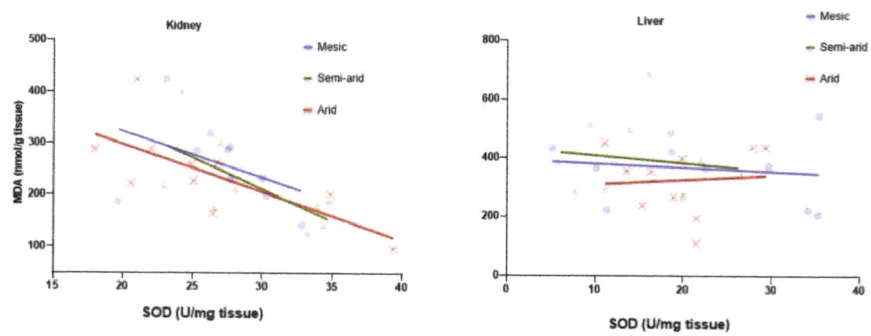

Figure 5. Correlations between malondialdehyde (MDA—nmol/g tissue) and superoxide dismutase (SOD—U/mg tissue) activity in the kidney and liver of the common mole-rat (*Cryptomys hottentotus hottentotus*) for each population along an aridity gradient (arid—red (circles), semi-arid—moss (triangle), mesic—blue (x)). Kidney—arid: r = −0.71, p = 0.02; semi-arid: r = −0.69, p = 0.03; mesic: (r = −0.42, p = 0.22. Liver—arid: r = 0.08, p = 0.84; semi-arid: r = −0.13, p = 0.71; mesic: r = −0.14, p = 0.71).

4. Discussion

In this study, we investigate for the first time (while controlling for age) the oxidative ecology of a wild subterranean rodent along an aridity gradient. In contrast to our initial predictions, we found that aridity did affect oxidative stress, but this was tissue-specific and driven by both the access to food and different types of food available rather than the production of ROS or resultant oxidative damage. Although the current study did not determine group sizes between the populations, our study investigated the same populations as Spinks, Bennett and Jarvis [21], where tendencies toward larger group sizes were observed in the arid population as compared to the mesic one.

The cost of digging in an arid environment may expose mole-rats to more frequent metabolic exertion that subsequently results in increased oxidative stress, but the social lifestyle of mole-rats may prevent over-exertion during digging, preventing oxidative stress [17,66]. As predicted, TOS levels for each tissue were similar regardless of aridity, where TOS levels are an indirect measure of the total ROS production [106]. Despite the increased energetic costs and the consequent predicted increase of ROS production from increased metabolism associated with digging in drier and harder soils [15–17], individuals in arid environments did not appear to exhibit elevated ROS from exercise, dehydration and/or heat stress generated as a consequence of digging. This implies that individuals from the arid environment consisting of larger groups may burrow less often compared to their counterparts in the mesic populations, that would burrow more frequently due to the increased ease of digging. Support for this includes field activity rates that were higher in a mesic-dwelling social African mole-rat species, the Natal mole-rat (*C. h. natalensis*) (average colony size of seven) [68], compared to an arid-dwelling social mole-rat species, the Damaraland mole-rat (*Fukomys damarensis*) (average colony size of 16) [114]. However, Hart et al. [60] found no significant difference when comparing the daily energy expenditure of the same two species. This may be a consequence of exposure to different soil moisture profiles when digging and colony size differences between the species. These factors promote social living and may help protect individuals from exercise-induced free radical production during digging [15–17], and further promote that mole-rats in arid environments are selectively digging during periods when the soil is workable [21,57,66]. Moreover, mole-rats may likely stop digging when at risk of hyperthermia [67].

In contrast to our initial predictions, TAC changes, and the concomitant OSI changes were significantly affected by aridity, which decreased with reduced aridity. A lower OSI would be observed, suggesting efficient ROS scavenging for individuals inhabiting arid environments in the kidneys. Non-enzymatic antioxidants are primarily obtained from food [115–117], with others created endogenously in an organism [118,119]. Interestingly, this trend is only observed in the kidneys, which may suggest that the food obtained may selectively enhance antioxidants in a tissue-specific manner [79,120,121] or prioritise kidney antioxidants through inter-organism transports of antioxidants [122,123]. Factors contributing to TAC differences can also be related to the different environments associated with aridity. Previous research has found a greater variety of geophytes available for arid mole-rats [21,49], suggesting that arid individuals have a wider dietary niche as a consequence of increased N and C isotopic values across all tissues [124]. Some geophytes with high antioxidants are only found in arid environments of Steinkopf [125,126]. Furthermore, alternative food sources such as clover and grass were also utilised [124]. As predicted by the arid food-distribution hypothesis, the reduced food density in arid environments may have relaxed the dietary specialization of arid dwelling mole-rats in order for them to adapt and survive in more arid habitats [124]. At the same time, geophytes in arid environments contain significantly more water [21], which may be a pre-emptive measure for plants to combat excessive ROS as a consequence of drought [125,127,128]. This pre-emptive measure to combat drought is also associated with elevated antioxidants in the plants and even their roots, which is likely to occur more regularly in arid environments [125]. It may be that food sources for the common mole-rat in arid environments are richer in antioxidants, compared to more mesic habitats, but this requires further investigation. Since group sizes were not accounted for in the current study, we cannot say with certainty that our data supports the aridity food-distribution hypothesis, however, access to the possibility of a variety of different food sources leading to higher TAC values compared to the semi-arid population suggests that arid animals have sufficient access to food.

The captivity effect on antioxidant activity also needs to be considered. Firstly, food is provided and no longer foraged for, which may decrease the metabolic rate, where a drop in resting metabolic rate was observed after 3 months of acclimation to captivity [58]. The observed acclimation was observed in mole-rats caught only from the most arid region, Steinkopf. The drop in metabolism may affect endogenous antioxidant production [129], likely affecting measured antioxidant levels. Furthermore, glutathione and other antioxidants are rapidly turned over [130] suggesting that observed antioxidant levels may result from captive feeding. Interestingly, if captive feeding was the only source of antioxidants and feeding acclimation occurred, mole-rats from all populations would have demonstrated the same TAC. This suggests that even if dietary acclimation occurred, arid dwelling mole-rats still possess higher non-enzymatic antioxidants in their kidneys and selectively protect them. Other variables which can influence observed levels include the rate of consumption of food by mole-rats which was not controlled for. Furthermore, if additional food consumption took place, the excess antioxidant could result in oxidative damage [131,132], but oxidative damage did not occur, suggesting the mole-rats incorporated the same amount of antioxidants into their system from exogenous sources.

Mammalian oxidative ecology demonstrates that generally, at the basal level, SOD activity is higher in the liver as compared to the kidneys [115,133]. The current study found that kidney SOD activity was slightly higher than that of the liver, and this pattern was consistent regardless of aridity. This is in contrast to other subterranean species which followed the general mammalian trend of higher SOD activity in the liver, such as Brandt's vole [10] and the Damaraland mole-rat [134]. It is not unusual for kidney SOD activity to be higher than liver SOD activity, as this trend has previously been observed in Wistar rats [135,136], broiler chickens [137] and North American beavers [115]. The common mole-rat may represent one of the species with inherently higher SOD activity in the kidneys as compared to the liver. Alternative ecological explanations for this phenomenon may include hormesis, the process of protection against an oxidative insult resulting in an overall benefit [138–141]. The kidneys play a critical role in the water balance and retention in mole-rats as they only obtain water from their food resources [62], and the inability of mole-rats to significantly concentrate urine [142], may necessitate protection of kidneys from oxidative insults to maintain optimal kidney function. Additionally, SOD activity levels in the kidneys demonstrated a near significant negative correlation with age, a likely factor in preserving the longevity of common mole-rats. Liver SOD activity also demonstrated a negative correlation, but not to the degree of the kidney. Lastly, an inverse relationship was observed for SOD in relation to MDA for both tissues, but its effect was significant only for the kidneys. Furthermore, this effect was only significant for the arid and semi-arid population and not for the mesic population. This trend being only significant for mole-rats occupying arid and semi-arid habitats suggests prioritised protection of the kidney as compared to the liver in an arid environment, likely due to its involvement in water retention [142–144] and possibly its susceptibility to hypoxia [145], a lifestyle associated with subterranean and/or fossorial animals [53,54]. Water is likely still a precious commodity despite its apparent abundance for social mole-rats, and emphasises the importance of kidneys in the common mole-rat and likely other mole-rats for future tissue oxidative status considerations.

One point of note of the current study is that all three sites have experienced aridification, although this effect was most profound in the mesic and arid sites. Aridification can influence vegetation, precipitation and soil properties [6–10], all ecological constraints which explain why mole-rats have become social and increased in group size as predicted by the arid food-distribution hypothesis [21,51]. Despite the mesic site being still considered mesic, this phenomenon highlights the concern about the rate of aridification of this site. It is uncertain whether current colony sizes may have contributed to differences observed in kidney TAC values due to reduced efficiency in group foraging and/or the possible differences in vegetation, and vegetation as a food source and antioxidant content. If the aridity food-distribution hypothesis holds true, if the mesic population does not increase group size with increasing aridity, it may limit the amount of food that is foraged and likely force increased cooperation in the mesic-dwelling social mole-rats. Since TOS and MDA did not change regardless of aridity in each tissue, our results suggest that aridification did not significantly affect oxidative damage at any degree of aridity.

There were some limitations to the current study where future studies may want to direct their focus. Our first concern is the possible acclimation of animals to captivity conditions, which could confound the mechanism behind our observation. As such, future studies may benefit from in-field sampling; however, this requires preserving samples in remote locations. Alternatively, animals may need to be sacrificed sooner to prevent captivity acclimation. Furthermore, the physiological mechanism can be investigated between populations focusing on the kidneys, such as the diuretic hormone vasopressin [3] and more in-depth exogenous antioxidants used by the kidneys, such as targeting glutathione, vitamins C and E, carotenoids etc. [146]. Also, mole-rat food (e.g., geophytes), as suggested, may differ in exogenous antioxidants along an aridity gradient, thus, future studies investigating aridification should collect food items and determine antioxidant content. Lastly, although we observed that the common mole-rat was unaffected by oxidative stress as a consequence of aridification, this observation may only be true for the social species, as aridification may pose a significant risk for solitary mole-rats. Hart et al. [60] found that the arid-dwelling solitary mole-rat species, the Namaqua dune mole-rat (*Bathergus janetta*), is close to local extinction, possibly due to the aridification of their habitat. The Namaqua dune mole-rat and other solitary mole-rats have been observed to possess the lowest plasma TAC when compared to their social counterparts, with the lowest observed for the Namaqua dune mole-rat, the only arid-dwelling solitary mole-rat species (Jacobs unpublish. data). This may suggest that solitary mole-rats may be more adversely affected by aridification and should be considered for future investigation for conservation purposes.

5. Conclusions

This study found that wild, social, common mole-rats in arid environments value and selectively protect their kidneys with higher antioxidants. The protection of the kidneys may have longevity consequences, but as it stands, social mole-rats do not suffer oxidative stress from aridification in the liver or kidney. An imbalance between oxidative damage and defence (OSI) can have detrimental effects on a mammal's health and life span. It is clear that aridity affects this balance, and in light of climate change, it is critical to address the dearth of knowledge regarding the effect of aridity on a mammal's oxidative ecology.

Supplementary Materials: The following supporting information can be downloaded at: https://www.mdpi.com/article/10.3390/antiox11112290/s1, Supplementary S1: The liver and kidney oxidant and antioxidant status difference between tissues; Table S2: The full backwards elimination models of all statistical outputs for total oxidant status (TOS), malondialdehyde (MDA), total antioxidant capacity (TAC), superoxide dismutase (SOD) and oxidative stress index (OSI) for kidney and liver; Table S3: The raw data for common mole-rats *Cryptomys hottentotus* hottentotus, consisting of total oxidant status (TOS), malondialdehyde (MDA), total antioxidant capacity (TAC), superoxide dismutase (SOD), oxidative stress index (OSI), age, body mass, sex and aridity for kidney and liver [147–156].

Author Contributions: Conceptualization, P.J.J., D.W.H. and N.C.B.; methodology, P.J.J., D.W.H. and H.N.M.; formal analysis, P.J.J., D.W.H. and B.V.J.; investigation, P.J.J., D.W.H., H.N.M. and A.K.J.v.V.; resources, N.C.B.; writing—original draft preparation, P.J.J. and D.W.H.; writing—review and editing, P.J.J., D.W.H., N.C.B., H.N.M., A.K.J.v.V., C.G.F., S.J.P., B.V.J. and N.C.B.; visualization, P.J.J. and D.W.H.; project administration, D.W.H. and N.C.B.; funding acquisition, N.C.B., S.J.P. and C.G.F. All authors have read and agreed to the published version of the manuscript.

Funding: NCB acknowledges funding from the SARChI chair of Mammalian Behavioural Ecology and Physiology from the DST-NRF South Africa, the National Research Foundation (grant no. 64756), the Natural Environment Research Council under grant NE\L002485\1 and the University of Pretoria.

Institutional Review Board Statement: Permission to capture common mole-rats was obtained from all landowners. A collecting permit was obtained from the relevant nature conservation authorities (permit No. FAUNA 0419/2021, FAUNA 042/2021, CPB6,-1161, CPB6-1163, CNN44-87-17699).

The Animal Use and Care Committee of the University of Pretoria evaluated and approved the experimental protocol (ethics clearance No. NAS016/2021) and DAFF section 20 approval (SDAH-Epi-21031811071).

Informed Consent Statement: Not applicable.

Data Availability Statement: Data is contained within the article or supplementary materials.

Conflicts of Interest: The authors declare no conflict of interest.

References

1. Huey, R.B.; Kearney, M.R.; Krockenberger, A.; Holtum, J.A.; Jess, M.; Williams, S.E. Predicting organismal vulnerability to climate warming: Roles of behaviour, physiology and adaptation. *Philos. Trans. R. Soc. B Biol. Sci.* **2012**, *367*, 1665–1679. [CrossRef]
2. Degen, A.A. *Ecophysiology of Small Desert Mammals*; Springer Science & Business Media: Berlin/Heidelberg, Germany, 2012.
3. Schwimmer, H.; Haim, A. Physiological adaptations of small mammals to desert ecosystems. *Integr. Zool.* **2009**, *4*, 357–366. [CrossRef]
4. Gardner, J.L.; Peters, A.; Kearney, M.R.; Joseph, L.; Heinsohn, R. Declining body size: A third universal response to warming? *Trends Ecol. Evol.* **2011**, *26*, 285–291. [CrossRef]
5. Sahin, S. An aridity index defined by precipitation and specific humidity. *J. Hydrol.* **2012**, *444*, 199–208. [CrossRef]
6. Park, C.-E.; Jeong, S.-J.; Joshi, M.; Osborn, T.J.; Ho, C.-H.; Piao, S.; Chen, D.; Liu, J.; Yang, H.; Park, H. Keeping global warming within 1.5 °C constrains emergence of aridification. *Nat. Clim. Chang.* **2018**, *8*, 70–74. [CrossRef]
7. Mosley, L.M. Drought impacts on the water quality of freshwater systems; review and integration. *Earth-Sci. Rev.* **2015**, *140*, 203–214. [CrossRef]
8. Huang, J.; Yu, H.; Dai, A.; Wei, Y.; Kang, L. Drylands face potential threat under 2 °C global warming target. *Nat. Clim. Chang.* **2017**, *7*, 417–422. [CrossRef]
9. Hu, W.; Ran, J.; Dong, L.; Du, Q.; Ji, M.; Yao, S.; Sun, Y.; Gong, C.; Hou, Q.; Gong, H. Aridity-driven shift in biodiversity–soil multifunctionality relationships. *Nat. Commun.* **2021**, *12*, 5350. [CrossRef]
10. Shi, L.; Chen, B.; Wang, X.; Huang, M.; Qiao, C.; Wang, J.; Wang, Z. Antioxidant response to severe hypoxia in Brandt's vole *Lasiopodomys brandtii*. *Integr. Zool.* **2022**, *17*, 581–595. [CrossRef]
11. Perrino, E.; Signorile, G. Costa di Monopoli (Puglia): Check-list della flora vascolare. *Inf. Bot. Ital.* **2009**, *41*, 263–279.
12. Galasso, G.; Domina, G.; Adorni, M.; Ardenghi, N.M.G.; Banfi, E.; Bedini, G.; Bertolli, A.; Brundu, G.; Calbi, M.; Cecchi, L. Notulae to the Italian alien vascular flora: 1. *Ital. Bot.* **2016**, *1*, 17–37. [CrossRef]
13. Pokorny, L.; Riina, R.; Mairal, M.; Meseguer, A.S.; Culshaw, V.; Cendoya, J.; Serrano, M.; Carbajal, R.; Ortiz, S.; Heuertz, M. Living on the edge: Timing of Rand Flora disjunctions congruent with ongoing aridification in Africa. *Front. Genet.* **2015**, *6*, 154. [CrossRef]
14. Fernández-González, F.; Loidi, J.; Moreno, J.C.; Del Arco, M.; Férnández-Cancio, A.; Galán, C.; García-Mozo, H.; Muñoz, J.; Pérez-Badia, R.; Sardinero, S. Impacts on plant biodiversity. In *Impacts on Climatic Change in Spain*; OCCE, Ministerio de Medio Ambiente: Madrid, Spain, 2005; pp. 183–248.
15. Vleck, D. The energy cost of burrowing by the pocket gopher Thomomys bottae. *Physiol. Zool.* **1979**, *52*, 122–136. [CrossRef]
16. Lovegrove, B. The cost of burrowing by the social mole rats (Bathyergidae) *Cryptomys damarensis* and *Heterocephalus glaber*: The role of soil moisture. *Physiol. Zool.* **1989**, *62*, 449–469. [CrossRef]
17. Zelová, J.; Šumbera, R.; Okrouhlík, J.; Burda, H. Cost of digging is determined by intrinsic factors rather than by substrate quality in two subterranean rodent species. *Physiol. Behav.* **2010**, *99*, 54–58. [CrossRef]
18. Van Jaarsveld, B.; Bennett, N.C.; Hart, D.W.; Oosthuizen, M.K. Locomotor activity and body temperature rhythms in the Mahali mole-rat (*C. h. mahali*): The effect of light and ambient temperature variations. *J. Therm. Biol.* **2019**, *79*, 24–32. [CrossRef]
19. McAnulty, S.; McAnulty, L.; Pascoe, D.; Gropper, S.; Keith, R.; Morrow, J.; Gladden, L. Hyperthermia increases exercise-induced oxidative stress. *Int. J. Sport. Med.* **2005**, *26*, 188–192. [CrossRef]
20. Belviranli, M.; Gökbel, H. Acute exercise induced oxidative stress and antioxidant changes. *Eur. J. Gen. Med.* **2006**, *3*, 126–131. [CrossRef]
21. Spinks, A.C.; Bennett, N.C.; Jarvis, J.U. A comparison of the ecology of two populations of the common mole-rat, *Cryptomys hottentotus hottentotus*: The effect of aridity on food, foraging and body mass. *Oecologia* **2000**, *125*, 341–349. [CrossRef]
22. Jacobs, P.J.; Oosthuizen, M.; Mitchell, C.; Blount, J.D.; Bennett, N.C. Heat and dehydration induced oxidative damage and antioxidant defenses following incubator heat stress and a simulated heat wave in wild caught four-striped field mice Rhabdomys dilectus. *PLoS ONE* **2020**, *15*, e0242279. [CrossRef]
23. Slimen, I.B.; Najar, T.; Ghram, A.; Dabbebi, H.; Ben Mrad, M.; Abdrabbah, M. Reactive oxygen species, heat stress and oxidative-induced mitochondrial damage. A review. *Int. J. Hyperth.* **2014**, *30*, 513–523. [CrossRef]
24. Habashy, W.S.; Milfort, M.C.; Rekaya, R.; Aggrey, S.E. Cellular antioxidant enzyme activity and biomarkers for oxidative stress are affected by heat stress. *Int. J. Biometeorol.* **2019**, *63*, 1569–1584. [CrossRef]
25. França, M.; Panek, A.; Eleutherio, E. Oxidative stress and its effects during dehydration. *Comp. Biochem. Physiol. Part A Mol. Integr. Physiol.* **2007**, *146*, 621–631. [CrossRef]

26. Altan, Ö.; Pabuçcuoğlu, A.; Altan, A.; Konyalioğlu, S.; Bayraktar, H. Effect of heat stress on oxidative stress, lipid peroxidation and some stress parameters in broilers. *Br. Poult. Sci.* **2003**, *44*, 545–550. [CrossRef]
27. Georgescu, V.P.; de Souza Junior, T.P.; Behrens, C.; Barros, M.P.; Bueno, C.A.; Utter, A.C.; McAnulty, L.S.; McAnulty, S.R. Effect of exercise-induced dehydration on circulatory markers of oxidative damage and antioxidant capacity. *Appl. Physiol. Nutr. Metab.* **2017**, *42*, 694–699. [CrossRef]
28. Hillman, A.R.; Vince, R.V.; Taylor, L.; McNaughton, L.; Mitchell, N.; Siegler, J. Exercise-induced dehydration with and without environmental heat stress results in increased oxidative stress. *Appl. Physiol. Nutr. Metab.* **2011**, *36*, 698–706. [CrossRef]
29. King, M.A.; Clanton, T.L.; Laitano, O. Hyperthermia, dehydration, and osmotic stress: Unconventional sources of exercise-induced reactive oxygen species. *Am. J. Physiol.-Regul. Integr. Comp. Physiol.* **2016**, *310*, R105–R114. [CrossRef]
30. Powers, S.K.; Radak, Z.; Ji, L.L. Exercise-induced oxidative stress: Past, present and future. *J. Physiol.* **2016**, *594*, 5081–5092. [CrossRef]
31. Vollaard, N.B.; Shearman, J.P.; Cooper, C.E. Exercise-induced oxidative stress. *Sport. Med.* **2005**, *35*, 1045–1062. [CrossRef]
32. Davies, K.J. Oxidative stress, antioxidant defenses, and damage removal, repair, and replacement systems. *IUBMB Life* **2000**, *50*, 279–289. [CrossRef]
33. Sies, H. Oxidative stress: A concept in redox biology and medicine. *Redox. Biol.* **2015**, *4*, 180–183. [CrossRef]
34. Costantini, D. Understanding diversity in oxidative status and oxidative stress: The opportunities and challenges ahead. *J. Exp. Biol.* **2019**, *222*, jeb194688. [CrossRef]
35. Sindhi, V.; Gupta, V.; Sharma, K.; Bhatnagar, S.; Kumari, R.; Dhaka, N. Potential applications of antioxidants–A review. *J. Pharm. Res.* **2013**, *7*, 828–835. [CrossRef]
36. Halliwell, B.; Gutteridge, J. *Free Radical in Biology and Medicine*, 2nd ed.; Oxford Science Publications: Oxford, UK, 1989.
37. Jones, D.P. Redefining oxidative stress. *Antioxid. Redox Signal.* **2006**, *8*, 1865–1879. [CrossRef]
38. Halliwell, B. Biochemistry of oxidative stress. *Biochem. Soc. Trans.* **2007**, *35*, 1147–1150. [CrossRef]
39. Coleman, J.C.; Downs, C.T. Variation in urine concentrating ability and water balance of the Black-tailed Tree Rat *Thallomys nigricauda*, along an aridity gradient. *Comp. Biochem. Physiol. Part A Mol. Integr. Physiol.* **2009**, *154*, 508–513. [CrossRef]
40. Coleman, J.C.; Downs, C.T. Daily rhythms of body temperature and activity in free-living Black-tailed Tree Rats (*Thallomys nigricauda*) along an aridity gradient. *Physiol. Behav.* **2010**, *99*, 22–32. [CrossRef]
41. Bozinovic, F.; Rojas, J.M.; Broitman, B.R.; Vásquez, R.A. Basal metabolism is correlated with habitat productivity among populations of degus (*Octodon degus*). *Comp. Biochem. Physiol. Part A Mol. Integr. Physiol.* **2009**, *152*, 560–564. [CrossRef]
42. Bozinovic, F.; Rojas, J.; Gallardo, P.; Palma, R.; Gianoli, E. Body mass and water economy in the South American olivaceous field mouse along a latitudinal gradient: Implications for climate change. *J. Arid Environ.* **2011**, *75*, 411–415. [CrossRef]
43. van Jaarsveld, B.; Bennett, N.C.; Kemp, R.; Czenze, Z.J.; McKechnie, A.E. Heat tolerance in desert rodents is correlated with microclimate at inter-and intraspecific levels. *J. Comp. Physiol. B* **2021**, *191*, 575–588. [CrossRef]
44. Conenna, I.; Santini, L.; Rocha, R.; Monadjem, A.; Cabeza, M.; Russo, D. Global patterns of functional trait variation along aridity gradients in bats. *Glob. Ecol. Biogeogr.* **2021**, *30*, 1014–1029. [CrossRef]
45. Faulkes, C.G.; Bennett, N.C. Plasticity and constraints on social evolution in African mole-rats: Ultimate and proximate factors. *Phil. Trans. R. Soc. B.* **2013**, *368*, 20120347. [CrossRef]
46. Faulkes, C.; Bennett, N.; Bruford, M.W.; O'brien, H.; Aguilar, G.; Jarvis, J. Ecological constraints drive social evolution in the African mole–rats. *Proc. R. Soc. Lond. Ser. B Biol. Sci.* **1997**, *264*, 1619–1627. [CrossRef]
47. Faulkes, C.; Verheyen, E.; Verheyen, W.; Jarvis, J.; Bennett, N. Phylogeographical patterns of genetic divergence and speciation in African mole-rats (Family: Bathyergidae). *Mol. Ecol.* **2004**, *13*, 613–629. [CrossRef]
48. McGowan, N.E.; Scantlebury, D.M.; Bennett, N.C.; Maule, A.G.; Marks, N.J. Thermoregulatory differences in African mole-rat species from disparate habitats: Responses and limitations. *J. Therm. Biol.* **2020**, *88*, 102495. [CrossRef]
49. Bennett, N.C.; Faulkes, C.G. *African Mole-Rats: Ecology and Eusociality*; Cambridge University Press: Cambridge, UK, 2000.
50. Faulkes, C.G.; Bennett, N.C. Social evolution in African mole-rats–a comparative overview. In *The Extraordinary Biology of the Naked Mole-Rat*; Springer: Cham, Switzerland, 2021; pp. 1–33. [CrossRef]
51. Jarvis, J.U.; O'Riain, M.J.; Bennett, N.C.; Sherman, P.W. Mammalian eusociality: A family affair. *Trends Ecol. Evol.* **1994**, *9*, 47–51. [CrossRef]
52. Roper, T.; Bennett, N.; Conradt, L.; Molteno, A. Environmental conditions in burrows of two species of African mole-rat, *Georhychus capensis* and *Cryptomys damarensis*. *J. Zool.* **2001**, *254*, 101–107. [CrossRef]
53. Ivy, C.M.; Sprenger, R.J.; Bennett, N.C.; van Jaarsveld, B.; Hart, D.W.; Kirby, A.M.; Yaghoubi, D.; Storey, K.B.; Milsom, W.K.; Pamenter, M.E. The hypoxia tolerance of eight related African mole-rat species rivals that of naked mole-rats, despite divergent ventilatory and metabolic strategies in severe hypoxia. *Acta Physiol.* **2020**, *228*, e13436. [CrossRef]
54. Logan, S.M.; Szereszewski, K.E.; Bennett, N.C.; Hart, D.W.; Van Jaarsveld, B.; Pamenter, M.E.; Storey, K.B. The brains of six African mole-rat species show divergent responses to hypoxia. *J. Exp. Biol.* **2020**, *223*, jeb215905. [CrossRef]
55. Burda, H.; Šumbera, R.; Begall, S. Microclimate in burrows of subterranean rodents—Revisited. In *Subterranean Rodents*; Springer: Berlin/Heidelberg, Germany, 2007; pp. 21–33.
56. Pamenter, M.E. Adaptations to a hypoxic lifestyle in naked mole-rats. *J. Exp. Biol.* **2022**, *225*, jeb196725. [CrossRef]
57. Šumbera, R. Thermal biology of a strictly subterranean mammalian family, the African mole-rats (Bathyergidae, Rodentia)-a review. *J. Therm. Biol.* **2019**, *79*, 166–189. [CrossRef] [PubMed]

58. Bennett, N.; Clarke, B.; Jarvis, J. A comparison of metabolic acclimation in two species of social mole-rats (Rodentia, Bathyergidae) in southern Africa. *J. Arid Environ.* **1992**, *23*, 189–198. [CrossRef]
59. Wallace, K.M.; van Jaarsveld, B.; Bennett, N.C.; Hart, D.W. The joint effect of micro-and macro-climate on the thermoregulation and heat dissipation of two African mole-rat (Bathyergidae) sub-species, *Cryptomys hottentotus mahali* and *C. h. pretoriae*. *J. Therm. Biol.* **2021**, *99*, 103025. [CrossRef] [PubMed]
60. Hart, D.; Bennett, N.; Oosthuizen, M.; Waterman, J.; Hambly, C.; Scantlebury, D. Energetics and water flux in the subterranean rodent family Bathyergidae. *Front. Ecol. Evol.* **2022**, *10*, 867350. [CrossRef]
61. Bennett, N.; Aguilar, G.; Jarvis, J.; Faulkes, C. Thermoregulation in three species of Afrotropical subterranean mole-rats (Rodentia: Bathyergidae) from Zambia and Angola and scaling within the genus Cryptomys. *Oecologia* **1994**, *97*, 222–227. [CrossRef]
62. Bennett, N.; Jarvis, J. Coefficients of digestibility and nutritional values of geophytes and tubers eaten by southern African mole-rats (Rodentia: Bathyergidae). *J. Zool.* **1995**, *236*, 189–198. [CrossRef]
63. Spinks, A.C.; Branch, T.A.; Croeser, S.; Bennett, N.C.; Jarvis, J.U. Foraging in wild and captive colonies of the common mole-rat *Cryptomys hottentotus hottentotus* (Rodentia: Bathyergidae). *J. Zool.* **1999**, *249*, 143–152. [CrossRef]
64. Jarvis, J.U.; Bennett, N.C.; Spinks, A.C. Food availability and foraging by wild colonies of Damaraland mole-rats (*Cryptomys damarensis*): Implications for sociality. *Oecologia* **1998**, *113*, 290–298. [CrossRef]
65. Lovegrove, B.; Painting, S. Variations in the foraging behaviour and burrow structures of the Damara molerat *Cryptomys damarensis* in the Kalahari Gemsbok National Park. *Koedoe* **1987**, *30*, 149–163. [CrossRef]
66. Okrouhlík, J.; Burda, H.; Kunc, P.; Knížková, I.; Šumbera, R. Surprisingly low risk of overheating during digging in two subterranean rodents. *Physiol. Behav.* **2015**, *138*, 236–241. [CrossRef]
67. Hart, D.W.; van Jaarsveld, B.; Lasch, K.G.; Grenfell, K.L.; Oosthuizen, M.K.; Bennett, N.C. Ambient Temperature as a Strong Zeitgeber of Circadian Rhythms in Response to Temperature Sensitivity and Poor Heat Dissipation Abilities in Subterranean African Mole-Rats. *J. Biol. Rhythm.* **2021**, *36*, 461–469. [CrossRef] [PubMed]
68. Finn, K.T.; Janse van Vuuren, A.K.; Hart, D.W.; Süess, T.; Zöttl, M.; Bennett, N. Seasonal changes in locomotor activity patterns of wild social Natal mole-rats (*Cryptomys hottentotus natalensis*). *Front. Ecol. Evol.* **2022**, *10*, 819393. [CrossRef]
69. McNab, B.K. The metabolism of fossorial rodents: A study of convergence. *Ecology* **1966**, *47*, 712–733. [CrossRef]
70. Contreras, L.; McNab, B. Thermoregulation and energetics in subterranean mammals. *Prog. Clin. Biol. Res.* **1990**, *335*, 231–250.
71. Lovegrove, B. Thermoregulation in the subterranean rodent *Georychus capensis* (Rodentia: Bathyergidae). *Physiol. Zool.* **1987**, *60*, 174–180. [CrossRef]
72. Thirumalai, T.; Therasa, S.V.; Elumalai, E.; David, E. Intense and exhaustive exercise induce oxidative stress in skeletal muscle. *Asian Pac. J. Trop. Dis.* **2011**, *1*, 63–66. [CrossRef]
73. Gündüz, F.; Şentürk, Ü.K. The effect of reactive oxidant generation in acute exercise-induced proteinuria in trained and untrained rats. *Eur. J. Appl. Physiol.* **2003**, *90*, 526–532. [CrossRef]
74. Huang, C.-C.; Lin, T.-J.; Lu, Y.-F.; Chen, C.-C.; Huang, C.-Y.; Lin, W.-T. Protective effects of L-arginine supplementation against exhaustive exercise-induced oxidative stress in young rat tissues. *Chin. J. Physiol.* **2009**, *52*, 306–315. [CrossRef]
75. Ji, L.L.; Fu, R. Responses of glutathione system and antioxidant enzymes to exhaustive exercise and hydroperoxide. *J. Appl. Physiol.* **1992**, *72*, 549–554. [CrossRef]
76. Liu, J.; Yeo, H.C.; Overvik-Douki, E.; Hagen, T.; Doniger, S.J.; Chu, D.W.; Brooks, G.A.; Ames, B.N. Chronically and acutely exercised rats: Biomarkers of oxidative stress and endogenous antioxidants. *J. Appl. Physiol.* **2000**, *89*, 21–28. [CrossRef]
77. Di Meo, S.; Venditti, P. Mitochondria in exercise-induced oxidative stress. *Neurosignals* **2001**, *10*, 125–140. [CrossRef] [PubMed]
78. Nayanatara, A.; Nagaraja, H.; Anupama, B. The effect of repeated swimming stress on organ weights and lipid peroxidation in rats. *Thai J. Pharm. Sci.* **2005**, *18*, 3–9.
79. Alessio, H.M.; Hagerman, A.E.; Romanello, M.; Carando, S.; Threlkeld, M.S.; Rogers, J.; Dimitrova, Y.; Muhammed, S.; Wiley, R.L. Consumption of green tea protects rats from exercise-induced oxidative stress in kidney and liver. *Nutr. Res.* **2002**, *22*, 1177–1188. [CrossRef]
80. Chapman, C.L.; Johnson, B.D.; Vargas, N.T.; Hostler, D.; Parker, M.D.; Schlader, Z.J. Both hyperthermia and dehydration during physical work in the heat contribute to the risk of acute kidney injury. *J. Appl. Physiol.* **2020**, *128*, 715–728. [CrossRef]
81. Pryor, R.R.; Pryor, J.L.; Vandermark, L.W.; Adams, E.L.; Brodeur, R.M.; Schlader, Z.J.; Armstrong, L.E.; Lee, E.C.; Maresh, C.M.; Casa, D.J. Acute kidney injury biomarker responses to short-term heat acclimation. *Int. J. Environ. Res. Public Health* **2020**, *17*, 1325. [CrossRef]
82. Chapman, C.L.; Johnson, B.D.; Parker, M.D.; Hostler, D.; Pryor, R.R.; Schlader, Z. Kidney physiology and pathophysiology during heat stress and the modification by exercise, dehydration, heat acclimation and aging. *Temperature* **2021**, *8*, 108–159. [CrossRef]
83. Bongers, C.C.; Alsady, M.; Nijenhuis, T.; Tulp, A.D.; Eijsvogels, T.M.; Deen, P.M.; Hopman, M.T. Impact of acute versus prolonged exercise and dehydration on kidney function and injury. *Physiol. Rep.* **2018**, *6*, e13734. [CrossRef]
84. Davies, K.J.; Quintanilha, A.T.; Brooks, G.A.; Packer, L. Free radicals and tissue damage produced by exercise. *Biochem. Biophys. Res. Commun.* **1982**, *107*, 1198–1205. [CrossRef]
85. Salo, D.C.; Donovan, C.M.; Davies, K.J. HSP70 and other possible heat shock or oxidative stress proteins are induced in skeletal muscle, heart, and liver during exercise. *Free Radic. Biol. Med.* **1991**, *11*, 239–246. [CrossRef]
86. King, M.A.; Leon, L.R.; Mustico, D.L.; Haines, J.M.; Clanton, T.L. Biomarkers of multiorgan injury in a preclinical model of exertional heat stroke. *J. Appl. Physiol.* **2015**, *118*, 1207–1220. [CrossRef]

87. Flanagan, S.; Ryan, A.; Gisolfi, C.; Moseley, P. Tissue-specific HSP70 response in animals undergoing heat stress. *Am. J. Physiol.-Regul. Integr. Comp. Physiol.* **1995**, *268*, R28–R32. [CrossRef] [PubMed]
88. Nielsen, S.; Frøkiær, J.; Marples, D.; Kwon, T.-H.; Agre, P.; Knepper, M.A. Aquaporins in the kidney: From molecules to medicine. *Physiol. Rev.* **2002**, *82*, 205–244. [CrossRef] [PubMed]
89. Segar, W.E. Chronic Hyperosmolality: A Condition Resulting From Absence of Thirst, Defective Osmoregulation, and Limited Ability to Concentrate Urine. *Am. J. Dis. Child.* **1966**, *112*, 318–327. [CrossRef] [PubMed]
90. Rosas-Rodríguez, J.A.; Valenzuela-Soto, E.M. Enzymes involved in osmolyte synthesis: How does oxidative stress affect osmoregulation in renal cells? *Life Sci.* **2010**, *87*, 515–520. [CrossRef]
91. Kerksick, C.; Willoughby, D. The antioxidant role of glutathione and N-acetyl-cysteine supplements and exercise-induced oxidative stress. *J. Int. Soc. Sport. Nutr.* **2005**, *2*, 38. [CrossRef]
92. Colantoni, A.; Delfanti, L.; Cossio, F.; Baciotti, B.; Salvati, L.; Perini, L.; Lord, R. Soil aridity under climate change and implications for agriculture in Italy. *Appl. Math. Sci.* **2015**, *9*, 2467–2475. [CrossRef]
93. Muñoz-Sabater, J.; Dutra, E.; Agustí-Panareda, A.; Albergel, C.; Arduini, G.; Balsamo, G.; Boussetta, S.; Choulga, M.; Harrigan, S.; Hersbach, H. ERA5-Land: A state-of-the-art global reanalysis dataset for land applications. *Earth Syst. Sci. Data* **2021**, *13*, 4349–4383. [CrossRef]
94. Romanenko, V. Computation of the autumn soil moisture using a universal relationship for a large area. *Proc. Ukr. Hydrometeorol. Res. Inst.* **1961**, *3*, 12–25.
95. Hickman, G. A live-trap and trapping technique for fossorial mammals. *S. Afr. J. Zool.* **1979**, *14*, 9–12. [CrossRef]
96. Bennett, N.C.; Faulkes, C.G.; Molteno, A.J. Reproductive suppression in subordinate, non-breeding female Damaraland mole-rats: Two components to a lifetime of socially induced infertility. *Proc. Biol. Sci.* **1996**, *263*, 1599–1603. [CrossRef]
97. Bennett, N.C.; Ganswindt, A.; Ganswindt, S.B.; Jarvis, J.; Zöttl, M.; Faulkes, C. Evidence for contrasting roles for prolactin in eusocial naked mole-rats, *Heterocephalus glaber* and Damaraland mole-rats, *Fukomys damarensis*. *Biol. Lett.* **2018**, *14*, 20180150. [CrossRef] [PubMed]
98. Hart, D.W.; Medger, K.; van Jaarsveld, B.; Bennett, N.C. Filling in the holes: The reproductive biology of the understudied Mahali mole-rat (*Cryptomys hottentotus mahali*). *Can. J. Zool.* **2021**, *99*, 801–811. [CrossRef]
99. Jacobs, P.J.; Finn, K.T.; van Vuuren, A.K.J.; Suess, T.; Hart, D.W.; Bennett, N.C. Defining the link between oxidative stress, behavioural reproductive suppression and heterothermy in the Natal mole-rat (*Cryptomys hottentotus natalensis*). *Comp. Biochem. Physiol. Part B Biochem. Mol. Biol.* **2022**, *261*, 110753. [CrossRef] [PubMed]
100. Jacobs, P.J.; Hart, D.W.; Bennett, N.C. Plasma oxidative stress in reproduction of two eusocial African mole-rat species, the naked mole-rat and the Damaraland mole-rat. *Front. Zool.* **2021**, *18*, 45. [CrossRef] [PubMed]
101. Bennett, N.; Jarvis, J.; Wallace, D. The relative age structure and body masses of complete wild-captured colonies of two social mole-rats, the common mole-rat, *Cryptomys hottentotus hottentotus* and the Damaraland mole-rat, *Cryptomys damarensis*. *J. Zool.* **1990**, *220*, 469–485. [CrossRef]
102. Golden, T.R.; Hinerfeld, D.A.; Melov, S. Oxidative stress and aging: Beyond correlation. *Aging Cell* **2002**, *1*, 117–123. [CrossRef]
103. Liguori, I.; Russo, G.; Curcio, F.; Bulli, G.; Aran, L.; Della-Morte, D.; Gargiulo, G.; Testa, G.; Cacciatore, F.; Bonaduce, D. Oxidative stress, aging, and diseases. *Clin. Interv. Aging* **2018**, *13*, 757. [CrossRef]
104. Shigenaga, M.K.; Hagen, T.M.; Ames, B.N. Oxidative damage and mitochondrial decay in aging. *Proc. Natl. Acad. Sci. USA* **1994**, *91*, 10771–10778. [CrossRef]
105. Finkel, T.; Holbrook, N.J. Oxidants, oxidative stress and the biology of ageing. *Nature* **2000**, *408*, 239. [CrossRef]
106. Erel, O. A new automated colorimetric method for measuring total oxidant status. *Clin. Biochem.* **2005**, *38*, 1103–1111. [CrossRef]
107. Halliwell, B.; Chirico, S. Lipid peroxidation: Its mechanism, measurement, and significance. *Am. J. Clin. Nutr.* **1993**, *57*, 715S–725S. [CrossRef] [PubMed]
108. Tunca, U.; Saygin, M.; Ozmen, O.; Aslankoc, R.; Yalcin, A. The impact of moderate-intensity swimming exercise on learning and memory in aged rats: The role of Sirtuin-1. *Iran. J. Basic Med. Sci.* **2021**, *24*, 1413. [CrossRef] [PubMed]
109. Kalyoncu, Ş.; Yilmaz, B.; Demir, M.; Tuncer, M.; Bozdağ, Z.; İnce, O.; Bozdayi, M.A.; Ulusal, H.; Taysi, S. Melatonin attenuates ovarian ischemia reperfusion injury in rats by decreasing oxidative stress index and peroxynitrite. *Turk. J. Med. Sci.* **2020**, *50*, 1513–1522. [CrossRef] [PubMed]
110. R Core Team. R: A language and environment for statistical computing. R Foundation for Statistical Computing. 2021. Available online: https://www.R-project.org/ (accessed on 10 November 2022).
111. Bates, D.; Maechler, M.; Bolker, B.; Walker, S. Fitting linear mixed-effects models using lme4. *J. Stat. Softw.* **2015**, *67*, 1–48. [CrossRef]
112. Kuznetsova, A.; Brockhoff, P.B.; Christensen, R.H.B. Package 'lmertest'; R Package Version. 2015; Volume 2, p. 734. Available online: http://cran.uib.no/web/packages/lmerTest/lmerTest.pdf (accessed on 10 November 2022).
113. Lenth, R.; Singmann, H.; Love, J.; Buerkner, P.; Herve, M. Emmeans: Estimated Marginal Means, Aka Least-Squares Means; R Package Version. 2018, Volume 1, p. 3. Available online: https://CRAN.R-project.org/package=emmeans (accessed on 10 November 2022).
114. Francioli, Y.; Thorley, J.; Finn, K.; Clutton-Brock, T.; Zöttl, M. Breeders are less active foragers than non-breeders in wild Damaraland mole-rats. *Biol. Lett.* **2020**, *16*, 20200475. [CrossRef] [PubMed]

115. Il'ina, T.; Ilyukha, V.; Baishnikova, I.; Belkin, V.; Sergina, S.; Antonova, E. Antioxidant defense system in tissues of semiaquatic mammals. *J. Evol. Biochem. Physiol.* **2017**, *53*, 282–288. [CrossRef]
116. Kumar, S. Free radicals and antioxidants: Human and food system. *Adv. Appl. Sci. Res.* **2011**, *2*, 129–135.
117. Song, W.; Derito, C.M.; Liu, M.K.; He, X.; Dong, M.; Liu, R.H. Cellular antioxidant activity of common vegetables. *J. Agric. Food Chem.* **2010**, *58*, 6621–6629. [CrossRef]
118. Mirończuk-Chodakowska, I.; Witkowska, A.M.; Zujko, M.E. Endogenous non-enzymatic antioxidants in the human body. *Adv. Med. Sci.* **2018**, *63*, 68–78. [CrossRef]
119. Sachdeva, M.; Karan, M.; Singh, T.; Dhingra, S. Oxidants and antioxidants in complementary and alternative medicine: A review. *Spatula DD* **2014**, *4*, 1–16. [CrossRef]
120. Limón-Pacheco, J.; Gonsebatt, M.E. The role of antioxidants and antioxidant-related enzymes in protective responses to environmentally induced oxidative stress. *Mutat. Res./Genet. Toxicol. Environ. Mutagen.* **2009**, *674*, 137–147. [CrossRef] [PubMed]
121. Fu, X.-J.; Liu, H.-B.; Wang, P.; Guan, H.-S. A study on the antioxidant activity and tissues selective inhibition of lipid peroxidation by saponins from the roots of *Platycodon grandiflorum*. *Am. J. Chin. Med.* **2009**, *37*, 967–975. [CrossRef] [PubMed]
122. Banerjee, A.K.; Mandal, A.; Chanda, D.; Chakraborti, S. Oxidant, antioxidant and physical exercise. *Mol. Cell. Biochem.* **2003**, *253*, 307–312. [CrossRef]
123. Ji, L.L. Exercise and oxidative stress: Role of the cellular antioxidant systems. *Exerc. Sport Sci. Rev.* **1995**, *23*, 135–166. [CrossRef] [PubMed]
124. Robb, G.; Harrison, A.; Woodborne, S.; Bennett, N. Diet composition of two common mole-rat populations in arid and mesic environments in South Africa as determined by stable isotope analysis. *J. Zool.* **2016**, *300*, 257–264. [CrossRef]
125. Daniels, C.; Rautenbach, F.; Marnewick, J.L.; Valentine, A.; Babajide, O.; Mabusela, W. Environmental stress effect on the phytochemistry and antioxidant activity of a South African bulbous geophyte, *Gethyllis multifolia* L. Bolus. *S. Afr. J. Bot.* **2015**, *96*, 29–36. [CrossRef]
126. Snijman, D. Gethyllis L. (Amaryllidaceae). Available online: http://opus.sanbi.org/bitstream/20.500.12143/3456/1/Gethyllis_PlantzAfrica.pdf (accessed on 3 November 2022).
127. Noctor, G.; Mhamdi, A.; Foyer, C.H. The roles of reactive oxygen metabolism in drought: Not so cut and dried. *Plant Physiol.* **2014**, *164*, 1636–1648. [CrossRef]
128. Talbi, S.; Romero-Puertas, M.C.; Hernández, A.; Terrón, L.; Ferchichi, A.; Sandalio, L.M. Drought tolerance in a Saharian plant *Oudneya africana*: Role of antioxidant defences. *Environ. Exp. Bot.* **2015**, *111*, 114–126. [CrossRef]
129. Reed, M.C.; Thomas, R.L.; Pavisic, J.; James, S.J.; Ulrich, C.M.; Nijhout, H.F. A mathematical model of glutathione metabolism. *Theor. Biol. Med. Model.* **2008**, *5*, 8. [CrossRef]
130. Wu, G.; Fang, Y.-Z.; Yang, S.; Lupton, J.R.; Turner, N.D. Glutathione metabolism and its implications for health. *J. Nutr.* **2004**, *134*, 489–492. [CrossRef]
131. Costantini, D.; Verhulst, S. Does high antioxidant capacity indicate low oxidative stress? *Funct. Ecol.* **2009**, *23*, 506–509. [CrossRef]
132. Fang, Y.-Z.; Yang, S.; Wu, G. Free radicals, antioxidants, and nutrition. *Nutrition* **2002**, *18*, 872–879. [CrossRef]
133. Marklund, S.L. Extracellular superoxide dismutase and other superoxide dismutase isoenzymes in tissues from nine mammalian species. *Biochem. J.* **1984**, *222*, 649–655. [CrossRef] [PubMed]
134. Schmidt, C.M.; Blount, J.D.; Bennett, N.C. Reproduction is associated with a tissue-dependent reduction of oxidative stress in eusocial female Damaraland mole-rats (*Fukomys damarensis*). *PLoS ONE* **2014**, *9*, e103286. [CrossRef] [PubMed]
135. Iman, M.M. Effect of aspartame on some oxidative stress parameters in liver and kidney of rats. *Afr. J. Pharm. Pharm.* **2011**, *5*, 678–682. [CrossRef]
136. Kaushik, S.; Kaur, J. Chronic cold exposure affects the antioxidant defense system in various rat tissues. *Clin. Chim. Acta* **2003**, *333*, 69–77. [CrossRef]
137. Liu, J.; Cui, H.; Liu, X.; Peng, X.; Deng, J.; Zuo, Z.; Cui, W.; Deng, Y.; Wang, K. Dietary high vanadium causes oxidative damage-induced renal and hepatic toxicity in broilers. *Biol. Trace Elem. Res.* **2012**, *145*, 189–200. [CrossRef]
138. Oliveira, M.F.; Geihs, M.A.; França, T.F.; Moreira, D.C.; Hermes-Lima, M. Is "preparation for oxidative stress" a case of physiological conditioning hormesis? *Front. Physiol.* **2018**, *9*, 945. [CrossRef]
139. Hood, W.; Zhang, Y.; Mowry, A.; Hyatt, H.; Kavazis, A. Life history trade-offs within the context of mitochondrial hormesis. *Integr. Comp. Biol.* **2018**, *58*, 567–577. [CrossRef]
140. Costantini, D. Variation in oxidative stress threats and hormesis across environments. In *Oxidative Stress and Hormesis in Evolutionary Ecology and Physiology*; Springer: Berlin/Heidelberg, Germany, 2014; pp. 75–109.
141. Luna-López, A.; González-Puertos, V.Y.; López-Diazguerrero, N.E.; Königsberg, M. New considerations on hormetic response against oxidative stress. *J. Cell Commun. Signal.* **2014**, *8*, 323–331. [CrossRef]
142. Urison, N.; Buffenstein, R. Kidney concentrating ability of a subterranean xeric rodent, the naked mole-rat (*Heterocephalus glaber*). *J. Comp. Physiol. B* **1994**, *163*, 676–681. [CrossRef] [PubMed]
143. Johnson, R.J.; Sánchez-Lozada, L.G.; Newman, L.S.; Lanaspa, M.A.; Diaz, H.F.; Lemery, J.; Rodriguez-Iturbe, B.; Tolan, D.R.; Butler-Dawson, J.; Sato, Y. Climate change and the kidney. *Ann. Nutr. Metab.* **2019**, *74*, 38–44. [CrossRef]
144. Kültz, D. Hyperosmolality triggers oxidative damage in kidney cells. *Proc. Natl. Acad. Sci. USA* **2004**, *101*, 9177–9178. [CrossRef] [PubMed]

145. Evans, R.G.; Smith, D.W.; Lee, C.J.; Ngo, J.P.; Gardiner, B.S. What makes the kidney susceptible to hypoxia? *Anat. Rec.* **2020**, *303*, 2544–2552. [CrossRef]
146. Bouayed, J.; Bohn, T. Exogenous antioxidants—Double-edged swords in cellular redox state: Health beneficial effects at physiologic doses versus deleterious effects at high doses. *Oxidative Med. Cell. Longev.* **2010**, *3*, 228–237. [CrossRef] [PubMed]
147. Balcerczyk, A.; Bartosz, G. Thiols are main determinants of total antioxidant capacity of cellular homogenates. *Free Radic. Res.* **2003**, *37*, 537–541. [CrossRef] [PubMed]
148. Bharti, V.K.; Srivastava, R.; Kumar, H.; Bag, S.; Majumdar, A.; Singh, G.; Pandi-Perumal, S.; Brown, G.M. Effects of melatonin and epiphyseal proteins on fluoride-induced adverse changes in antioxidant status of heart, liver, and kidney of rats. *Adv. Pharmacol. Sci.* **2014**, *2014*, 532969. [CrossRef]
149. Lauterburg, B.H.; Adams, J.D.; Mitchell, J.R. Hepatic glutathione homeostasis in the rat: Efflux accounts for glutathione turnover. *Hepatology* **1984**, *4*, 586–590. [CrossRef]
150. Hellsten, Y.; Svensson, M.; Sjödin, B.; Smith, S.; Christensen, A.; Richter, E.; Bangsbo, J. Allantoin formation and urate and glutathione exchange in human muscle during submaximal exercise. *Free. Radic. Biol. Med.* **2001**, *31*, 1313–1322. [CrossRef]
151. Kasapoglu, M.; Özben, T. Alterations of antioxidant enzymes and oxidative stress markers in aging. *Exp. Gerontol.* **2001**, *36*, 209–220. [CrossRef]
152. Tiana, L.; Caib, Q.; Wei, H. Alterations of antioxidant enzymes and oxidative damage to macromolecules in different organs of rats during aging. *Free. Radic. Biol. Med.* **1998**, *24*, 1477–1484. [CrossRef]
153. Sani, M.; Sebaï, H.; Gadacha, W.; Boughattas, N.A.; Reinberg, A.; Mossadok, B.A. Catalase activity and rhythmic patterns in mouse brain, kidney and liver. *Comp. Biochem. Physiol. Part B Biochem. Mol. Biol.* **2006**, *145*, 331–337. [CrossRef] [PubMed]
154. Hulbert, A.J.; Turner, N.; Hinde, J.; Else, P.; Guderley, H. How might you compare mitochondria from different tissues and different species? *J. Comp. Physiol. B* **2006**, *176*, 93–105. [CrossRef] [PubMed]
155. Apel, K.; Hirt, H. Reactive oxygen species: Metabolism, oxidative stress, and signaling transduction. *Annu. Rev. Plant Biol.* **2004**, *55*, 373. [CrossRef] [PubMed]
156. Sani, M.; Ghanem-Boughanmi, N.; Gadacha, W.; Sebai, H.; Boughattas, N.A.; Reinberg, A.; Ben-Attia, M. Malondialdehyde content and circadian variations in brain, kidney, liver, and plasma of mice. *Chronobiol. Int.* **2007**, *24*, 671–685. [CrossRef] [PubMed]

antioxidants

Review

The Eco-Immunological Relevance of the Anti-Oxidant Response in Invasive Molluscs

Davide Malagoli [1,2,*], Nicola Franchi [1] and Sandro Sacchi [1]

[1] Department of Life Sciences, University of Modena and Reggio Emilia, 41125 Modena, Italy; nicola.franchi@unimore.it (N.F.); sandro.sacchi@unimore.it (S.S.)
[2] NBFC, National Biodiversity Future Center, 90133 Palermo, Italy
* Correspondence: davide.malagoli@unimore.it

Abstract: Reactive oxygen species (ROS) are volatile and short-lived molecules playing important roles in several physiological functions, including immunity and physiological adaptation to unsuitable environmental conditions. In an eco-immunological view, the energetic costs associated with an advantageous metabolic apparatus able to cope with wide changes in environmental parameters, e.g., temperature range, water salinity or drought, could be further balanced by the advantages that this apparatus may also represent in other situations, e.g., during the immune response. This review provides an overview of molluscs included in the IUCN list of the worst invasive species, highlighting how their relevant capacity to manage ROS production during physiologically challenging situations can also be advantageously employed during the immune response. Current evidence suggests that a relevant capacity to buffer ROS action and their damaging consequences is advantageous in the face of both environmental and immunological challenges, and this may represent a trait for potential invasiveness. This should be considered in order to obtain or update information when investigating the potential of the invasiveness of emerging alien species, and also in view of ongoing climate changes.

Keywords: ecoimmunology; stress; immunity; haemocytes; biodiversity; *Pomacea canaliculata; Achatina fulica; Mytilus galloprovincialis; Dreissena polymorpha*; alien species

Citation: Malagoli, D.; Franchi, N.; Sacchi, S. The Eco-Immunological Relevance of the Anti-Oxidant Response in Invasive Molluscs. *Antioxidants* **2023**, *12*, 1266. https://doi.org/10.3390/antiox12061266

Academic Editors: Marcelo Hermes-Lima, Daniel Carneiro Moreira and Tania Zenteno-Savín

Received: 7 May 2023
Revised: 1 June 2023
Accepted: 3 June 2023
Published: 13 June 2023

1. Introduction

Climate change and invasive species are among the most relevant threats to biodiversity [1]. Invasive alien species are demonstrating an outstanding capacity to adapt to the new environment they colonize. Beside the adaptations to new physical-chemical environmental aspects, invasive species must also present a high degree of metabolic plasticity and an outstanding capacity of adaptation to new antigenic ecospaces [2,3]. In the IUCN list of the worst invasive species [4], numerous invertebrate species are retrievable, presenting hypervariable defence-related molecules and specific anti-microbial peptides [5,6]. In addition to the immune features and mediators that evolved or differentiated along specific evolutionary lineages, one widespread function of the immune anti-pathogen responses is represented by the controlled production of reactive oxygen species (ROS) [7,8]. These volatile and short-lived molecules are of extreme importance both in invertebrate and vertebrate immune response and, thanks to their non-specificity, they can be directed against a plethora of potential pathogens. Beside this, environmental conditions can also create oxidative stress in organisms [1,9]. In this last case, the ROS become a menace for the organism itself, and oxidative stress must be managed in order to avoid permanent damage.

Oxidative stress is a ubiquitous phenomenon that can be studied at cellular, tissue, organ and organism level. It occurs when the production or presence of ROS exceeds antioxidant defence capabilities [10,11]. ROS-mediated oxidative stress can be promoted either from the environment [12,13] or from the activity of endogenous enzymes, such as those involved in cellular respiration or various forms of oxydase [14]. The main cellular effects of oxidative stress are unspecific and include lipid peroxidation, protein misfolding and nucleotide alteration [15] although other more complex and less investigated effects cannot be excluded [16].

As cellular respiration and metabolism are natural and constitutive sources of ROS [14], all cells have evolved mechanisms of defence that include enzymes, e.g., catalases, peroxidases and superoxide dismutase and molecular scavengers, e.g., glutathione. More complex molecular chaperones, such as heat shock proteins (HSPs), can also intervene in recovering the protein misfolding promoted by oxidative stress [17]. While the molecular bases of oxidative stress and their association with the immune defences are well conserved, differences exist among organisms in the ability to manage ROS production during the immune response [18] and physiologically challenging situations such as those experienced as a consequence of a sudden change in oxygen availability, e.g., changes in water level or arousal after estivation/hibernation periods [12,17,19,20].

2. The Oxidative Stress Response against Immune or Environmental Challenges in Highly Invasive Mollusc Species

The resistance to oxidative stress introduces a further level of complexity in the proper management of invasive species. It has already been documented that on some occasions the usage of biocides may positively affect the spread of invasive species. An experiment performed by exposing native and invasive ant species to sublethal doses of a neonicotinoid, a class of highly diffused chemical biocides, demonstrated that it could either increase or decrease the probability of invasive ant survival according to the exposure status of the native ants [21]. Neonicotinoids may have important effects in non-target organisms. These effects include DNA damage, protein oxidation and lipid peroxidation that are largely the consequence of the increased ROS concentration that follows the altered mitochondrial Ca^{2+} homeostasis and the hindered mitochondrial respiration [22]. The invasive species endowed with a high resistance to ROS-mediated insults are less likely to suffer consequences from those pesticides provoking ROS-mediated oxidative stress; either the invasive species are the direct target of the treatment or they are non-target species. This further restricts the number of the compounds available for pest control, pushing researchers to investigate the effects of biological control [23] or biopesticides [24]. However, as is described in more detail below, biopesticide effects can also be limited by highly efficient ROS detoxification.

The 2013 IUCN list of the worst invasive species [4] includes bivalves, e.g., *Mytilus galloprovincialis*, *Dreissena polymorpha* and *Potamocorbula amurensis* and gastropods, e.g., *Achatina fulica*, *Euglandina rosea* and *Pomacea canaliculata*. Although a different level of information is available for these species, the retrievable data point towards a common trait of an efficient immune response and a relevant resistance to oxidative stress.

The Mediterranean mussel *Mytilus galloprovincialis* presents circulating haemocytes (immunocytes) that support efficient cell-mediated innate immune functions [25–28]. ROS species have been demonstrated to intervene during immune response and after the tissue injury of *M. galloprovincialis* [29,30]. As a filter feeder and a species of economic relevance, several studies have investigated the effects of environmental pollutants on Mediterranean mussel health and anti-oxidant stress response [13,30]. As the increase in water temperature may represent a threat for mussel cultivation, the possibility to improve *M. galloprovincialis* heat stress response has been studied at physiological and molecular levels. It has been demonstrated that a brief pre-exposure to thermal stress (a procedure labelled as "hardening") significantly increases the capability of mussels to cope with a further and more prolonged exposure to high temperatures [31]. Among the numerous parameters taken into account, oxidative stress and redox signalling were studied by measuring the expression of different superoxide dismutases, glutathione-S-transferase and catalase, the enzymatic activities of superoxide dismutases, catalase and glutathione reductase and the protein levels of HSP70 isoforms (i.e., HSP72 and HSP73).

Superoxide dismutase and catalase expression during the prolonged exposure to high temperatures were increased, and they were significantly higher in hardened mussels with respect to stressed but non-hardened animals. For some molecules, the increase in enzymatic activities followed a similar temporal profile in hardened and non-hardened mussels, but it was higher in hardened ones. Glutathione reductase acted similarly to superoxide dismutases and catalase. Consistently, HSP72 and HSP73 protein levels were higher in hardened animals, and the increased expression of electron transport system elements seemed synchronized with that of the other components of the anti-oxidant response [31]. The complex and adaptive anti-oxidant response of hardened mussels to the experimental thermal stresses suggested that hardened mussels are able to manage and coordinate both the metabolic and the oxidative stress responses in the face of sudden temperature increase.

The zebra mussel *D. polymorpha* is an invasive species that can also be studied as a sentinel organism in eco-physiological studies. As frequently reported in molluscs [32], the haemocytes can be divided into two main morphologies, namely agranular and granular haemocytes. Agranular haemocytes may include cells with different morphologies and, possibly, functions, i.e., blast-like cells and hyalinocytes. In the zebra mussel, the latter exhibited the highest oxidative activity, also in comparison with granular haemocytes [33]. In ex vivo experiments, *D. polymorpha* haemocytes exposed to chemical, physical and biological stresses elicited a complex response that included the expression of anti-oxidant enzymes and they changed significantly on the basis of the stressor considered [34]. Numerous studies included the effects of cadmium, as the zebra mussel has been proposed as a sentinel organism in freshwater environments for the evaluation of water pollutant effects [35]. The oxidative activity of *D. polymorha* haemocytes changed in different haemocyte sub-populations, as the granular cells were the less affected by the treatment and the hyalinocytes seemed to be influenced only at the highest doses of cadmium ions, indicating an important stability of mitochondrial activity and ROS management for these two cell populations [36], even in the presence of relevant stresses.

To our knowledge, no information comparable to that reported for *M. galloprovincialis* and *D. polymorpha* is available for the Asian clam *P. amurensis*. This notwithstanding, experiments correlating the HSP protein levels to water salinity values, as well as studies comparing the aerobic-fermentative metabolism rates in relation to water salinity and seasonal temperature changes, have been presented [37–39], suggesting that the Asian clam could also finely adjust stress response and metabolism as a consequence of environmental changes.

The African giant snail *A. fulica* has been labelled as the most widely distributed invasive pest land snail [40]. Its large haemocytes were described a long time ago; they displayed phagocytic activity, were able to synthesise superoxide anion radical, but did not present endogenous peroxidase activity [41]. As the intermediate vector of the human pathogenic nematodes of the genus *Angiostrongylus*, the immune response of *A. fulica* to these parasites has been investigated. In snails infected with *Angiostrongylus vasorum* larvae, increased phenoloxidase (PO) activity and consequently a ROS-mediated immune response were observed, with a PO activity peak in the immediacy of the infection. The activation of PO was also followed by melanisation. Beside PO activation, the immediate increase in nitric oxide (NO) production was also observed via Griess reaction, confirming that the immune response against the nematode included a significant ROS-mediated component, especially during the first phases of the infection [42]. Ex vivo experiments performed by confronting withdrawn haemocytes with the axenic larvae of diverse metastrongyloid lungworms also demonstrated that *A. fulica* haemocytes can release their nuclear content, thus forming extracellular traps (ETs). Beside containing histones that exert a well-known antimicrobial function [43], the *A. fulica* ETs also contained molecules similar to myeloperoxidase, an enzyme involved in the production of hypohalous acids and exerting a cytotoxic function especially by means of oxidative stress [44]. The exposure of *A. fulica* to the causative agent of human eosinophilic encephalitis, *Angiostrongylus cantonensis*, promoted the metabolic shift towards oxidative activity by increasing the glycolytic pathway and the activity of lactate dehydrogenase in order to keep the redox balance, at least in the haemolymph [45]. As an invasive snail, *A. fulica* has also been the subject of experiments aimed at controlling its diffusion without damaging other species. In this regard, the efficacy of a nematode, *Phasmarhabditis hermaphrodita*, also used as a component of a bio-based molluscicide, has been assessed on the giant African snail. Juvenile snails are highly resistant to the nematodes as they are able to entrap them into the inner layer of the shell, apparently without further metabolic or immune responses [46] (Figure 1).

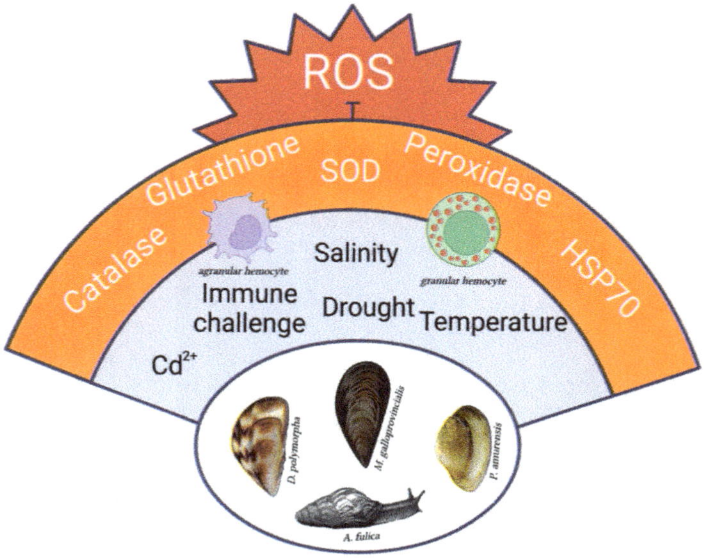

Figure 1. Highly invasive molluscs of different Classes present common traits of resistance versus oxidative stress that could be originated either from environmental stimuli or from immune challenges. The advantage of managing ROS increase is shared among diverse physiological responses, thus limiting the associated energy expenditure and allowing efficient energy trade-offs [47]. Created with BioRender.com (accessed on 2 May 2023).

While no specific information is retrievable about the immune defences and oxidative responses of the rosy wolfsnail that, however, can be vehicle of *A. cantonesis* [48], more data can be retrieved for the snail *P. canaliculata*. The circulating haemocytes of *P. canaliculata* have been described [49–51] and their proteome is available [52]. In addition to the haemolymph, the haemocytes have also been recognized within organs and in regenerating tissues [53,54], increasing the number of potential sites where the haemocyte-mediated immune response can take place and haemocytes can replicate. The immune system of *P. canaliculata* has been the target of numerous studies focused on the control of its spread. The vegetal pesticide pedunsaponin A modified the haemocyte number, membrane potential and morphology and promoted ciliary loss in ciliated tissues, thus affecting snail respiration and excretion [55,56]. The effects of the pesticide were increased after silencing the expression of HSP70 by RNAi [57], suggesting a role for this chaperone in buffering pedunsaponin A-mediated damage. As already reported above for *A. fulica*, the immune system of *P. canaliculata* was also targeted by using a commercially available molluscicide based on the nematode *P. hermaphrodita*. *P. canaliculata* exhibited a significant resistance to this molluscicide. The recommended concentration of *P. hermaphrodita* determined an overall low mortality and reduced the synthesis of an antimicrobial peptide alternatively in two immune-related organs of the snails, i.e., the gills and the anterior kidney, in dependence of the temperature of the treatment [58]. The ultrastructural observation and proteomic analysis of an organ associated with the oxidative stress resistance and potentially involved in the immune response, namely the aortic ampulla [59,60], revealed that *P. canaliculata* systemic reaction to the immune challenge, represented by the pathogenic nematode, involved a diffused oxidative stress response, as suggested by the increased expression in the ampulla of enzymes associated with ROS detoxification, such as Cu^{2+}–Zn^{2+} superoxide dismutase, catalase-like isoform X1, glutathione peroxidase-like protein and peroxiredoxin. As the nematodes were not present in the ampulla of exposed snails, and no nematode proteins were isolated or sequenced among the ampulla proteome, the increase in anti-oxidant enzymes allowed the speculation that the organ was responding to a systemic increase in ROS. Consistently, the snail self-protection from ROS-mediated damage was also associated with the increased expression of HSP60, 70 and 90 [61].

3. The Eco-Immunological Advantage of Managing Oxidative Stress Response

While significant differences exist in the anatomical organization and physiology of the invasive species introduced above, their high tolerance versus oxidative stress represents a common trait in their immune response towards pathogens, pesticides and environmental or experimental physico-chemical stressors. The invasiveness of a species relies on different physiological features [62,63], not least the capacity to overcome or tolerate the new potential pathogens. Invertebrate innate-only immune response has been progressively described as anticipatory and highly specific, especially in light of the numerous hypervariable molecules discovered in invertebrate taxa [64–68] and the increasing evidence of circulating microbiotas [69–71] that some invertebrate species are able to manage together with the intestinal microbiota. However, a specific and anticipatory immune response could also miss new pathogens that may be encountered during the colonization of new environments by the invasive species. In this respect, well-conserved and unspecific responses, such as the increased synthesis of ROS, could become essential for the adaptation to new environments. As an unspecific response based on short-lived and volatile molecules, ROS-mediated immune defence is self-limited by the capability of the host to manage the potential harm that these volatile molecules could determine; in the majority of the invasive species mentioned here, a marked resilience versus oxidative stress has been reported.

The mussel *M. galloprovincialis* can populate subtidal and intertidal environments. It displays tolerance to temperature fluctuations [72] and its metabolic plasticity allows the Mediterranean mussel to adapt and to react differently to environmental challenges on the basis of the colonized habitat [73]. In *D. polymorpha*, an exceptional tolerance to variable environmental conditions associated with strong anti-oxidant defences has been reported [74], maybe in consequence of a high genetic diversity [75]. *P. amurensis* can colonize subtidal and intertidal habitats and present a high tolerance to wide ranges of water salinity and temperature [76,77], conditions that are connected to changes in oxygen availability and require efficient oxidative stress response. Among invasive freshwater gastropods, both *A. fulica* and *P. canaliculata* can undergo estivation [78,79], a physiological adaptative response to unsuitable environmental conditions that relies on the ability to manage the significant increase in ROS during the arousal, when oxygen concentration quickly increases. In the African giant snail, mitochondrial or cytosolic superoxide dismutase levels did not change during estivation, and the same held true for other anti-oxidant enzymes such as glutathione peroxidase, glutathione reductase or catalase. While their amount and activity are stable during estivation, during the arousal from estivation the antioxidant enzymes increase their detoxifying activities with a time- and tissue-specific distribution. Consistently, no signs of protein damage or lipid peroxidation were detected in the heart tissues of African giant snails after four weeks of experimental estivation [78]. During estivation, *A. fulica* can accumulate urea, a nitrogen waste product less harmful than ammonia. Evidence collected in *A. fulica* has demonstrated an astonishing capacity to detoxify exogenous and injected ammonia via a significant increase in the urea synthesis rate, suggesting that urea could have other metabolic roles beside the excretory one [80]. In *P. canaliculata*, the involvement of nitrogen compounds in the anti-oxidant response associated with the arousal from estivation has been documented in detail. In agreement with the observations from *A. fulica*, the *P. canaliculata* presented a complex pattern of anti-oxidant responses that involved enzymatic and non-enzymatic components which fluctuated in an organ-specific fashion, especially during the arousal [81]. The nitrogen compound uric acid is likely to play an important role in buffering the ROS concentration increase during the arousal [82]. A specific tissue, labelled as urate tissue, is distributed among numerous organs, i.e., the lung, the digestive system, the anterior kidney (also identified as pallial ureter) and the ampulla, which are considered relevant organs in managing ROS levels during the arousal [60,79] (Figure 2).

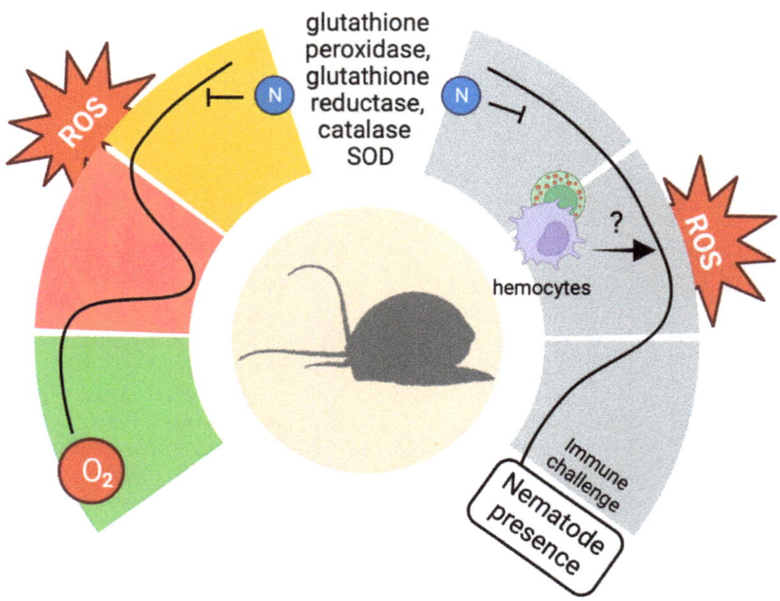

Figure 2. The invasive gastropods *A. fulica* and *P. canaliculata* can cope with ROS peaks in different situations. Left: when the water and oxygen availability are sufficient (green), the snails do not display a significant activity in detoxifying enzymes. When the water availability is reduced, e.g., as a consequence of drought, the snails can enter estivation, reducing their metabolism, as a consequence of reduced oxygen availability (red). Once the environmental conditions become more suitable for their survival, the snails face a peak of ROS that follows the increase in the oxygen availability and in the metabolic activity. This peak is associated with an increased enzymatic activity and, at least in *P. canaliculata*, is sustained by nitrogen-containing compounds. Right: during the immune response against pathogenic nematodes (grey), a similar trend of ROS production is observed, although in this case it is also possible that immune-related components, e.g., haemocytes, can contribute to ROS synthesis and increase. The protective elements employed during the arousal can also be utilized in response to the immune challenge, allowing the snails to express a powerful ROS-mediated immune response, avoiding self-damage. Created with BioRender.com (accessed 2 May 2023).

Recently, the response of *P. canaliculata* circulating haemocytes has also been investigated, and differences have emerged in the mediators involved in the anti-oxidant response versus experimental hibernation or estivation [83], confirming the high plasticity and adaptability of the oxidative stress response in this invasive snail. The proteomic analysis of ampullae collected from snails challenged with the pathogenic nematode *P. hermaphrodita* evidenced a metabolic response that paralleled the adaptation to estivation, with a significant increase in anti-oxidant defences and an increase in aerobic metabolism, witnessed by the increase in mitochondrial enzymes related to ATP synthesis [61].

The ability to cope with the environmental conditions that require an increased capability of managing oxidative stress response represents a significant advantage that could help the adaptivity and invasiveness of some species. Nonetheless, that capacity is associated with the metabolic costs that the maintenance of a prompt anti-oxidant defence system requires. In the eco-immunological perspective of efficient energy expenditure and management [84,85], the advantage of controlling the self-damaging consequences of oxidative stress response could be more relevant if extended to other physiological functions such as the immune response against pathogens. While the physiological immune functions are deeply interconnected with metabolism, neural functions and development [86], and their energetic demand is balanced together with that of the other physiological systems, during the response against pathogens the energy demand of the immune system suddenly increases and must be managed [84]. In this context, the possibility to take advantage of a pre-set and dynamic anti-oxidant defence system such as those described above for invasive molluscs may allow the organisms to express a potent and efficacious ROS-mediated immune response without incurring the self-damage that ROS may determine for the host. In organisms adapted to sustain the oxidant stress response derived from changes in water temperature or salinity, or the arousal from estivation or hibernation, the self-limitations that must be applied to the ROS-mediated immune response [87] could be less marked. This would prove particularly functional for *A. fulica* and *P. canaliculata* in the presence of large pathogens, such as the nematode *P. hermaphrodita*. From this perspective, the metabolic costs associated with the adaptation to specific environmental conditions could be counterbalanced by the advantages also reflecting on the immune functions, thus giving the species endowed with this capacity a special advantage in the face of both environmental and immunological challenges [1].

4. Conclusions

The updated determination of alien species' potential of invasiveness and the consequent polices for alien species control are becoming more urgent in view of the increasing consequences of ongoing climate change. For instance, in 2012, the EU Parliament classified *Pomacea* as a pest and invading genus and requested a comprehensive risk assessment on *Pomacea* species [88] that included the potential establishment of these snails in EU territory. Similar studies highlighted regions of India considered as more prone to invasion by the African giant snail, *A. fulica* [89]. The different scenarios regarding the areas potentially invaded and prospected in those risk assessments almost ten years ago, which considered specific patterns of air, water temperature and rainfall fluctuations, might now need some adjustments or updates. At the same time, climate change has also promoted an unanticipated reduction of sea water pH. This acidification has been demonstrated to modify the freeze tolerance of subtidal species, such as for instance *M. galloprovincialis*, and it will likely modify the area of distribution of subtidal and intertidal *Mytilus* species [90].

Current evidence suggests that most invasive molluscs present a relevant capacity to manage ROS increase, either derived from environmental cues or from immune stimuli. From an eco-immunological perspective, the possibility of taking advantage of the same metabolic ability in diverse functional contexts allows a better use of resources and energy trade-offs. This shared capacity may likely represent a trait of potential invasiveness which, associated with more species-specific characteristics, could justify the success of species such as *M. galloprovincialis*, *D. polymorpha*, *A. fulica* and *P. canaliculata* to invade and adapt to new environments and niches. From this perspective, the resistance to experimental oxidative stress of emerging alien species could be further investigated in order to gain more complete information on their potential of invasiveness and to formulate consistent policies for their control.

Author Contributions: Conceptualization, D.M.; investigation, D.M., N.F. and S.S.; resources, D.M.; writing—original draft preparation, D.M.; writing—review and editing, D.M., N.F. and S.S.; visualization, N.F.; supervision, D.M.; project administration, D.M.; funding acquisition, D.M. All authors have read and agreed to the published version of the manuscript.

Funding: This research was funded by the National Recovery and Resilience Plan (NRRP), Mission 4, Component 2 Investment 1.4—Call for tender No. 3138 of 16 December 2021, rectified by Decree n.3175 of 18 December 2021 of the Italian Ministry of University and Research funded by the European Union—NextGenerationEU, grant number Project code CN_00000033, Concession Decree No. 1034 of 17 June 2022 adopted by the Italian Ministry of University and Research, CUP E93C22001090001, Project title "National Biodiversity Future Center—NBFC" and by the Department of Life Sciences, University of Modena and Reggio Emilia, Bando FAR2020. The APC was waived by the Editor.

Institutional Review Board Statement: Ethical review and approval were waived for this study, as the use of bivalves and gastropods as experimental animals is neither detailed in the Directive 2010/63/EU of the European Parliament and of the Council (https://eur-lex.europa.eu/legal-content/EN/TXT/HTML/?uri=CELEX:32010L0063&from=EN, accessed on 2 June 2023) nor in the Italian legislative decree n. 26/2014 (https://www.gazzettaufficiale.it/eli/id/2014/03/14/14G00036/sg, accessed on 2 June 2023). Experiments have been performed and presented in accordance with the ARRIVE guidelines (https://arriveguidelines.org/arrive-guidelines, accessed on 2 June 2023).

Informed Consent Statement: Not applicable.

Data Availability Statement: No new data were created or analysed in this study. Data sharing is not applicable to this article.

Conflicts of Interest: The authors declare no conflict of interest.

References

1. Byers, J.E.; Blaze, J.A.; Dodd, A.C.; Hall, H.L.; Gribben, P.E. Exotic Asphyxiation: Interactions between Invasive Species and Hypoxia. *Biol. Rev.* **2023**, *98*, 150–167. [CrossRef] [PubMed]
2. Stock, A.; Murray, C.C.; Gregr, E.J.; Steenbeek, J.; Woodburn, E.; Micheli, F.; Christensen, V.; Chan, K.M.A. Exploring Multiple Stressor Effects with Ecopath, Ecosim, and Ecospace: Research Designs, Modeling Techniques, and Future Directions. *Sci. Total Environ.* **2023**, *869*, 161719. [CrossRef] [PubMed]
3. Capri, M.; Salvioli, S.; Monti, D.; Bucci, L.; Garagnani, P.; Ottaviani, E.; Franceschi, C. *Eco-Immunology, Evolutive Aspects and Future Perspectives*; Springer: Heidelberg, Germany, 2014; pp. 125–144. [CrossRef]
4. Lowe, S.; Browne, M.; Boudjelas, S.; De Poorter, M. *100 of the World's Worst Invasive Alien Species: A Selection from the Global Invasive Species Database*; The Invasive Species Specialist Group (ISSG): Auckland, New Zealand; p. 12. First published as special lift-out in Aliens 12, December 2000. Updated and reprinted version: November 2004.
5. Ng, T.H.; Kurtz, J. Dscam in Immunity: A Question of Diversity in Insects and Crustaceans. *Dev. Comp. Immunol.* **2020**, *105*, 103539. [CrossRef] [PubMed]
6. Gerdol, M.; Moro, G.D.; Manfrin, C.; Venier, P.; Pallavicini, A. Big Defensins and Mytimacins, New AMP Families of the Mediterranean Mussel *Mytilus galloprovincialis*. *Dev. Comp. Immunol.* **2012**, *36*, 390–399. [CrossRef]
7. Morris, G.; Gevezova, M.; Sarafian, V.; Maes, M. Redox Regulation of the Immune Response. *Cell. Mol. Immunol.* **2022**, *19*, 1079–1101. [CrossRef] [PubMed]
8. Staerck, C.; Gastebois, A.; Vandeputte, P.; Calenda, A.; Larcher, G.; Gillmann, L.; Papon, N.; Bouchara, J.-P.; Fleury, M.J.J. Microbial Antioxidant Defense Enzymes. *Microb. Pathog.* **2017**, *110*, 56–65. [CrossRef]
9. Sokolova, I. Bioenergetics in Environmental Adaptation and Stress Tolerance of Aquatic Ectotherms: Linking Physiology and Ecology in a Multi-Stressor Landscape. *J. Exp. Biol.* **2021**, *224*, jeb236802. [CrossRef]
10. Betteridge, D.J. What Is Oxidative Stress? *Metabolis* **2000**, *49*, 3–8. [CrossRef]
11. Pizzino, G.; Irrera, N.; Cucinotta, M.; Pallio, G.; Mannino, F.; Arcoraci, V.; Squadrito, F.; Altavilla, D.; Bitto, A. Oxidative Stress: Harms and Benefits for Human Health. *Oxid. Med. Cell. Longev.* **2017**, *2017*, 8416763. [CrossRef]
12. Gostyukhina, O.L.; Yu, A.A.; Chelebieva, E.S.; Vodiasova, E.A.; Lantushenko, A.O.; Kladchenko, E.S. Adaptive Potential of the Mediterranean Mussel *Mytilus galloprovincialis* to Short-Term Environmental Hypoxia. *Fish Shellfish Immun.* **2022**, *131*, 654–661. [CrossRef]
13. Sun, C.; Teng, J.; Wang, D.; Zhao, J.; Shan, E.; Wang, Q. The Adverse Impact of Microplastics and Their Attached Pathogen on Hemocyte Function and Antioxidative Response in the Mussel *Mytilus galloprovincialis*. *Chemosphere* **2023**, *325*, 138381. [CrossRef] [PubMed]
14. Ramzan, R.; Vogt, S.; Kadenbach, B. Stress-Mediated Generation of Deleterious ROS in Healthy Individuals—Role of Cytochrome c Oxidase. *J. Mol. Med.* **2020**, *98*, 651–657. [CrossRef]

15. Melis, J.P.M.; van Steeg, H.; Luijten, M. Oxidative DNA Damage and Nucleotide Excision Repair. *Antioxid. Redox Signal.* **2013**, *18*, 2409–2419. [CrossRef] [PubMed]
16. Tauffenberger, A.; Magistretti, P.J. Reactive Oxygen Species: Beyond Their Reactive Behavior. *Neurochem. Res.* **2021**, *46*, 77–87. [CrossRef] [PubMed]
17. Eaton, L.; Pamenter, M.E. What to Do with Low O2: Redox Adaptations in Vertebrates Native to Hypoxic Environments. *Comp. Biochem. Physiol. Part A Mol. Integr. Physiol.* **2022**, *271*, 111259. [CrossRef]
18. Dickson, K.B.; Zhou, J. Role of Reactive Oxygen Species and Iron in Host Defense against Infection. *Front. Biosci.* **2020**, *25*, 1600–1616. [CrossRef]
19. Ferreira-Cravo, M.; Welker, A.F.; Hermes-Lima, M. Aestivation, Molecular and Physiological Aspects. *Prog. Mol. Subcell Biol.* **2009**, *49*, 47–61. [CrossRef]
20. Nowakowska, A.; Świderska-Kołacz, G.; Rogalska, J.; Caputa, M. Antioxidants and Oxidative Stress in *Helix pomatia* Snails during Estivation. *Comp. Biochem. Physiol. Part C Toxicol. Pharmacol.* **2009**, *150*, 481–486. [CrossRef]
21. Barbieri, R.F.; Lester, P.J.; Miller, A.S.; Ryan, K.G. A Neurotoxic Pesticide Changes the Outcome of Aggressive Interactions between Native and Invasive Ants. *Proc. R. Soc. B Biol. Sci.* **2013**, *280*, 20132157. [CrossRef]
22. Xu, X.; Wang, X.; Yang, Y.; Ares, I.; Martínez, M.; Lopez-Torres, B.; Martínez-Larrañaga, M.-R.; Wang, X.; Anadón, A.; Martinez, M.-A. Neonicotinoids: Mechanisms of Systemic Toxicity Based on Oxidative Stress-Mitochondrial Damage. *Arch. Toxicol.* **2022**, *96*, 1493–1520. [CrossRef]
23. McLaughlin, G.M.; Dearden, P.K. Invasive Insects: Management Methods Explored. *J. Insect. Sci.* **2019**, *19*, 17. [CrossRef] [PubMed]
24. Kumar, J.; Ramlal, A.; Mallick, D.; Mishra, V. An Overview of Some Biopesticides and Their Importance in Plant Protection for Commercial Acceptance. *Plants* **2021**, *10*, 1185. [CrossRef] [PubMed]
25. Malagoli, D.; Mandrioli, M.; Tascedda, F.; Ottaviani, E. Circulating Phagocytes: The Ancient and Conserved Interface between Immune and Neuroendocrine Function. *Biol. Rev.* **2017**, *92*, 369–377. [CrossRef] [PubMed]
26. Malagoli, D.; Casarini, L.; Sacchi, S.; Ottaviani, E. Stress and Immune Response in the Mussel *Mytilus galloprovincialis*. *Fish Shellfish Immun.* **2007**, *23*, 171–177. [CrossRef]
27. Malagoli, D.; Gobba, F.; Ottaviani, E. Effects of 50-Hz Magnetic Fields on the Signalling Pathways of FMLP-Induced Shape Changes in Invertebrate Immunocytes: The Activation of an Alternative "Stress Pathway" . *Biochim. Biophys. Acta BBA-Gen. Subj.* **2003**, *1620*, 185–190. [CrossRef]
28. Malagoli, D.; Ottaviani, E. Yessotoxin Affects FMLP-induced Cell Shape Changes in *Mytilus galloprovincialis* Immunocytes. *Cell. Biol. Int.* **2004**, *28*, 57–61. [CrossRef]
29. Franco-Martínez, L.; Martínez-Subiela, S.; Escribano, D.; Schlosser, S.; Nöbauer, K.; Razzazi-Fazeli, E.; Romero, D.; Cerón, J.J.; Tvarijonaviciute, A. Alterations in Haemolymph Proteome of *Mytilus galloprovincialis* Mussel after an Induced Injury. *Fish Shellfish Immun.* **2018**, *75*, 41–47. [CrossRef]
30. Canesi, L.; Ciacci, C.; Fabbri, R.; Marcomini, A.; Pojana, G.; Gallo, G. Bivalve Molluscs as a Unique Target Group for Nanoparticle Toxicity. *Mar. Environ. Res.* **2012**, *76*, 16–21. [CrossRef]
31. Georgoulis, I.; Feidantsis, K.; Giantsis, I.A.; Kakale, A.; Bock, C.; Pörtner, H.O.; Sokolova, I.M.; Michaelidis, B. Heat Hardening Enhances Mitochondrial Potential for Respiration and Oxidative Defence Capacity in the Mantle of Thermally Stressed *Mytilus galloprovincialis*. *Sci. Rep.* **2021**, *11*, 17098. [CrossRef]
32. Smith, V.J.; Accorsi, A.; Malagoli, D. *The Evolution of the Immune System*; Academic Press: London, UK, 2016; pp. 1–28. [CrossRef]
33. Evariste, L.; Auffret, M.; Audonnet, S.; Geffard, A.; David, E.; Brousseau, P.; Fournier, M.; Betoulle, S. Functional Features of Hemocyte Subpopulations of the Invasive Mollusk Species *Dreissena polymorpha*. *Fish Shellfish Immun.* **2016**, *56*, 144–154. [CrossRef]
34. Guernic, A.L.; Geffard, A.; Rioult, D.; Bigot-Clivot, A.; Leprêtre, M.; Ladeiro, M.P. Cellular and Molecular Complementary Immune Stress Markers for the Model Species *Dreissena polymorpha*. *Fish Shellfish Immun.* **2020**, *107*, 452–462. [CrossRef] [PubMed]
35. Binelli, A.; Torre, C.D.; Magni, S.; Parolini, M. Does Zebra Mussel (*Dreissena polymorpha*) Represent the Freshwater Counterpart of Mytilus in Ecotoxicological Studies? A Critical Review. *Environ. Pollut.* **2015**, *196*, 386–403. [CrossRef] [PubMed]
36. Evariste, L.; Rioult, D.; Brousseau, P.; Geffard, A.; David, E.; Auffret, M.; Fournier, M.; Betoulle, S. Differential Sensitivity to Cadmium of Immunomarkers Measured in Hemocyte Subpopulations of Zebra Mussel *Dreissena polymorpha*. *Ecotoxicol. Environ. Saf.* **2017**, *137*, 78–85. [CrossRef] [PubMed]
37. Miller, N.A.; Chen, X.; Stillman, J.H. Metabolic Physiology of the Invasive Clam, *Potamocorbula amurensis*: The Interactive Role of Temperature, Salinity, and Food Availability. *PLoS ONE* **2014**, *9*, e91064. [CrossRef] [PubMed]
38. Werner, I. The Influence of Salinity on the Heat-Shock Protein Response of *Potamocorbula amurensis* (Bivalvia). *Mar. Environ. Res.* **2004**, *58*, 803–807. [CrossRef]
39. Werner, I.; Hinton, D.E. Spatial Profiles of Hsp70 Proteins in Asian Clam (*Potamocorbula amurensis*) in Northern San Francisco Bay May Be Linked to Natural Rather than Anthropogenic Stressors. *Mar. Environ. Res.* **2000**, *50*, 379–384. [CrossRef]
40. Lima, M.G.; Augusto, R.D.C.; Pinheiro, J.; Thiengo, S.C. Physiology and Immunity of the Invasive Giant African Snail, *Achatina (Lissachatina) fulica*, Intermediate Host of *Angiostrongylus cantonensis*. *Dev. Comp. Immunol.* **2020**, *105*, 103579. [CrossRef]
41. Adema, C.M.; Harris, R.A.; van Deutekom-Mulder, E.C. A Comparative Study of Hemocytes from Six Different Snails: Morphology and Functional Aspects. *J. Invertebr. Pathol.* **1992**, *59*, 24–32. [CrossRef]

42. Coaglio, A.L.; Ferreira, M.A.N.D.; dos Santos Lima, W.; de Jesus Pereira, C.A. Identification of a Phenoloxidase- and Melanin-Dependent Defence Mechanism in *Achatina fulica* Infected with *Angiostrongylus vasorum*. *Parasites Vectors* **2018**, *11*, 113. [CrossRef]
43. Brinkmann, V.; Reichard, U.; Goosmann, C.; Fauler, B.; Uhlemann, Y.; Weiss, D.S.; Weinrauch, Y.; Zychlinsky, A. Neutrophil Extracellular Traps Kill Bacteria. *Science* **2004**, *303*, 1532–1535. [CrossRef]
44. Lange, M.K.; Penagos-Tabares, F.; Muñoz-Caro, T.; Gärtner, U.; Mejer, H.; Schaper, R.; Hermosilla, C.; Taubert, A. Gastropod-Derived Haemocyte Extracellular Traps Entrap Metastrongyloid Larval Stages of *Angiostrongylus vasorum*, *Aelurostrongylus abstrusus* and *Troglostrongylus brevior*. *Parasites Vectors* **2017**, *10*, 50. [CrossRef] [PubMed]
45. Tunholi-Alves, V.M.; Tunholi, V.M.; Garcia, J.; Mota, E.M.; Castro, R.N.; Pontes, E.G.; Pinheiro, J. Unveiling the Oxidative Metabolism of *Achatina fulica* (Mollusca: Gastropoda) Experimentally Infected to *Angiostrongylus cantonensis* (Nematoda: Metastrongylidae). *Parasitol. Res.* **2018**, *117*, 1773–1781. [CrossRef]
46. Williams, A.J.; Rae, R. Susceptibility of the Giant African Snail (*Achatina fulica*) Exposed to the Gastropod Parasitic Nematode *Phasmarhabditis hermaphrodita*. *J. Invertebr. Pathol.* **2015**, *127*, 122–126. [CrossRef] [PubMed]
47. Ottaviani, E.; Malagoli, D.; Capri, M.; Franceschi, C. Ecoimmunology: Is There Any Room for the Neuroendocrine System? *Bioessays* **2008**, *30*, 868–874. [CrossRef] [PubMed]
48. Campbell, B.G.; Little, M.D. The Finding of *Angiostrongylus cantonensis* in Rats in New Orleans. *Am. J. Trop. Med. Hyg.* **1988**, *38*, 568–573. [CrossRef]
49. Shozawa, A.; Suto, C. Hemocytes of *Pomacea canaliculata*: I. Reversible Aggregation Induced by Ca^{2+}. *Dev. Comp. Immunol.* **1990**, *14*, 175–184. [CrossRef] [PubMed]
50. Accorsi, A.; Bucci, L.; de Eguileor, M.; Ottaviani, E.; Malagoli, D. Comparative Analysis of Circulating Hemocytes of the Freshwater Snail *Pomacea canaliculata*. *Fish Shellfish Immun.* **2013**, *34*, 1260–1268. [CrossRef]
51. Cueto, J.A.; Rodriguez, C.; Vega, I.A.; Castro-Vazquez, A. Immune Defenses of the Invasive Apple Snail *Pomacea canaliculata* (Caenogastropoda, Ampullariidae): Phagocytic Hemocytes in the Circulation and the Kidney. *PLoS ONE* **2015**, *10*, e0123964. [CrossRef]
52. Boraldi, F.; Lofaro, F.D.; Accorsi, A.; Ross, E.; Malagoli, D.; Boraldi, F.; Lofaro, F.D. Toward the Molecular Deciphering of *Pomacea canaliculata* Immunity: First Proteomic Analysis of Circulating Hemocytes. *Proteomics* **2019**, *19*, 1800314. [CrossRef]
53. Rodriguez, C.; Prieto, G.I.; Vega, I.A.; Castro-Vazquez, A. Assessment of the Kidney and Lung as Immune Barriers and Hematopoietic Sites in the Invasive Apple Snail *Pomacea canaliculata*. *PeerJ* **2018**, *6*, e5789. [CrossRef]
54. Bergamini, G.; Ahmad, M.; Cocchi, M.; Malagoli, D. A New Protocol of Computer-Assisted Image Analysis Highlights the Presence of Hemocytes in the Regenerating Cephalic Tentacles of Adult *Pomacea canaliculata*. *Int. J. Mol. Sci.* **2021**, *22*, 5023. [CrossRef] [PubMed]
55. Yang, C.; Tian, Y.; Lv, T.; Chang, X.; Zhang, M.; Gong, G.; Zhao, L.; Yang, S.; Chen, H. Histopathological Effects of Pedunsaponin A on *Pomacea canaliculata*. *Pestic Biochem. Phys.* **2018**, *148*, 151–158. [CrossRef] [PubMed]
56. Yang, C.; Lv, T.; Wang, B.; Qiu, X.; Luo, L.; Zhang, M.; Yue, G.; Qin, G.; Xie, D.; Chen, H. The Damaging Effects of Pedunsaponin A on *Pomacea canaliculata* Hemocytes. *Toxins* **2019**, *11*, 390. [CrossRef] [PubMed]
57. Yang, C.; Ran, X.; Zhou, Y.; Huang, Y.; Yue, G.; Zhang, M.; Gong, G.; Chang, X.; Qiu, X.; Chen, H. Study on the Relationship of Hsp70 with the Temperature Sensitivity of Pedunsaponin a Poisoning *Pomacea canaliculata*. *Pestic Biochem. Phys.* **2022**, *188*, 105243. [CrossRef] [PubMed]
58. Montanari, A.; Bergamini, G.; Ferrari, A.; Ferri, A.; Nasi, M.; Simonini, R.; Malagoli, D. The Immune Response of the Invasive Golden Apple Snail to a Nematode-Based Molluscicide Involves Different Organs. *Biology* **2020**, *9*, 371. [CrossRef]
59. Accorsi, A.; Benatti, S.; Ross, E.; Nasi, M.; Malagoli, D. A Prokineticin-like Protein Responds to Immune Challenges in the Gastropod Pest *Pomacea canaliculata*. *Dev. Comp. Immunol.* **2017**, *72*, 37–43. [CrossRef] [PubMed]
60. Giraud-Billoud, M.; Koch, E.; Vega, I.A.; Gamarra-Luques, C.; Castro-Vazquez, A. Urate Cells and Tissues in the South American Apple Snail *Pomacea canaliculata*. *J. Mollus Stud.* **2008**, *74*, 259–266. [CrossRef]
61. Boraldi, F.; Lofaro, F.D.; Bergamini, G.; Ferrari, A.; Malagoli, D. *Pomacea canaliculata* Ampullar Proteome: A Nematode-Based Bio-Pesticide Induces Changes in Metabolic and Stress-Related Pathways. *Biology* **2021**, *10*, 1049. [CrossRef]
62. Lennox, R.; Choi, K.; Harrison, P.M.; Paterson, J.E.; Peat, T.B.; Ward, T.D.; Cooke, S.J. Improving Science-Based Invasive Species Management with Physiological Knowledge, Concepts, and Tools. *Biol. Invasions* **2015**, *17*, 2213–2227. [CrossRef]
63. Mooney, H.A.; Cleland, E.E. The Evolutionary Impact of Invasive Species. *Proc. Natl. Acad. Sci. USA* **2001**, *98*, 5446–5451. [CrossRef]
64. Armitage, S.A.O.; Peuß, R.; Kurtz, J. Dscam and Pancrustacean Immune Memory—A Review of the Evidence. *Dev. Comp. Immunol.* **2015**, *48*, 315–323. [CrossRef] [PubMed]
65. Yakovenko, I.; Donnyo, A.; Ioscovich, O.; Rosental, B.; Oren, M. The Diverse Transformer (Trf) Protein Family in the Sea Urchin *Paracentrotus lividus* Acts through a Collaboration between Cellular and Humoral Immune Effector Arms. *Int. J. Mol. Sci.* **2021**, *22*, 6639. [CrossRef] [PubMed]
66. Schultz, J.H.; Bu, L.; Adema, C.M. Comparative Immunological Study of the Snail *Physella acuta* (Hygrophila, Pulmonata) Reveals Shared and Unique Aspects of Gastropod Immunobiology. *Mol. Immunol.* **2018**, *101*, 108–119. [CrossRef] [PubMed]
67. Liu, D.; Yi, Q.; Wu, Y.; Lu, G.; Gong, C.; Song, X.; Sun, J.; Qu, C.; Liu, C.; Wang, L.; et al. A Hypervariable Immunoglobulin Superfamily Member from *Crassostrea gigas* Functions as Pattern Recognition Receptor with Opsonic Activity. *Dev. Comp. Immunol.* **2018**, *86*, 96–108. [CrossRef]

68. Cantet, F.; Toubiana, M.; Parisi, M.-G.; Sonthi, M.; Cammarata, M.; Roch, P. Individual Variability of Mytimycin Gene Expression in Mussel. *Fish Shellfish Immun.* **2012**, *33*, 641–644. [CrossRef]
69. Desriac, F.; Chevalier, P.; Brillet, B.; Leguerinel, I.; Thuillier, B.; Paillard, C.; Fleury, Y. Exploring the Hologenome Concept in Marine Bivalvia: Haemolymph Microbiota as a Pertinent Source of Probiotics for Aquaculture. *FEMS Microbiol. Lett.* **2014**, *350*, 107–116. [CrossRef]
70. Liu, H.; Zha, S.; Yang, Z.; Zhang, W.; Lin, Z.; Wang, S.; Bao, Y. Acute Sulfide Exposure Induces Hemocyte Toxicity and Microbiota Dysbiosis in Blood Clam *Tegillarca granosa*. *Aquat. Toxicol.* **2022**, *249*, 106224. [CrossRef]
71. Destoumieux-Garzón, D.; Canesi, L.; Oyanedel, D.; Travers, M.; Charrière, G.M.; Pruzzo, C.; Vezzulli, L. Vibrio–Bivalve Interactions in Health and Disease. *Environ. Microbiol.* **2020**, *22*, 4323–4341. [CrossRef]
72. Vasquez, M.C.; Martinez, D.A.; Tomanek, L. Multiple Stressor Responses Are Regulated by Sirtuins in Mytilus Congeners. *Comp. Biochem. Physiol. Part A Mol. Integr. Physiol.* **2020**, *246*, 110719. [CrossRef]
73. Collins, C.L.; Burnett, N.P.; Ramsey, M.J.; Wagner, K.; Zippay, M.L. Physiological Responses to Heat Stress in an Invasive Mussel *Mytilus galloprovincialis* Depend on Tidal Habitat. *Mar. Environ. Res.* **2020**, *154*, 104849. [CrossRef]
74. Wojtal-Frankiewicz, A.; Bernasińska, J.; Frankiewicz, P.; Gwoździński, K.; Jurczak, T. The Role of Environmental Factors in the Induction of Oxidative Stress in Zebra Mussel (*Dreissena polymorpha*). *Aquat. Ecol.* **2017**, *51*, 289–306. [CrossRef]
75. Ventura, L.D.; Sarpe, D.; Kopp, K.; Jokela, J. Variability in Phenotypic Tolerance to Low Oxygen in Invasive Populations of Quagga and Zebra Mussels. *Aquat. Invasions* **2016**, *11*, 267–276. [CrossRef]
76. Carlton, J.; Tompson, J.; Schemel, L.; Nichols, F. Remarkable Invasion of San Francisco Bay (California, USA), by the Asian Clam *Potamocorbula amurensis*. I. Introduction and Dispersal. *Mar. Ecol. Prog. Ser.* **1990**, *66*, 81–94. [CrossRef]
77. Paganini, A.; Kimmerer, W.; Stillman, J. Metabolic Responses to Environmental Salinity in the Invasive Clam *Corbula amurensis*. *Aquat. Biol.* **2010**, *11*, 139–147. [CrossRef]
78. Salway, K.D.; Tattersall, G.J.; Stuart, J.A. Rapid Upregulation of Heart Antioxidant Enzymes during Arousal from Estivation in the Giant African Snail (*Achatina fulica*). *Comp. Biochem. Physiol. Part A Mol. Integr. Physiol.* **2010**, *157*, 229–236. [CrossRef]
79. Giraud-Billoud, M.; Abud, M.A.; Cueto, J.A.; Vega, I.A.; Castro-Vazquez, A. Uric Acid Deposits and Estivation in the Invasive Apple-Snail, *Pomacea canaliculata*. *Comp. Biochem. Physiol. Part A Mol. Integr. Physiol.* **2011**, *158*, 506–512. [CrossRef] [PubMed]
80. Hiong, K.C.; Loong, A.M.; Chew, S.F.; Ip, Y.K. Increases in Urea Synthesis and the Ornithine–Urea Cycle Capacity in the Giant African Snail, *Achatina fulica*, during Fasting or Aestivation, or after the Injection with Ammonium Chloride. *J. Exp. Zool. Part A Comp. Exp. Biol.* **2005**, *303*, 1040–1053. [CrossRef]
81. Giraud-Billoud, M.; Campoy-Diaz, A.D.; Dellagnola, F.A.; Rodriguez, C.; Vega, I.A. Antioxidant Responses Induced by Short-Term Activity–Estivation–Arousal Cycle in *Pomacea canaliculata*. *Front. Physiol.* **2022**, *13*, 805168. [CrossRef]
82. Giraud-Billoud, M.; Vega, I.A.; Tosi, M.E.R.; Abud, M.A.; Calderón, M.L.; Castro-Vazquez, A. Antioxidant and Molecular Chaperone Defences during Estivation and Arousal in the South American Apple Snail *Pomacea canaliculata*. *J. Exp. Biol.* **2012**, *216*, 614–622. [CrossRef]
83. Rodriguez, C.; Campoy-Diaz, A.D.; Giraud-Billoud, M. Short-Term Estivation and Hibernation Induce Changes in the Blood and Circulating Hemocytes of the Apple Snail *Pomacea canaliculata*. *Metabolites* **2023**, *13*, 289. [CrossRef]
84. Lochmiller, R.L.; Deerenberg, C. Trade-offs in Evolutionary Immunology: Just What Is the Cost of Immunity? *Oikos* **2000**, *88*, 87–98. [CrossRef]
85. Malagoli, D.; Ottaviani, E. Life Is a Huge Compromise: Is the Complexity of the Vertebrate Immune-Neuroendocrine System an Advantage or the Price to Pay? *Comp. Biochem. Physiol. Part A Mol. Integr. Physiol.* **2010**, *155*, 134–138. [CrossRef] [PubMed]
86. Gordon, S.; Martinez-Pomares, L. Physiological Roles of Macrophages. *Pflügers Arch.-Eur. J. Physiol.* **2017**, *469*, 365–374. [CrossRef]
87. Molina-Cruz, A.; DeJong, R.J.; Charles, B.; Gupta, L.; Kumar, S.; Jaramillo-Gutierrez, G.; Barillas-Mury, C. Reactive Oxygen Species Modulate *Anopheles gambiae* Immunity against Bacteria and Plasmodium. *J. Biol. Chem.* **2008**, *283*, 3217–3223. [CrossRef]
88. EFSA Panel on Plant Health (PLH). Scientific Opinion on the Assessment of the Potential Establishment of the Apple Snail in the EU. *EFSA J.* **2013**, *11*, 3487. [CrossRef]
89. Sarma, R.R.; Munsi, M.; Ananthram, A.N. Effect of Climate Change on Invasion Risk of Giant African Snail (*Achatina fulica* Férussac, 1821: Achatinidae) in India. *PLoS ONE* **2015**, *10*, e0143724. [CrossRef]
90. Thyrring, J.; Macleod, C.D.; Marshall, K.E.; Kennedy, J.; Tremblay, R.; Harley, C.D. Ocean Acidification Increases Susceptibility to Sub-Zero Air Temperatures in Ecosystem Engineers and Limits Poleward Range Shifts. *Elife* **2023**, *12*, e81080. [CrossRef]

antioxidants

MDPI

Article

The Effects of Lipoic Acid on Yolk Nutrient Utilization, Energy Metabolism, and Redox Balance over Time in *Artemia* sp.

Juan Rafael Buitrago Ramírez [1,2,*], Robson Matheus Marreiro Gomes [1,2], Alan Carvalho de Sousa Araujo [1,2], Sonia Astrid Muñoz Buitrago [1,2], Jean Piraine Souza [1,2] and José María Monserrat [1,2,3,*]

[1] Programa de Pós Graduação em Aquicultura, Instituto de Oceanografia (IO), Universidade Federal do Rio Grande—FURG, Rua do Hotel, n° 2, Cassino, Rio Grande 96210-030, RS, Brazil; roobinho_matheus@furg.br (R.M.M.G.); alancsa@furg.br (A.C.d.S.A.); samunozb@unal.edu.co (S.A.M.B.); jeanpiraine@furg.br (J.P.S.)

[2] Laboratório de Bioquímica Funcional de Organismos Aquáticos (BIFOA), Instituto de Oceanografia (IO), Universidade Federal do Rio Grande—FURG, Rua do Hotel, n° 2, Cassino, Rio Grande 96210-030, RS, Brazil

[3] Instituto of Ciências Biológicas (ICB), Universidade Federal do Rio Grande—FURG, Av. Itália, Km 08, Rio Grande 96201-900, RS, Brazil

* Correspondence: juanrafaelb25@gmail.com (J.R.B.R.); josemmonserrat@gmail.com (J.M.M.)

Abstract: Lipoic acid (LA) is a mitochondrial coenzyme that, depending on the concentration and exposure time, can behave as an antioxidant or pro-oxidant agent and has a proven ability to modulate metabolism by promoting lipid and glucose oxidation for energy production. To assess the effects of LA on energy metabolism and redox balance over time, *Artemia* sp. nauplii was used as an animal model. The administered concentrations of the antioxidant were 0.05, 0.1, 0.5, 1.0, 5.0, and 10.0 μM. Therefore, possible differences in protein, triglyceride, glucose, and lactate concentrations in the artemia samples and total ammoniacal nitrogen (TAN) in the culture water were evaluated. We also measured the effects of LA on in vivo activity of the electron transport system (ETS), antioxidant capacity, and production of reactive oxygen species (ROS) at 6, 12, 18, and 24 h post-hatching. There was a decrease in glucose concentration in the LA-treated animals, and a decrease in ammonia production was observed in the 0.5 μM LA treatment. ETS activity was positively regulated by the addition of LA, with the most significant effects at concentrations of 5.0 and 10.0 μM at 12 and 24 h. For ETS activity, treatments with LA presented the highest values at 24 h, a period when ROS production decreased significantly, for the treatment with 10.0 μM. LA showed positive regulation of energy metabolism together with a decrease in ROS and TAN excretion.

Keywords: antioxidants; energy metabolism; nutritional supplements; nitrogen compounds

Citation: Buitrago Ramírez, J.R.; Marreiro Gomes, R.M.; de Sousa Araujo, A.C.; Muñoz Buitrago, S.A.; Piraine Souza, J.; Monserrat, J.M. The Effects of Lipoic Acid on Yolk Nutrient Utilization, Energy Metabolism, and Redox Balance over Time in *Artemia* sp.. *Antioxidants* **2023**, *12*, 1439. https://doi.org/10.3390/antiox12071439

Academic Editors: Marcelo Hermes-Lima, Daniel Carneiro Moreira, Tania Zenteno-Savín and José Eduardo Serrão

Received: 28 April 2023
Revised: 30 June 2023
Accepted: 13 July 2023
Published: 18 July 2023

1. Introduction

The manner in which animals transform matter and the energy provided in their diet varies widely depending on genotypic and environmental characteristics [1–3]. In general, animals in aquatic environments rely primarily on proteins as an energy source; thus, their utilization of protein synthesis and growth depends considerably on their energy requirements [4]. Consequently, the lipids and carbohydrates supplied by the diet play secondary roles as energy substrates. Thus, it is estimated that approximately 80% of the required ATP is synthesized by amino acid catabolism in fish [4,5]. Although lipids are important energetic substrates, they are primarily used during the preprandial period, especially when prolonged [6]. As for carbohydrates, there is a consensus that their contribution to ATP synthesis in aquatic organisms is low, although their use also increases during preprandial periods or prolonged fasting. Lipid mobilization is important for energy production [6,7], and this may be due to the low availability of digestible carbohydrates in natural aquatic environments [8]. There is a special interest in minimizing the protein used as an energy source in aquaculture because it comes from raw materials in the diet with a considerable economic cost, as it is a non-renewable ingredient extracted from the natural environment [9]. Thus, optimizing the use of amino acids from proteins for anabolic processes linked to development and growth through various strategies, such as supplementation with compounds capable of modulating metabolism, is an alternative approach to explore in the search for more sustainable aquaculture activities [10,11]. Supplemental feed within aquaculture is increasingly being used to improve the physiological parameters that influence production performance [12–17]. This requires greater importance to be placed on new trends, such as precision nutritional regulation under principles such as the optimization of feeding and cycling of generated organic matter to decrease pollutant emissions [18]. Precise nutritional regulation requires optimizing feed, products, and technical and technological support for waste and metabolic management [18]. Thus, nutritional supplements, which are understood as metabolic modulators, could become tools with which to couple physiology with the zootechnical demands of a species of productive interest [19–21]. There are compounds with the proven ability to significantly modulate the energy metabolism of aquatic species, including an increase in β-oxidation and glycolysis activity that can induce a protein-sparing effect [10,20,22]. Compounds with potential for this purpose include lipoic acid (LA), a well-known antioxidant with physiological benefits, as well as the ability to contain or decrease the impact of pro-oxidant events and modulate the metabolism of the body by prioritizing metabolic pathways for mitochondrial biogenesis, mobilization of energy substrates, and energy production [23,24]. However, lipoic acid undergoes a reductive reaction and produces dihydro lipoic acid, which consumes NADPH, thereby lowering the reductive potential in the cell. Thus, depending on dose and time, lipoic acid can also act as a pro-oxidant [25]. It has been noted that the mobilization of energy reserves caused by LA leads to a decrease in protein oxidation for energy production through the positive regulation of glycolysis and β-oxidation [11,26,27]. In aquaculture, LA can be used as a promoter of protein efficiency and as a potential supplement with which to decrease the protein requirements of cultured organisms [11,26]. Although there is evidence that LA promotes protein efficiency in aquatic and terrestrial animals, the results of Terjesen et al. (2004) [28] diverge from this idea, which may point to the possible interactions of LA with the metabolism of the species and diet provided [22,29,30]. However, the possible effects of a diet that is not adjusted to the animal's requirements could lead to inconclusive results regarding how LA functions at the metabolic level. Thus, a strategy is needed at the experimental level to avoid possible misunderstandings in the assessment of the metabolic effects of LA, such as using animal models at stages where they depend on yolk reserves and adjusting to the requirements of animals at the predetermined stage of development [31,32]. Therefore, this study aimed to evaluate the metabolic changes in the use of nutrients in newly hatched *Artemia* sp. nauplii treated with LA and the effect of this compound on the redox balance of the animals. The choice of artemia nauplii as an animal model is due to the fact that this animal depends on yolk reserves, meaning that there is a supply of LA within a nutritional context where the species' requirements are met.

2. Materials and Methods

2.1. *Artemia sp.*

Artemia cysts were incubated in Imhoff cones to hatch at a ratio of 1 g L^{-1} cysts at 28 °C, 28 g L^{-1} salinity, aeration, and constant light. After 24 h of incubation, the newly hatched nauplii were collected in a single beaker, from which two aliquots of 1 mL were diluted in 99 mL seawater. From these diluted suspensions, 1 mL was taken with a glass pipette and the number of organisms was counted in duplicate. The average counts from each 100 mL beaker were calculated. The result was multiplied by 100 to determine the number of artemia mL^{-1} in the beaker. Thus, the density of artemia was adjusted to 250 mL^{-1} by replacing 100% of the water in the collection beaker with water treated with 15% sodium hypochlorite (final concentration 0.015%) and dechlorinated with sodium thiosulfate.

2.2. Standardization of the In Vivo Electron Transport System (ETS) Activity Protocol

This protocol was partially based on the protocol described by [33] Reid et al. (2018) using *Danio rerio*. This procedure uses resazurin as a fluorophore, which, when reduced by ETS mitochondrial proteins, is transformed into resorufin, a fluorescent substance (excitation: 530 nm; emission: 590 nm). The variation in fluorescence per minute was considered an indicator of the ETS activity. To standardize the artemia protocol, the influence of the resazurin concentration (0.5, 0.75, and 1.00 mg mL^{-1} final concentrations) and animal density (60, 120, and 240 artemia nauplii mL^{-1}) on fluorescence readings was tested using a full factorial design with three replicates per treatment. Readings were performed on white 96-well plates at 28 °C for 1 hour with a reading frequency of 2 min using a Synergy HT spectrofluorimeter (BioTek, São Paulo, Brazil). To assess the sensitivity of the protocol to changes in ETS activity, the effect of potassium cyanide (KCN), an inhibitor of ETS complex IV, on fluorescence kinetics was tested. Thus, final KCN concentrations of 50, 100, 150, and 200 µM were used. Immediately after KCN was added to the nauplii arranged in 96 white plate wells (240 nauplii mL^{-1}), resazurin (0.01 mg mL^{-1}) was added to nine wells of each KCN concentration and nine wells that received only distilled water as the KCN vehicle. Additionally, dichlorofluorescein diacetate (H_2DCF-DA) was used at a final concentration of 8.3 µM [34] and read at 485 nm (excitation) and 520 nm (emission) in nine additional wells exposed to different KCN concentrations or distilled water to verify ROS production as an additional marker of mitochondrial activity to confirm the effect of KCN on mitochondria using a standardized method. It is worth emphasizing that the nauplii remained alive after ETS and ROS measurements, including those exposed to KCN.

2.2.1. Isolation of the Mitochondrial Fraction

Mitochondria were extracted from artemia to estimate their capacity to reduce resazurin in vivo. This analysis was designed to quantitatively determine the contribution of the mitochondrial fraction of artemia to in vivo ETS activity. Mitochondrial fractions were isolated in triplicate from three samples of artemia incubated separately under the same conditions as those previously reported. Additionally, non-mitochondrial fractions were recovered to evaluate their contribution to the fluorescence observed in the live artemia specimens.

Two g of newly hatched artemia nauplii were gently homogenized using a Teflon homogenizer in 16 mL of buffer 1 (0.125 M sucrose, 0.375 M sorbitol, 1 mM EGTA, 150 mM KCl, 0.5% bovine serum albumin free of fatty acids, and 20 mM HEPES KOH, pH 7.5) which had been cooled with ice [35]. Subsequently, the homogenate was centrifuged for 10 s at 3026× *g* and 4 °C. The supernatant was recovered and centrifuged for 15 min at 17,409× *g* and 4 °C. After discarding the supernatant from the previous centrifugation, the pellet was resuspended in 32 mL of buffer 2 (0.125 M sucrose, 0.375 M sorbitol, 0.025 mM EGTA, 150 mM KCl, 0.5% acid-free bovine serum albumin fatty acids, and 20 mM HEPES KOH, pH 7.5) and centrifuged at 1082× *g* and 4 °C for 5 min. After centrifugation, the supernatant was recovered and centrifuged for 15 min at 17,409× *g* and 4 °C. The generated pellet (mitochondrial fraction) was resuspended in 300 µL of buffer 2 [35].

2.2.2. Measurement of the Reductive Capacity of Artemia Mitochondria

From an aliquot of the mitochondrial fraction, the protein concentration of the mitochondrial fraction was determined using the Biuret method. Subsequently, in white 96-well plates, the reaction medium (5 mM sodium succinate; 0.125 M sucrose; 0.065 M KCl; 0.002 M K_2HPO_4; and 0.01 M KOH-HEPES, pH 7.5) plus distilled water, adenosine diphosphate (ADP; final concentration: 103 µM), or ADP 103 µM + KCN as inhibitor (final concentration: 103 µM) along with resazurin (final concentration: 32.5 µM) were added in sequence. Finally, buffer 2 was added to measure the blank for the analysis: fraction 1 (pellet resulting from the first centrifugation), fraction 2 (fractions discarded from subsequent centrifugations), and mitochondrial fraction, all of which had a final concentration of 1 mg of protein mL^{-1}. The generated fluorescence was used to measure the reducing mitochondrial capacity of resazurin (530 nm excitation and 590 nm emission) for 10 min at 28 °C, with readings taken every 1 min [35,36]. The fluorescence data were multiplied by the amount of protein in each fraction, as follows:

$$mT_{mit} = m_{mit} \times prot_{mit} \cdot g^{-1} \text{ of brine shrimp}$$

$$mT_{f1} = m_{f1} \times prot_{f1} \cdot g^{-1} \text{ of brine shrimp}$$

$$mT_{f2} = m_{f2} \times prot_{f2} \cdot g^{-1} \text{ of brine shrimp}$$

where mT_{mit}, mT_{f1}, and mT_{f2} represent the slopes of the mitochondrial fraction, fraction 1, and fraction 2, respectively, for the total amount of protein present in each fraction for 1 g of brine shrimp. In turn, m_{mit}, m_{f1}, and m_{f2} represent the slopes of the fluorescence units obtained for the mitochondrial fraction, fraction 1, and fraction 2 exposed to succinate + ADP, respectively. Multiplying the values of m_{mit}, m_{f1}, and m_{f2} by the amount of total protein in each fraction present in 1 g of brine shrimp ($prot_{mit} \cdot g^{-1}$ of *Artemia* sp. nauplii, $prot_{f1} \cdot g^{-1}$ of *Artemia* sp. nauplii, and $prot_{f2} \cdot g^{-1}$ of *Artemia* sp. nauplii, respectively), the values of mT_{mit}, mT_{f1}, and mT_{f2} were estimated. Based on the mT_{mit}, mT_{f1}, and mT_{f2} values, the percentage of participation of each fraction in the resazurin reduction rate was calculated. Additionally, statistical differences between the slopes of the fluorescence lines for each substrate or inhibitor (succinate, succinate + ADP, and succinate + ADP + KCN) for each fraction were evaluated.

2.3. *In Vivo Exposure of Artemia to Lipoic Acid (LA)*

Two experiments were performed to evaluate the effects of LA on artemia nauplii. In the first experiment, we exposed the animals to LA to assess changes in nutrient reserve consumption in the yolk, antioxidant capacity, and ETS for 24 h post-hatching. This duration was selected because, according to [37] (1967), 30 h after hatching, artemia salina had nearly consumed their yolk and obtained their first food. In the second experiment, artemia nauplii were exposed to LA, and the ETS and ROS concentrations were measured after 18 and 24 h. This is because, in animals at 12 h or less post-hatching, the ETS kinetic readings showed determination coefficients below 40% (see Section 4).

2.3.1. Experiment 1

In 8 24-well plates, 2 mL of artemia nauplii was stocked (24 h post-rehydration) per well at a density of 250 artemia nauplii mL^{-1}, and 6 plates were treated with final concentrations of 0.05, 0.1, 0.5, 1.0, 5.0, and 10.0 μM. One of the plates was used as a control for the LA vehicle, which was treated only with dimethyl sulfoxide (DMSO) at a final concentration of 0.005%. The remaining plate, to which only distilled water was added, was used as the experimental control. Nauplii were collected from three wells of each treatment every six hours for 24 h, filtered with 60 μm pore size screens, deposited into previously weighed 2 mL microtubes, quickly submerged in liquid nitrogen, and stored at −80 °C for further analysis. The water from each well was used to measure the total ammonia nitrogen (TAN) concentration by means of the phenol-hypochlorite method [38]. A calibration curve with ammonium chloride obtained after the analysis of the water samples was used to quantify the TAN. Owing to interference from DMSO in the reaction, DMSO was added to the standard curve at the concentration used for the experiment.

2.3.2. Experiment 2

Artemia nauplii were stocked in 4 96-well plates 24 h post-rehydration at 240 organisms mL^{-1}. Twelve wells per plate were exposed to 0.05, 0.1, 0.5, 1.0, 5.0, and 10.0 μM lipoic acid or the LA vehicle dimethyl sulfoxide (DMSO) at a final concentration of 0.005%. In each case, only 20 μL of each solution was added to the wells to avoid extreme dilution of the water salinity (the dilution factor was 1%). Because the metabolic rate of artemia nauplii remained low at 6 and 12 h, ETS activity and ROS generation were only evaluated at 18 and 24 h. Thus, the fluorophores resazurin and H_2DCF-DA were added to the plates, with six wells for each fluorophore and each LA concentration or control. The fluorescence of each fluorophore was read every 2 min at the previously mentioned lengths for 3 h to obtain the fluorescence variations per minute and per well, following the study of Rodrigues et al. (2021) [34].

2.4. Biochemical Analysis

2.4.1. Sample Processing

The microtubes with artemia nauplii were weighed, and the microtubes' weights were recorded to determine the weights of the collected samples. The samples were diluted five times with buffer (0.09 M Na_2HPO_4, 0.09 M $KHPO_4$, 0.45 mg mL^{-1} polyvinylpyrrolidone, 22.5 μM $MgSO_4$, and 0.16% Triton X-100) and sonicated at 40 kHz with a 3 mm diameter tip in 30 s pulses for 3 min while kept permanently on ice. After homogenization, samples were centrifuged at 2500× *g* for 10 min at 4 °C. The supernatants were stored in 500 μL microtubes at 80 °C for further analysis.

2.4.2. Determination of Protein Concentration

For protein analysis, the Bioclin kit for total protein was used. In 1.5 mL microtubes, 7.5 μL of sample supernatant, the homogenization buffer (blank), or a solution of 40 mg of albumin mL^{-1} (protein standard), as well as 375 μL of Biuret reagent, was added [39]. The resulting solutions were then homogenized using a vortex and incubated for ten minutes before being transferred to 96 transparent flat-bottom microplates using two wells per sample, including blank and standard. The absorbance was measured at 550 nm using a spectrofluorometer, and the protein concentration of the samples was calculated based on the absorbance obtained from the standard protein solution. The protein results are expressed in mg g^{-1} of artemia nauplii.

2.4.3. Determination of Glucose Concentration

The dosage was determined using the Bioclin Monoreagent kit (Porto Alegre, RS, Brazil) according to the manufacturer's instructions. Homogenates of artemia nauplii samples, homogenization buffer as an analysis blank, and a standard glucose solution at 1 mg mL^{-1} were used. The procedure was performed in 96-well transparent plates, and after adding the monoreagent for the glucose samples and aliquots in duplicate, the plates were incubated at 37 °C for 10 min and read at 505 nm. The glucose concentration in the samples was calculated based on the absorbance of the standard glucose solution. The glucose results are expressed in mg g^{-1} of artemia.

2.4.4. Determination of Lactate Concentration

This was aided by a commercial kit for lactate (Bioclin, Porto Alegre, RS, Brazil) based on the production of NADH after the lactate and the NAD^+ reaction to generate pyruvate + NADH, catalyzed by the enzyme lactate dehydrogenase. NADH production was measured fluorometrically for each sample in duplicate at 340 and 440 nm. A standard lactate solution (0.3 mg mL^{-1}) was used, from which six serial dilutions were made up to a concentration of 0.004 mg mL^{-1}, with which a calibration curve was constructed. The lactate concentrations of the samples were measured five minutes after the addition of the lactate dehydrogenase enzyme and NAD^+ solution, and were then calculated against the fluorescence obtained from the lactate calibration curve. Readings were taken on 96-well white plates and each sample was replicated three times. The lactate results are expressed in mg g^{-1} of artemia nauplii.

2.4.5. Determination of Triglyceride Concentration

The analysis was performed with the initial undiluted artemia nauplii homogenates using a commercial monoreagent kit from Bioclin (Porto Alegre, RS, Brazil). The analysis was performed in 96-well transparent plates. The samples, blank, and triglyceride standard

(1 mg mL^{-1}) were analyzed in duplicate. Next, the plates were incubated at 37 °C for five minutes, and the absorbance was measured at 500 nm. The triglyceride concentrations in the samples were calculated based on the absorbance of the standard triglyceride solution. The results are expressed as artemia mg triglycerides g^{-1}.

2.4.6. Determination of Total Antioxidant Capacity

The analysis of antioxidant capacity against peroxyl radicals was performed with samples adjusted to a concentration of 0.5 mg of protein mL^{-1}. The results are expressed as the relative area of the fluorescence curves generated over time by the oxidation of H_2DCF-DA (Sigma-Aldrich, St. Louis, MO, USA) in the presence of the sample with or without the peroxyl radical generator, 2,2-azobis-2-methylpropionamidine dihydrochloride (ABAP Sigma-Aldrich). In this analysis, the larger the relative area obtained, the lower the antioxidant capacity of the sample, and vice versa [40].

2.5. Statistical Analysis

After checking the assumptions of normality, homoscedasticity, and independence of the variables, two-way analysis of variance (ANOVA) was performed (time and LA concentration) and Bonferroni's post hoc test was used to compare the means of each treatment. For the experimental control without a vehicle, a single mean was calculated for each evaluated response variable, grouping all collection times. Therefore, these data were not included in the comparisons between the treatments. The slopes of the reductive mitochondrial capacity isolated from artemia nauplii experiments were compared using ANOVA and Tukey's contrasts after verification of normality and variance homogeneity (using the Shapiro–Wilk and Levene tests, respectively). Finally, principal component analysis (PCA) was performed with the mean of each experimental group for each time, scaling the variables used with a mean of 0 and a standard deviation of 1. In the statistical tests, a significance level of 0.05 was adopted. In this sense, results with *p*-values lower than 0.05 were considered significant.

3. Results

3.1. Standardization of Activity Protocol for the In Vivo Electron Transport System (ETS)

Figure 1a shows the significant effect of animal density on the increase in fluorescence units per minute. However, the resazurin concentration did not have a significant effect ($p > 0.05$). KCN exposure, as seen in Figure 1b ($p < 0.050$), promoted a dependent response in terms of ETS determination using resazurin, whereas for ROS production, there were no differences between KCN concentrations (Figure 1c; $p > 0.05$).

3.2. Measurement of the Reductive Capacity of Artemia Nauplii Mitochondria

All fractions reduced resazurin. However, the reduction rates found in the mitochondrial fraction were, on average, approximately five times higher than those found in the other fractions. Percentage-wise, the mitochondrial fraction exposed to succinate and ADP represented 80.29 ± 12.43% of the total resazurin reduction slope, followed by fraction 2 (14.90 ± 0.5%) and fraction 1 (4.80 ± 0.8%) (Figure 2a). For the mitochondrial fraction, significant differences were observed between the fluorescence slopes per minute, depending on the type of substrate or inhibitor used. The mitochondrial fraction exposed to succinate + ADP showed the highest resazurin reduction rate, followed by that exposed only to succinate. The samples exposed to succinate + ADP + KCN showed the lowest fluorescence growth rates (Figure 2b).

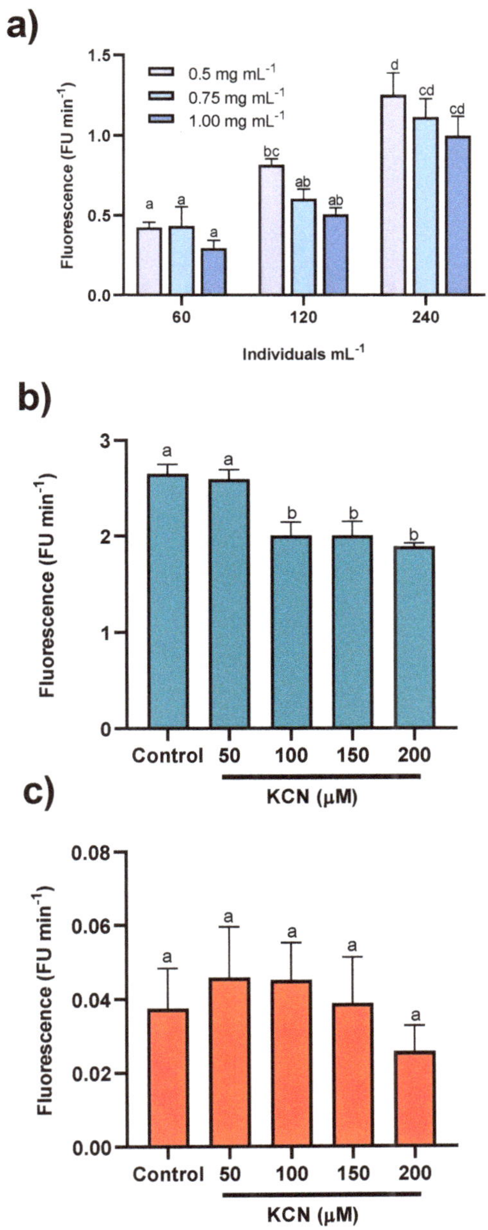

Figure 1. (**a**) Fluorescence increment per minute using different densities of artemia nauplii and different resazurin concentrations. Data are expressed as mean ± standard error (n = 3). (**b**) Effect of KCN on the electron transport system (ETS) activity. (**c**) Effect of KCN in reactive oxygen species (ROS) production. In (**b**,**c**), data are expressed as the mean ± standard error (n = 9). In all cases, different letters indicate statistically significant differences ($p < 0.05$) between the treatments.

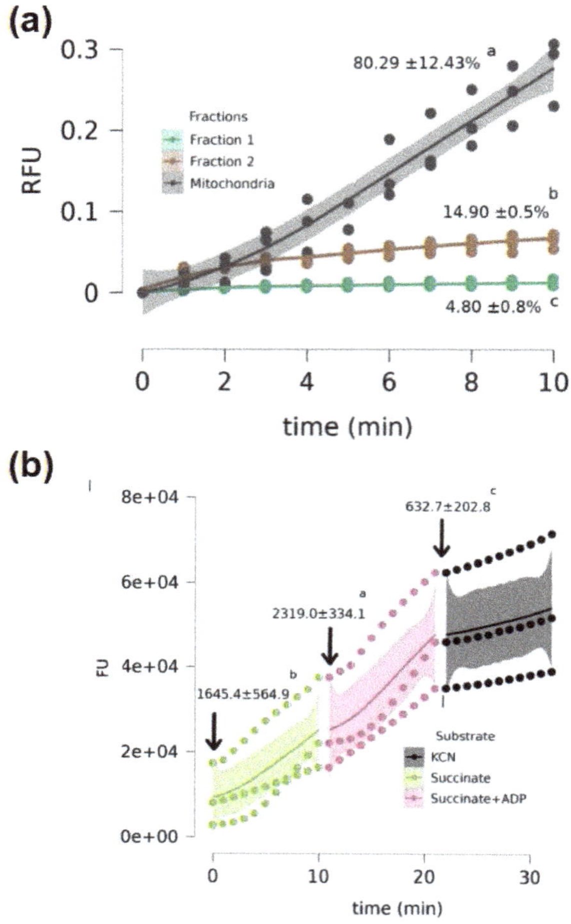

Figure 2. (**a**) Slopes of fluorescence readings (RFU) over time (min) estimated for fractions 1, 2, and the mitochondrial fraction exposed to succinate + ADP. The percentage values ± SD indicate the contribution of each fraction to the net fluorescence increment per min and per mg of protein contained in 1 g of artemia nauplii. Different letters indicate differences between the slopes of different fractions ($p < 0.05$). (**b**) Response of the mitochondrial fraction to different substrates. Values ± SD indicate the mean slopes of fluorescence increment per mg of protein contained in 1 g of artemia nauplii for each evaluated substrate. Colored bands represent the 95% confidence intervals calculated for each substrate. Different letters represent significant differences between the means of the slopes ($p < 0.05$).

3.3. *Experiment 1*

3.3.1. Protein Concentration

The results are shown in Figure 3a. At six hours, only the 10 μM treatment showed a lower protein concentration than the SCtrl treatment (control with DMSO) and the other LA treatments. At 12 h, all groups treated with LA showed lower protein levels than the SCtrl group. However, at 18 h, the treatment with 0.5 μM LA showed higher protein levels than SCtrl, which contrasted with the values obtained by the treatment with 0.05 μM, which were lower than those of SCtrl. After 24 h, no differences were found between the treatments.

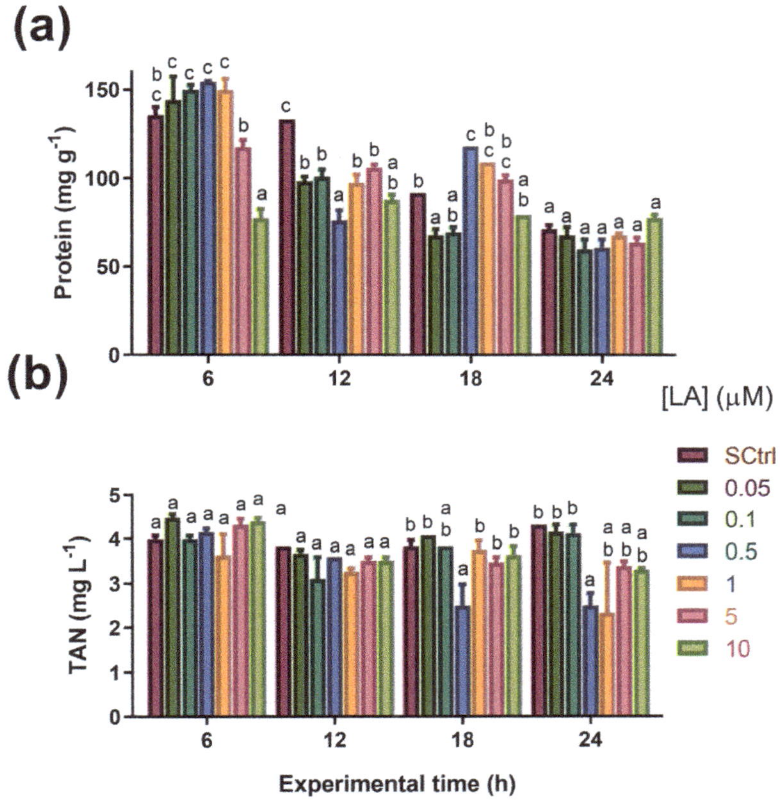

Figure 3. (**a**) Protein levels in artemia nauplii at different experimental times (h). (**b**) Total ammoniacal nitrogen (TAN) concentration in the water over time (h). Data are expressed as mean ± standard error (n = 3). Different letters indicate statistically significant differences ($p < 0.05$) between treatments. SCtrl: solvent control. Bars of different colors indicate exposure to different concentrations of lipoic acid (μM).

3.3.2. Total Ammoniacal Nitrogen (TAN)

At 18 and 24 h, there were decreases in the TAN concentration in the water of the treatment with 0.5 μM LA compared to SCtrl (Figure 3b).

3.3.3. Glucose Concentration

The results for glucose are shown in Figure 4a. For the six-hour duration, the treatment with 10 μM showed significantly lower glucose values than in SCtrl; therefore, for the twelve-hour duration, the treatment with 5 μM LA demonstrated the lowest glucose values. At 18 h, all groups that received LA showed lower glucose values than those found in SCtrl, and at 24 h, only treatments with 0.5, 1, and 5 μM LA showed differences compared to SCtrl.

3.3.4. Lactate Concentration

The lactate concentration showed differences between treatments. The treatments conducted for six hours presented the lowest lactate values, with 0.05 and 0.1 μM LA. At 12 h, there was an increase in lactate in the 1 and 10 μM treatments compared to that in SCtrl. At 18 h, the lactate values of the treatments with LA were generally lower than those presented by SCtrl, except for the treatments with 0.5 and 5 μM LA. Finally, at 24 h, the 0.05 μM LA treatment showed lactate values above those presented by SCtrl (Figure 4b).

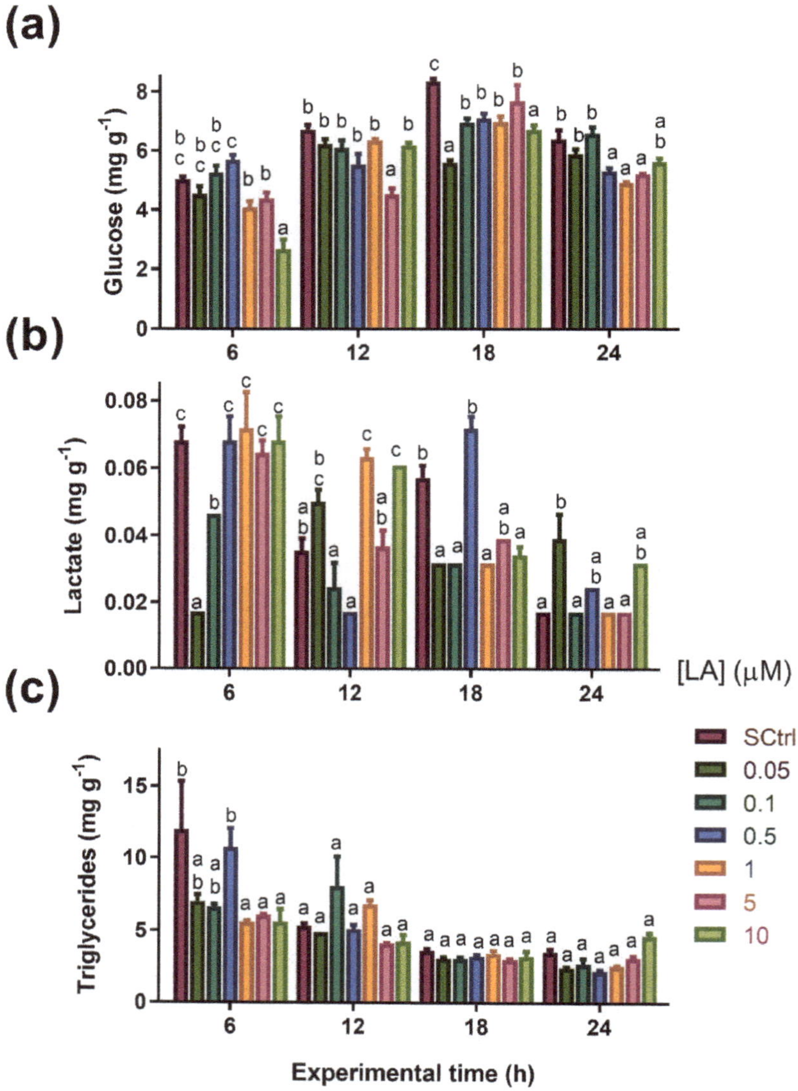

Figure 4. Glucose (**a**), lactate (**b**), and triglyceride (**c**) levels over time (h) in artemia nauplii. Data are expressed as means ± standard error (n = 3). Different letters indicate statistical differences ($p < 0.05$) between treatments. SCtrl: solvent control. Bars of different colors indicate exposure to different lipoic acid concentrations (in μM).

3.3.5. Triglyceride Content

The triglyceride content decreased at six hours in the groups treated with the highest LA concentrations (1, 5, and 10 μM LA) (Figure 4c).

3.3.6. Total Antioxidant Capacity

Overall, LA decreased the antioxidant capacity (ACAP). An increase in the antioxidant capacity was observed for the 10 μM treatment at the 24 h time point compared to the 1 and 5 μM treatments (Figure 5).

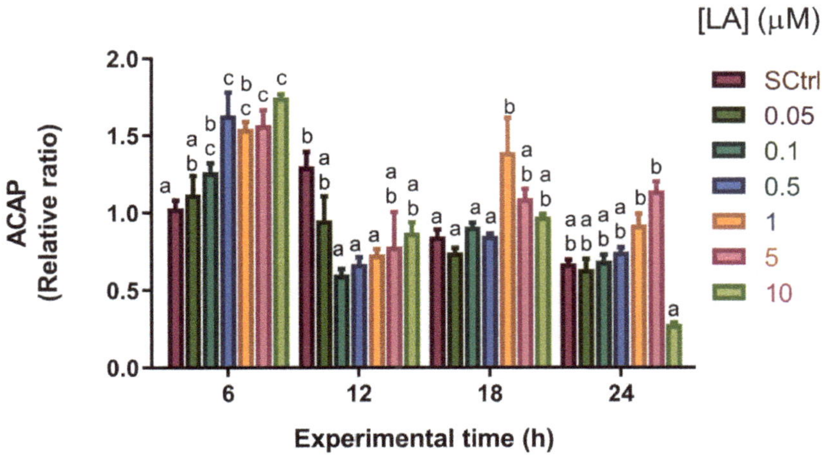

Figure 5. Total antioxidant capacity against peroxyl radicals (ACAP) in artemia nauplii over time (h). Data are expressed as means ± standard error (n = 3). Different letters indicate statistical differences ($p < 0.05$) between treatments. SCtrl: solvent control. Bars of different colors indicate exposure to different lipoic acid concentrations (in μM).

3.3.7. PCA Analysis

The first four PCA components explained 84.95% of the variance in the data matrix (PC1 45.65%, PC2 16.64%, PC3 13.69%, and PC4 8.95%). Positive correlations were found between PC1 and time (94%), as well as glucose (48%), and negative correlations were observed for protein (−87%), TAN (−43%), lactate (67%), triglycerides (−80%), and ACAP (76%). For PC2, two variables were strongly and negatively correlated with LA concentration (−81%) and positively correlated with glucose (61%). PC3 showed the highest correlation with TAN (73%), glucose, and lactate (−56%). PC4 showed the highest correlation with glucose (−49%) and ACAP (−40%). Protein levels correlated positively with lactate and triglycerides (53 and 70%, respectively) and negatively with time (77%). Triglycerides were negatively correlated with time (−76%) and glucose (−31%), and positively correlated with lactate (45%). ACAP was negatively correlated with time (−62%) and positively correlated with protein (57%), TAN (40%), lactate (48%), and triglycerides (34%) (Figure 6).

3.4. *Experiment 2*

The LA concentration and experimental time influenced the parameters of ETS activity and ROS production, and a significant interaction ($p < 0.05$) was observed between the LA treatments and exposure time for the variable ETS activity. Regarding this variable, at 18 h, no differences were detected ($p > 0.05$) among the treatments, whereas at 24 h, all treatments with LA showed higher ETS activity ($p < 0.05$) than SCtrl, with no differences among them. For ROS production at 18 h, it was not possible to observe differences between the experimental groups; however, after 24 h, there was a decrease in ROS production in the treatment with 10 μM compared to SCtrl ($p < 0.05$) (Figure 7).

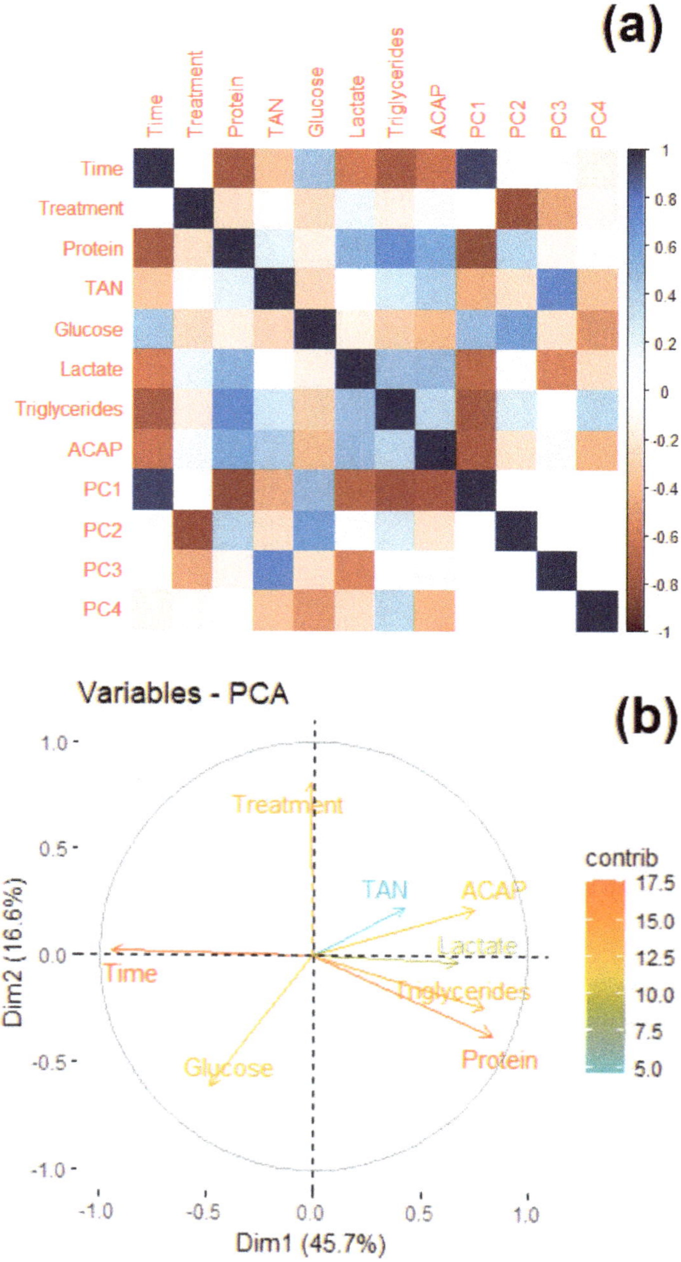

Figure 6. PCA results. (**a**) The correlation matrix is transformed into a color scale. (**b**) Contribution of each variable to the percentage of variance explained by the main components PC1 (x-axis) and PC2 (y-axis).

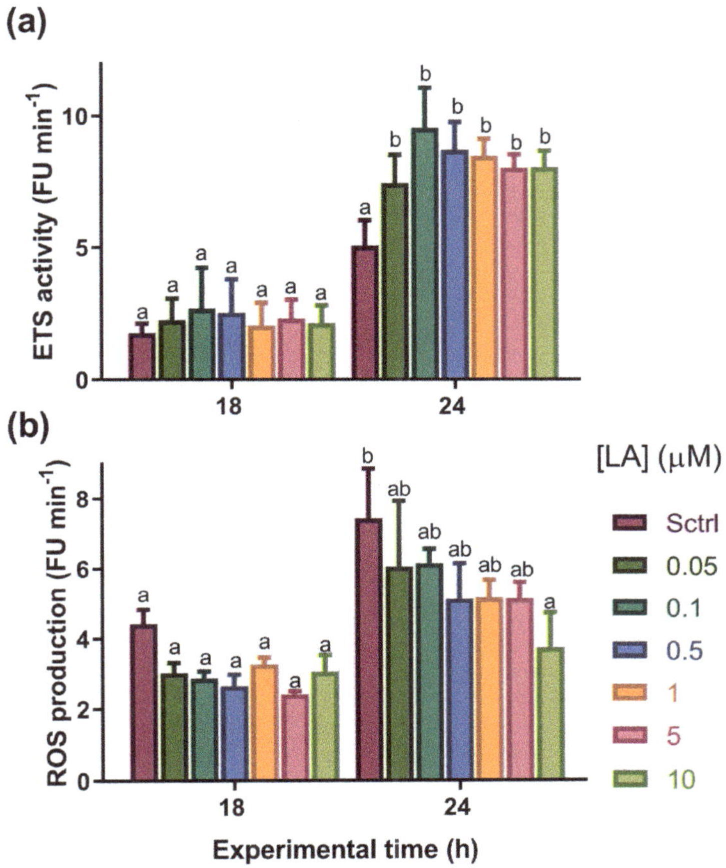

Figure 7. (**a**) Electron transport system activity and (**b**) reactive oxygen species (ROS) concentration at 18 and 24 h in artemia nauplii. Data are expressed as means ± standard error (n = 3). Different letters indicate statistical differences ($p < 0.05$) between treatments. SCtrl: solvent control. Bars of different colors indicate exposure to different lipoic acid concentrations (in µM).

4. Discussion

The standardization results of the resazurin protocol for kinetic analysis of ETS activity indicate that this protocol can be used to evaluate the ETS function of artemia nauplii. In animals 12 h or less post-hatching, the kinetic readings showed determination coefficients below 40%. Therefore, we do not recommend using the protocol during this period of life, as this phase should be associated with a low metabolic rate of the animals. At 16 h post-hatching, readings with determination coefficients above 90% were observed, which was the time taken to perform the ETS measurements. As expected, the density of the animals had a significant effect on the kinetics, indicating that the appearance of fluorescence is dependent on the metabolic activity of the animals contained within the well. However, the resazurin concentrations tested did not significantly affect the fluorescence generated when a final concentration of 0.005 mg mL^{-1} was used, which was four times lower than the concentration used by Reid et al. (2018) [33] for *Danio rerio* fish. This concentration did not produce toxic effects in either the larval or adult stage. Thus, it can be assumed that resazurin concentration allows for reliable estimation of the organism's metabolism without introducing technical artifacts. The lack of an effect of KCN on ROS levels (Figure 1c) could be associated with the existence of alternative oxidases (AOX) [35], as discussed below. The analysis performed after the isolation of the mitochondrial fraction of artemia nauplii showed that the highest net reduction of resazurin occurred in the mitochondrial fraction. This result indicates the feasibility of the method of quantifying in vivo mitochondrial activity in artemia nauplii, at least for the first hours of life, as previously described in *D. rerio* [33]. Regarding the mitochondrial fraction, the results indicated significant effects of the substrates on the resazurin reduction. By comparing the obtained results with those generally expected from mitochondrial respiratory activity, it is possible to explain the effects of each substrate on resazurin reduction. In the case of succinate, an ETS substrate that enters complex II to be oxidized as an electron donor increases the respiratory activity of the mitochondria. When this occurs in the absence of ADP, the rates of oxygen consumption, despite existing, are low (stage 4 of mitochondrial respiration), which leads to reduced mitochondrial capacity (Figure 2b). However, with the addition of ADP (stage 3 of mitochondrial respiration), an increase in oxygen consumption was expected to lead to an increase in the reducing capacity of mitochondria (Figure 2b). Thus, the obtained results were consistent with the bioenergetic mechanics of mitochondria, suggesting that resazurin is a reliable tool for measuring in vivo mitochondrial activity. Using KCN, the reductive mitochondrial capacity was not completely abolished. Rodriguez-Armenta et al. (2018) mentioned that the use of cyanide and octyl-gallate were necessary to induce a full inhibition of *A. salina* mitochondrial oxygen consumption, suggesting that an alternative oxidase may be present in this organism. In this study, the authors reported that when KCN was added first, it partially inhibited mitochondrial oxygen consumption (in our case, mitochondrial reductive capacity, Figure 2b), and was completely inhibited after adding octyl-gallate. As mentioned previously [41,42], oxyconformers present branched mitochondrial respiratory electron chains. Together with the well-known mitochondrial oxidative phosphorylation (OxPhos) components, other redox enzymes are present as alternative oxidases (AOX), which, as mentioned above, are inhibited by other molecules such as octyl-gallate [35]. It is also important to note that Talbot et al. (2008) [42] mentioned that cellular sites where resazurin is reduced include the mitochondrial matrix. A study by Springer et al. (1998) [43] showed that the inhibition of resazurin reduction occurs either by inhibiting complex I (using antimycin A) or complex II (using malonate). It is known that the reduction potential (E_0 at pH = 7.0 and temperature = 25 °C) of resazurin is +380 mV, which is reduced by molecules such as NAD(P)H (E_0 = +320 mV), $FADH_2$ (E_0 = +220 mV), and cytochromes (E_0 ranging from −80 to +290 mV) [44]. It should be noted that cyanide, as well as other poisons, like CO, arrest the whole process of mitochondrial respiration, reducing the oxidation of NADH or $FADH_2$ and, thus, decreasing the electron flow to reduce resazurin (Figure 2b). Regarding the protein values observed in Experiment 1, the greatest changes were observed over the exposure time, presenting a gradual decrease in their values. LA influenced the values of this parameter, although to a lesser extent than time, which can be explained by the fact that the animals were fasting. In this situation, the available protein is likely to be used as an energy source and reduced to amino acids for the resynthesis of protein or for the production of osmolytes such as taurine, which is synthesized from methionine [45–47]. Given that the protein concentration patterns of the LA-treated groups changed at different exposure times compared with SCtrl (Figure 3a), it is possible to consider the interactions between the developmental stage of artemia and LA as a metabolic modulator. Most of the mass contained in yolk platelets is in the form of proteins [48], which need to be degraded into amino acids that serve as the forming units of new proteins for nauplii [48]. This process is catalyzed in the nauplii phase, primarily by thiol proteases, such as cathepsin B-like [49]. There is evidence of redox regulation by cathepsin B because molecules with thiol groups, such as reduced glutathione (GSH), can elevate proteolytic activity in purified bovine cathepsin B [49]. In the case of reduced LA (dihydrolipoic acid or DHLA), concentrations between 1 and 10 μM have been observed to elevate proteolytic activity by up to 80% of the maximum enzyme activity in a solution with 2 mM GSH [49]. The induced effect of LA on yolk protein degradation could subsequently favor the protein degradation of yolk platelets for protein synthesis in the nauplii, as was observed at 18 h in the group supplemented with 0.5 μM (Figure 3a).

Regarding TAN concentration in water, the decrease observed in the treatment with 0.5 μM LA after 18 and 24 h may indicate a decrease in the deamination of amino acids for use as an energy source, possibly allowing greater efficiency to be achieved in protein utilization for artemia development. This hypothesis could explain why the concentration of 0.5 μM LA showed significantly lower TAN concentrations in water and higher protein concentrations at the 18 h time point. LA has been observed to negatively regulate the expression of proteins related to oxidative amino acid metabolism in experiments performed with other species. In carp *Ctenopharyngodon idellus*, decreased expression of the enzymes aspartate aminotransferase and alanine aminotransferase was observed when supplemented with 600 and 1200 mg LA Kg^{-1} in the diet [11]. Both proteins actively participate in amino acid catabolism and can be used as markers of the intensity of this process [11,50]. In aquaculture, lower rates of protein utilization as an energy substrate promote mass protein gain in animals, and can have considerable repercussions in terms of decreasing the protein requirement of the diet [9]. In addition, lower ammonia production influences the reduction in effluents rich in nitrogenous compounds from aquaculture and, thereby, the sustainability of the activity [51]. However, a disclaimer observation must be stated: no differences in protein levels at 24 h were observed between treatments (Figure 3a), even when the TAN level of the group exposed to 0.5 μM LA was lower than that of the control group (Figure 3b), a result that requires further confirmation.

In the case of triglycerides, it was not possible to observe a considerable effect of LA compared to SCtrl over the experimental period. Throughout the experiment, there was only one peak with a higher triglyceride concentration in the treatment with 0.5 μM LA in the first six hours (Figure 4c). In artemia, most lipid fractions are stored as neutral lipids (80%), with triglycerides representing approximately 70% [52]. Triglyceride utilization is mainly divided into the synthesis of structural lipids and the supply of free fatty acid for energy production [45]. Although these results do not show a significant effect of LA on triglyceride concentration, unlike other existing studies [11,53], similar responses to those found herein have been observed in other aquatic animal species, including crustaceans. For example, in the crab *Eriocheris sinensis*, two studies have reported that LA has no significant effect on the lipid content in this species [27,54].

In the case of glucose, there was an overall gradual increase in glucose levels until the 18 h time period, and from the 24 h time period (Figure 4a), there was a general decrease in the levels of this energy substrate. This behavior can be explained by the amount of trehalose in artemia cysts. Trehalose is a disaccharide composed of glucose monomers and is stored at high concentrations inside cysts, acting as a cellular protector during the cryptobiosis phase [55]. However, when the organism returns to development after rehydration, trehalose is hydrolyzed by the trehalase enzyme and transformed into glucose, which acts as an energy substrate to supply accelerated metabolic activity during development [56,57]. It is worth noting that glucose showed a negative correlation with the protein and triglycerides variables according to the results obtained by PCA (Figure 6), since, while the levels of protein and triglycerides decreased, glucose concentrations increased at least until the 18 h time point, perhaps due to increased gluconeogenesis. This could indicate that proteins and triglycerides participate as substrates for energy metabolism under normal physiological conditions. From this moment on, glucose has begun to occupy a place of greater importance as an energy substrate. In addition, it is particularly interesting to note that during the experiment, there were always groups treated with LA with glucose concentrations lower than those presented by the SCtrl group, corroborating the importance of LA as a promoter of carbohydrate metabolism [22,24]. These results seem to agree with those obtained for *Macrobrachium nipponense*, in which LA supplementation (700 and 1400 mg Kg^{-1} feed) was found to increase the expression of glycolysis, Krebs cycle, and oxidative phosphorylation enzymes [26]. In this same species, LA supplementation levels from 1000 to 5000 mg Kg^{-1} of feed resulted in biphasic behavior of the expression of the enzymes hexokinase, phosphofructokinase, and pyruvate kinase, with the highest expression at intermediate LA concentrations. This behavior was also observed for isocitrate dehydrogenase expression, indicating an LA-induced hormetic-type effect [58]. Throughout the experiment, notable increases in lactate concentration were observed in organisms exposed to lactic acid. Under certain conditions and concentrations, LA may increase energy metabolism above respiratory capacity, which could be compensated for by activating metabolic pathways such as anaerobic glycolysis [59]. However, this should be explored in future studies because of the lack of previous data indicating that LA increases lactate concentrations, even when considering studies on other crustaceans [26,58]. Evidence for an increase in metabolic activity caused by LA was found in the ETS capacity results. During the experiment, this parameter increased according to the LA concentration, as shown by the PCA results (Figure 6b), where a highly positive correlation between ETS capacity and LA inclusion in the medium was observed. However, the mechanisms by which this occurs remain unclear. Nevertheless, there is evidence for the participation of the reduced nucleotides NADH and NADPH in LA reduction. Several proteins, including thioredoxin reductase, lipoamide dehydrogenase, and the E3 subunit of α-ketoglutarate dehydrogenase, can catalyze this reduction [60]. This process could imply a decrease in cytoplasmic and mitochondrial $NAD(P)H/NAD(P)^+$ ratios. Mitochondrial ADP and AMP concentrations should increase under these conditions, favoring the consequent activation of proteins such as AMPK, which promotes the catabolic metabolism of organic macromolecules for the release of energy substrates, such as glucose or fatty acids, to restore ATP levels [25,53,61]. One effect observed after LA supplementation was increased sirtuin 1-SIRT1 expression [22]. This NAD^+-dependent deacetylase can also increase the activity of transcription factors such as peroxisome proliferation-activated receptor gamma (PPARγ), which acts as an agonist in mitochondrial biogenesis, with potential consequences in terms of increasing ETS capacity [62].

This study did not show a consistent effect of LA on the antioxidant capacity, unlike the results observed in several other animal models. States of lower antioxidant capacity were observed at 6 and 18 h, whereas others had higher antioxidant capacity at 12 and 24 h, although at the latter time, this occurred only at the concentration of 10 μM LA (Figure 5). These different responses reflect the dual role of LA as a pro-oxidant and antioxidant, which explains the gradual decrease in the antioxidant capacity observed at 6 h as the concentration of LA increased. It should be considered that the increments in ETS promoted by adding LA to the medium could have influenced the antioxidant capacity results. Thus, these results can be strongly linked to the effects of LA as a promoter of energy metabolism as an increase in oxidative phosphorylation can cause increased ROS production [63], which, in principle, should be intercepted as a protective mechanism by antioxidant defenses [64]. Alpha-lipoic acid, as an agent that modulates the redox state of a cell by influencing the concentration of reduced nucleotides NAD(P)H, has the ability to meddle with energy production routes by altering the equilibrium of $NAD^+/NADH$ concentrations [25]. A relative surge in NAD^+ concentrations can stimulate mitochondrial activity to replenish reduced nucleotides serving as energy transporters for the electron transport chain [65]. It is crucial to note that there is evidence of redox-type interactions between alpha-lipoic acid and insulin receptors, which in turn boost insulin signaling within cells [66].

The demand for endogenous antioxidants in situations of high ROS production can decrease the antioxidant capacity [67]. This could explain the metabolic increment associated with the energetic cost of antioxidant synthesis to cope with the ROS increase. This idea is based on evidence of the intimate relationship between AMPK activation and antioxidant gene expression, where the post-transcriptional changes performed by this protein on erythroid nuclear factor 2-related factor 2 (Nrf2) cause it to migrate to the nucleus, where it acts as a ligand on AREs to activate its expression [68]. LA induces metabolic changes (glucose and ETS output) in animals, which implies accelerated use of energy reserves that may interact with (or antagonize) normal physiological demands at this stage of life. In this situation, the organism could be forced to prioritize physiological functions differently than a more slowly developing organism and/or an organism with a constant nutrient intake. Further studies are needed to determine how LA acts in animals at an accelerated developmental stage with limited nutrient availability.

In the PCA analysis (Figure 6), owing to the high correlation of PC1 with time, it was possible to determine that this factor contributed the most to the variability observed within the data, where, in general, there was a decrease in nutrient concentration as the experiment progressed. The correlation between treatment with LA and PC2 provided evidence of the role of LA in data variability. Finally, it should be noted that the negative correlation between glucose and LA concentrations may indicate the influence of LA as a promoter of glucose catabolism (Figure 6a), possibly through positive regulation of mitochondrial energy activity. Additionally, at 24 h, a gradual decrease in ROS production was observed with the addition of LA, although the differences were only significant between the SCtrl and 10 μM treatment groups. LA did not promote antioxidant capacity, but paradoxically, it seemed to have had a sizeable effect on ROS production. This seems counterintuitive given its high ETS activity and low antioxidant capacity. However, considering that ROS production was lower than in SCtrl, the decrease in antioxidant capacity in LA-treated animals may have been the result of decreased ROS production as an effect of LA, which would be an adaptive response for the organism. Additionally, the observed increase in ETS activity may have influenced the decrease in the antioxidant capacity to contain ROS, a product of mitochondrial respiration.

To summarize the main findings observed in this study, we can mention that: (1) LA induced a higher aerobic metabolism, as indicated by the electron transport system (ETS) activity, pointing to its role as a metabolic regulator; (2) the in vivo ETS results can be associated mainly with the mitochondrial activity of artemia; (3) additionally, LA promoted glucose catabolism; (4) LA promoted a decrease in ROS production, as is consistent with its well-known antioxidant properties; (5) some combinations of LA concentrations (0.5 μM) and exposure time (18 h) induced higher protein levels and lowered TAN production.

Author Contributions: J.R.B.R., R.M.M.G., A.C.d.S.A., S.A.M.B. and J.P.S. carried out the experimental and analytical processes. J.R.B.R. and J.M.M. designed the experiment, conducted the statistical analyses, and wrote the main manuscript. All authors have read and agreed to the published version of the manuscript.

Funding: This study was financed in part by the Coordenação de Aperfeiçoamento de Pessoal de Nível Superior—Brasil (CAPES), Finance Code 001. José M. Monserrat is a research fellow of Conselho Nacional de Desenvolvimento Científico e Tecnológico (CNPq) (process number: 307888/2020-7). Juan Rafael Buitrago Ramírez acknowledges the fellowship from CNPq (Process number 152947/2022-1).

Institutional Review Board Statement: Not applicable.

Informed Consent Statement: Not applicable.

Data Availability Statement: The data that support the findings of this study are openly available in Figshare at https://doi.org/10.6084/m9.figshare.23690880.v1.

Conflicts of Interest: The authors declare no conflict of interest.

References

1. Cristóbal-Azkarate, J.; Maréchal, L.; Semple, S.; Majolo, B.; MacLarnon, A. Metabolic Strategies in Wild Male Barbary Macaques: Evidence from Faecal Measurement of Thyroid Hormone. *Biol. Lett.* **2016**, *12*, 20160168. [CrossRef] [PubMed]
2. Koyama, T.; Texada, M.J.; Halberg, K.A.; Rewitz, K. Metabolism and Growth Adaptation to Environmental Conditions in Drosophila. *Cell. Mol. Life Sci.* **2020**, *77*, 4523–4551. [CrossRef] [PubMed]
3. Seibel, B.A.; Drazen, J.C. The Rate of Metabolism in Marine Animals: Environmental Constraints, Ecological Demands and Energetic Opportunities. *Philos. Trans. R. Soc. B Biol. Sci.* **2007**, *362*, 2061–2078. [CrossRef] [PubMed]
4. Jia, S.; Li, X.; Zheng, S.; Wu, G. Amino Acids Are Major Energy Substrates for Tissues of Hybrid Striped Bass and Zebrafish. *Amino Acids* **2017**, *49*, 2053–2063. [CrossRef]

5. Li, X.; Han, T.; Zheng, S.; Wu, G. Nutrition and Functions of Amino Acids in Aquatic Crustaceans. In *Amino Acids in Nutrition and Health: Amino Acids in the Nutrition of Companion, Zoo and Farm Animals*; Wu, G., Ed.; Springer International Publishing: Cham, Switzerland, 2021; pp. 169–198; ISBN 978-3-030-54462-1.
6. Navarro, I.; Gutiérrez, J. Chapter 17 Fasting and Starvation. In *Metabolic Biochemistry*; Hochachka, P.W., Mommsen, T.P., Eds.; Biochemistry and Molecular Biology of Fishes; Elsevier: Amsterdam, The Netherlands, 1995; Volume 4, pp. 393–434.
7. Wang, X.; Li, E.; Chen, L. A Review of Carbohydrate Nutrition and Metabolism in Crustaceans. *N. Am. J. Aquac.* **2016**, *78*, 178–187. [CrossRef]
8. Polakof, S.; Panserat, S.; Soengas, J.L.; Moon, T.W. Glucose Metabolism in Fish: A Review. *J. Comp. Physiol. B* **2012**, *182*, 1015–1045. [CrossRef]
9. Hua, K.; Cobcroft, J.M.; Cole, A.; Condon, K.; Jerry, D.R.; Mangott, A.; Praeger, C.; Vucko, M.J.; Zeng, C.; Zenger, K.; et al. The Future of Aquatic Protein: Implications for Protein Sources in Aquaculture Diets. *One Earth* **2019**, *1*, 316–329. [CrossRef]
10. Mohseni, M.; Ozorio, R.O.A.; Pourkazemi, M.; Bai, S.C. Effects of Dietary L-Carnitine Supplements on Growth and Body Composition in Beluga Sturgeon (*Huso Huso*) Juveniles. *J. Appl. Ichthyol.* **2008**, *24*, 646–649. [CrossRef]
11. Shi, X.C.; Jin, A.; Sun, J.; Tian, J.J.; Ji, H.; Chen, L.Q.; Du, Z.Y. The Protein-Sparing Effect of α-Lipoic Acid in Juvenile Grass Carp, *Ctenopharyngodon Idellus*: Effects on Lipolysis, Fatty Acid β-Oxidation and Protein Synthesis. *Br. J. Nutr.* **2018**, *120*, 977–987. [CrossRef]
12. Carbone, D.; Faggio, C. Importance of Prebiotics in Aquaculture as Immunostimulants. Effects on Immune System of *Sparus Aurata* and Dicentrarchus Labrax. *Fish Shellfish Immunol.* **2016**, *54*, 172–178. [CrossRef]
13. Encarnação, P. Functional Feed Additives in Aquaculture Feeds. *Aquafeed Formul.* **2016**, 217–237. [CrossRef]
14. Fuchs, V.I.; Schmidt, J.; Slater, M.J.; Zentek, J.; Buck, B.H.; Steinhagen, D. The Effect of Supplementation with Polysaccharides, Nucleotides, Acidifiers and Bacillus Strains in Fish Meal and Soy Bean Based Diets on Growth Performance in Juvenile Turbot (*Scophthalmus maximus*). *Aquaculture* **2015**, *437*, 243–251. [CrossRef]
15. Hoseinifar, S.H.; Sun, Y.Z.; Wang, A.; Zhou, Z. Probiotics as Means of Diseases Control in Aquaculture, a Review of Current Knowledge and Future Perspectives. *Front. Microbiol.* **2018**, *9*, 2429. [CrossRef] [PubMed]
16. Kütter, M.T.; Monserrat, J.M.; Primel, E.G.; Caldas, S.S.; Tesser, M.B. Effects of Dietary α-Lipoic Acid on Growth, Body Composition and Antioxidant Status in the Plata Pompano *Trachinotus marginatus* (Pisces, Carangidae). *Aquaculture* **2012**, *368–369*, 29–35. [CrossRef]
17. Xu, N.; Fu, J.; Wang, H.; Lu, L. Quercetin Counteracts the Pro-Viral Effect of Heat Shock Response in Grass Carp Cells with Its Therapeutic Potential against Aquareovirus. *Aquac. Res.* **2021**, *52*, 3164–3173. [CrossRef]
18. Zhang, Y.; Lu, R.; Qin, C.; Nie, G. Precision Nutritional Regulation and Aquaculture. *Aquac. Rep.* **2020**, *18*, 100496. [CrossRef]
19. Lu, D.L.; Limbu, S.M.; Lv, H.B.; Ma, Q.; Chen, L.Q.; Zhang, M.L.; Du, Z.Y. The Comparisons in Protective Mechanisms and Efficiencies among Dietary α-Lipoic Acid, β-Glucan and l-Carnitine on Nile Tilapia Infected by Aeromonas Hydrophila. *Fish Shellfish Immunol.* **2019**, *86*, 785–793. [CrossRef]
20. Panserat, S.; Skiba-Cassy, S.; Seiliez, I.; Lansard, M.; Plagnes-Juan, E.; Vachot, C.; Aguirre, P.; Larroquet, L.; Chavernac, G.; Medale, F.; et al. Metformin Improves Postprandial Glucose Homeostasis in Rainbow Trout Fed Dietary Carbohydrates: A Link with the Induction of Hepatic Lipogenic Capacities? *Am. J. Physiol.-Regul. Integr. Comp. Physiol.* **2009**, *297*, R707–R715. [CrossRef]
21. Walsh, M.C.; Jacquier, V.; Schyns, G.; Claypool, J.; Tamburini, I.; Blokker, B.; Geremia, J.M. A Novel Microbiome Metabolic Modulator Improves the Growth Performance of Broiler Chickens in Multiple Trials and Modulates Targeted Energy and Amino Acid Metabolic Pathways in the Cecal Metagenome. *Poult. Sci.* **2021**, *100*, 100800. [CrossRef]
22. Huang, C.-C.; Sun, J.; Ji, H.; Kaneko, G.; Xie, X.-D.; Chang, Z.-G.; Deng, W. Systemic Effect of Dietary Lipid Levels and α-Lipoic Acid Supplementation on Nutritional Metabolism in Zebrafish (*Danio rerio*): Focusing on the Transcriptional Level. *Fish Physiol. Biochem.* **2020**, *46*, 1631–1644. [CrossRef]
23. Maczurek, A.; Hager, K.; Kenklies, M.; Sharman, M.; Martins, R.; Engel, J.; Carlson, D.A.; Münch, G. Lipoic Acid as an Anti-Inflammatory and Neuroprotective Treatment for Alzheimer's Disease. *Adv. Drug Deliv. Rev.* **2008**, *60*, 1463–1470. [CrossRef] [PubMed]
24. Sohaib, M.; Anjum, F.M.; Nasir, M.; Saeed, F.; Arshad, M.S.; Hussain, S. Alpha-Lipoic Acid: An Inimitable Feed Supplement for Poultry Nutrition. *J. Anim. Physiol. Anim. Nutr.* **2018**, *102*, 33–40. [CrossRef] [PubMed]
25. Chen, W.-L.; Kang, C.-H.; Wang, S.-G.; Lee, H.-M. α-Lipoic Acid Regulates Lipid Metabolism through Induction of Sirtuin 1 (SIRT1) and Activation of AMP-Activated Protein Kinase. *Diabetologia* **2012**, *55*, 1824–1835. [CrossRef] [PubMed]
26. Ding, Z.; Xiong, Y.; Zheng, J.; Zhou, D.; Kong, Y.; Qi, C.; Liu, Y.; Ye, J.; Limbu, S.M. Modulation of Growth, Antioxidant Status, Hepatopancreas Morphology, and Carbohydrate Metabolism Mediated by Alpha-Lipoic Acid in Juvenile Freshwater Prawns *Macrobrachium nipponense* under Two Dietary Carbohydrate Levels. *Aquaculture* **2022**, *546*, 737314. [CrossRef]
27. Xu, C.; Wang, X.; Han, F.; Qi, C.; Li, E.; Guo, J.; Qin, J.G.; Chen, L. α-Lipoic Acid Regulate Growth, Antioxidant Status and Lipid Metabolism of Chinese Mitten Crab *Eriocheir sinensis*: Optimum Supplement Level and Metabonomics Response. *Aquaculture* **2019**, *506*, 94–103. [CrossRef]
28. Terjesen, B.F.; Park, K.; Tesser, M.B.; Portella, M.C.; Zhang, Y.; Dabrowski, K. Biochemical and Molecular Actions of Nutrients Lipoic Acid and Ascorbic Acid Affect Plasma Free Amino Acids Selectively in the Teleost Fish Pacu (*Piaractus mesopotamicus*). *J. Nutr.* **2004**, *134*, 2930–2934. [CrossRef] [PubMed]

29. Castro, M.C.; Massa, M.L.; Schinella, G.; Gagliardino, J.J.; Francini, F. Lipoic Acid Prevents Liver Metabolic Changes Induced by Administration of a Fructose-Rich Diet. *Biochim. Biophys. Acta (BBA) Gen. Subj.* **2013**, *1830*, 2226–2232. [CrossRef] [PubMed]
30. Huang, D.; Lei, Y.; Xu, W.; Zhang, Y.; Zhou, H.; Zhang, W.; Mai, K. Protective Effects of Dietary α-Lipoic Acid on Abalone Haliotis Discus Hannai against the Oxidative Damage under Waterborne Cadmium Stress. *Aquac. Nutr.* **2019**, *25*, 263–270. [CrossRef]
31. Tong, X.; Yang, X.; Bao, C.; Wang, J.; Tang, X.; Jiang, D.; Yang, L. Changes of Biochemical Compositions during Development of Eggs and Yolk-Sac Larvae of Turbot *Scophthalmus maximus*. *Aquaculture* **2017**, *473*, 317–326. [CrossRef]
32. Vázquez, R.; González, S.; Rodríguez, A.; Mourente, G. Biochemical Composition and Fatty Acid Content of Fertilized Eggs, Yolk Sac Stage Larvae and First-Feeding Larvae of the Senegal Sole (*Solea senegalensis* Kaup). *Aquaculture* **1994**, *119*, 273–286. [CrossRef]
33. Reid, R.M.; D'Aquila, A.L.; Biga, P.R. The Validation of a Sensitive, Non-Toxic in Vivo Metabolic Assay Applicable across Zebrafish Life Stages. *Comp. Biochem. Physiol. Part C Toxicol. Pharmacol.* **2018**, *208*, 29–37. [CrossRef] [PubMed]
34. Rodrigues, R.V.; Maltez, L.C.; Ferreira, C.C.; Oliveira, T.P.A.P.; Sampaio, L.A.; Monserrat, J.M. ROS in Vivo Determination and Antioxidant Responses in Rotifers *Brachionus plicatilis* Fed with Commercial Yeast *Saccharomyces cerevisiae* or Microalgae *Nannochloropsis oculata*. *Aquac. Int.* **2021**, *29*, 1657–1667. [CrossRef]
35. Rodriguez-Armenta, C.; Uribe-Carvajal, S.; Rosas-Lemus, M.; Chiquete-Felix, N.; Huerta-Ocampo, J.A.; Muhlia-Almazan, A. Alternative Mitochondrial Respiratory Chains from Two Crustaceans: *Artemia franciscana* Nauplii and the White Shrimp, *Litopenaeus vannamei*. *J. Bioenerg. Biomembr.* **2018**, *50*, 143–152. [CrossRef] [PubMed]
36. Grey, J.F.E.; Townley, A.R.; Everitt, N.M.; Campbell-Ritchie, A.; Wheatley, S.P. A Cost-Effective, Analytical Method for Measuring Metabolic Load of Mitochondria. *Metab. Open* **2019**, *4*, 100020. [CrossRef] [PubMed]
37. Paffenhöfer, G.-A. Caloric Content of Larvae of the Brine Shrimp *Artemia salina*. *Helgoländer Wiss. Meeresunters.* **1967**, *16*, 130–135. [CrossRef]
38. Helder, W.; De Vries, R.T.P. An Automatic Phenol-Hypochlorite Method for the Determination of Ammonia in Sea- and Brackish Waters. *Neth. J. Sea Res.* **1979**, *13*, 154–160. [CrossRef]
39. Krohn, R.I. The Colorimetric Detection and Quantitation of Total Protein. *Curr. Protoc. Toxicol.* **2005**, *23*, A.3I.1–A.3I.28. [CrossRef]
40. Amado, L.L.; Garcia, M.L.; Ramos, P.B.; Freitas, R.F.; Zafalon, B.; Ferreira, J.L.R.; Yunes, J.S.; Monserrat, J.M. A Method to Measure Total Antioxidant Capacity against Peroxyl Radicals in Aquatic Organisms: Application to Evaluate Microcystins Toxicity. *Sci. Total Environ.* **2009**, *407*, 2115–2123. [CrossRef]
41. Mendez-Romero, O.; Ricardez-García, C.; Castañeda-Tamez, P.; Chiquete-Félix, N.; Uribe-Carvajal, S. Thriving in Oxygen While Preventing ROS Overproduction: No Two Systems Are Created Equal. *Front. Physiol.* **2022**, *13*, 582. [CrossRef]
42. Talbot, J.D.; Barrett, J.N.; Barrett, E.F.; David, G. Rapid, Stimulation-Induced Reduction of C12-Resorufin in Motor Nerve Terminals: Linkage to Mitochondrial Metabolism. *J. Neurochem.* **2008**, *105*, 807–819. [CrossRef]
43. Springer, J.E.; Azbill, R.D.; Carlson, S.L. A Rapid and Sensitive Assay for Measuring Mitochondrial Metabolic Activity in Isolated Neural Tissue. *Brain Res. Protoc.* **1998**, *2*, 259–263. [CrossRef] [PubMed]
44. Rampersad, S.N. Multiple Applications of Alamar Blue as an Indicator of Metabolic Function and Cellular Health in Cell Viability Bioassays. *Sensors* **2012**, *12*, 12347–12360. [CrossRef] [PubMed]
45. Evjemo, J.O.; Danielsen, T.L.; Olsen, Y. Losses of Lipid, Protein and N−3 Fatty Acids in Enriched *Artemia Franciscana* Starved at Different Temperatures. *Aquaculture* **2001**, *193*, 65–80. [CrossRef]
46. Helland, S.; Triantaphyllidis, G.V.; Fyhn, H.J.; Evjen, M.S.; Lavens, P.; Sorgeloos, P. Modulation of the Free Amino Acid Pool and Protein Content in Populations of the Brine Shrimp *Artemia* spp. *Mar. Biol.* **2000**, *137*, 1005–1016. [CrossRef]
47. Tanaka, K. The Proteasome: From Basic Mechanisms to Emerging Roles. *Keio J. Med.* **2013**, *62*, 1–12. [CrossRef]
48. Warner, A.H.; Puodziukas, J.G.; Finamore, F.J. Yolk Platelets in Brine Shrimp Embryos: Site of Biosynthesis and Storage of the Diguanosine Nucleotides. *Exp. Cell Res.* **1972**, *70*, 365–375. [CrossRef] [PubMed]
49. Warner, A.H.; Perz, M.J.; Osahan, J.K.; Zielinski, B.S. Potential Role in Development of the Major Cysteine Protease in Larvae of the Brine Shrimp *Artemia franciscana*. *Cell Tissue Res.* **1995**, *282*, 21–31. [CrossRef] [PubMed]
50. Jiang, J.; Feng, L.; Tang, L.; Liu, Y.; Jiang, W.; Zhou, X. Growth Rate, Body Composition, Digestive Enzymes and Transaminase Activities, and Plasma Ammonia Concentration of Different Weight Jian Carp (*Cyprinus Carpio* Var. Jian). *Anim. Nutr.* **2015**, *1*, 373–377. [CrossRef]
51. Herbeck, L.S.; Krumme, U.; Nordhaus, I.; Jennerjahn, T.C. Pond Aquaculture Effluents Feed an Anthropogenic Nitrogen Loop in a SE Asian Estuary. *Sci. Total Environ.* **2021**, *756*, 144083. [CrossRef] [PubMed]
52. Navarro, J.C.; Amat, F.; Sargent, J.R. The Lipids of the Cysts of Freshwater- and Marine-Type Artemia. *Aquaculture* **1993**, *109*, 327–336. [CrossRef]
53. Ruderman, N.B.; Julia Xu, X.; Nelson, L.; Cacicedo, J.M.; Saha, A.K.; Lan, F.; Ido, Y. AMPK and SIRT1: A Long-Standing Partnership? *Am. J. Physiol.-Endocrinol. Metab.* **2010**, *298*, E751–E760. [CrossRef] [PubMed]
54. Xu, C.; Li, E.; Liu, S.; Huang, Z.; Qin, J.G.; Chen, L. Effects of α-Lipoic Acid on Growth Performance, Body Composition, Antioxidant Status and Lipid Catabolism of Juvenile Chinese Mitten Crab *Eriocheir Sinensis* Fed Different Lipid Percentage. *Aquaculture* **2018**, *484*, 286–292. [CrossRef]
55. Clegg, J.S. The Origin of Threhalose and Its Significance during the Formation of Encysted Dormant Embryos of *Artemia salina*. *Comp. Biochem. Physiol.* **1965**, *14*, 135–143. [CrossRef] [PubMed]
56. Boulton, A.P.; Huggins, A.K. Biochemical Changes Occurring during Morphogenesis of the Brine Shrimp *Artemia Salina* and the Effect of Alterations in Salinity. *Comp. Biochem. Physiol. Part A Physiol.* **1977**, *57*, 17–22. [CrossRef]

57. Vallejo, C.G. Artemia Trehalase: Regulation by Factors That Also Control Resumption of Development. In *Cell and Molecular Biology of Artemia Development*; Warner, A.H., MacRae, T.H., Bagshaw, J.C., Eds.; Springer: Boston, MA, USA, 1989; pp. 173–189; ISBN 978-1-4757-0004-6.
58. Xiong, Y.; Li, Q.; Ding, Z.; Zheng, J.; Zhou, D.; Wei, S.; Han, X.; Cheng, X.; Li, X.; Xue, Y. Dietary α-Lipoic Acid Requirement and Its Effects on Antioxidant Status, Carbohydrate Metabolism, and Intestinal Microflora in Oriental River Prawn *Macrobrachium Nipponense* (De Haan). *Aquaculture* **2022**, *547*, 737531. [CrossRef]
59. Sokolova, I. Bioenergetics in Environmental Adaptation and Stress Tolerance of Aquatic Ectotherms: Linking Physiology and Ecology in a Multi-Stressor Landscape. *J. Exp. Biol.* **2021**, *224*, jeb236802. [CrossRef] [PubMed]
60. Ambrus, A.; Tretter, L.; Adam-Vizi, V. Inhibition of the Alpha-Ketoglutarate Dehydrogenase-Mediated Reactive Oxygen Species Generation by Lipoic Acid. *J. Neurochem.* **2009**, *109*, 222–229. [CrossRef] [PubMed]
61. Srivastava, S. Emerging Therapeutic Roles for NAD+ Metabolism in Mitochondrial and Age-Related Disorders. *Clin. Transl. Med.* **2016**, *5*, e25. [CrossRef]
62. Corona, J.C.; Duchen, M.R. PPARγ as a Therapeutic Target to Rescue Mitochondrial Function in Neurological Disease. *Free Radic. Biol. Med.* **2016**, *100*, 153. [CrossRef]
63. Hou, C.; Metcalfe, N.B.; Salin, K. Is Mitochondrial Reactive Oxygen Species Production Proportional to Oxygen Consumption? A Theoretical Consideration. *BioEssays* **2021**, *43*, 2000165. [CrossRef]
64. Zheng, J.-L.; Zeng, L.; Shen, B.; Xu, M.-Y.; Zhu, A.-Y.; Wu, C.-W. Antioxidant Defenses at Transcriptional and Enzymatic Levels and Gene Expression of Nrf2-Keap1 Signaling Molecules in Response to Acute Zinc Exposure in the Spleen of the Large Yellow Croaker *Pseudosciaena crocea*. *Fish Shellfish Immunol.* **2016**, *52*, 1–8. [CrossRef] [PubMed]
65. Stein, L.R.; Imai, S. The Dynamic Regulation of NAD Metabolism in Mitochondria. *Trends Endocrinol. Metab.* **2012**, *23*, 420–428. [CrossRef] [PubMed]
66. Moini, H.; Packer, L.; Saris, N.-E.L. Antioxidant and Prooxidant Activities of α-Lipoic Acid and Dihydrolipoic Acid. *Toxicol. Appl. Pharmacol.* **2002**, *182*, 84–90. [CrossRef] [PubMed]
67. Bolzan, L.P.; Barroso, D.C.; Souza, C.F.; Oliveira, F.C.; Wagner, R.; Baldisserotto, B.; Val, A.L.; Baldissera, M.D. Dietary Supplementation with Nerolidol Improves the Antioxidant Capacity and Muscle Fatty Acid Profile of *Brycon Amazonicus* Exposed to Acute Heat Stress. *J. Therm. Biol.* **2021**, *99*, 103003. [CrossRef]
68. Joo, M.S.; Kim, W.D.; Lee, K.Y.; Kim, J.H.; Koo, J.H.; Kim, S.G. AMPK Facilitates Nuclear Accumulation of Nrf2 by Phosphorylating at Serine 550. *Mol. Cell. Biol.* **2016**, *36*, 1931–1942. [CrossRef] [PubMed]

antioxidants

Article

Effects of Low Temperature on Antioxidant and Heat Shock Protein Expression Profiles and Transcriptomic Responses in Crayfish (*Cherax destructor*)

Ying Yang [1], Wenyue Xu [1], Qichen Jiang [2], Yucong Ye [1], Jiangtao Tian [1], Yingying Huang [1], Xinglin Du [1], Yiming Li [1], Yunlong Zhao [1,*] and Zhiquan Liu [3,4,*]

1 School of Life Science, East China Normal University, Shanghai 200241, China
2 Freshwater Fisheries Research Institute of Jiangsu Province, 79 Chating East Street, Nanjing 210017, China
3 School of Life and Environmental Sciences, Hangzhou Normal University, Hangzhou 311121, China
4 School of Engineering, Hangzhou Normal University, Hangzhou 310018, China
* Correspondence: ylzhao426@163.com (Y.Z.); liuzhiquan1024@163.com (Z.L.); Tel.: +86-21-54345387 (Y.Z.); Fax: +86-21-54341006 (Y.Z.)

Abstract: Low temperature is a critical factor restricting the growth and survival of aquatic animals, but research on the mechanism of response to low temperature in *Cherax destructor* is limited. *C. destructor* is one of the most important freshwater crustaceans with strong adaptability in Australia, and it has been commercialized gradually in recent years. Here, growth indicators, antioxidant parameters, anti-stress gene expression, and transcriptome sequencing were used on crayfish following 8 weeks of low-temperature acclimation. The results showed that weight gain, length gain, and molting rates decreased as the temperature decreased. The activity of antioxidant enzymes decreased, while the content of antioxidant substances and the expression of anti-stress genes increased. Transcriptome sequencing identified 589 differentially expressed genes, 279 of which were upregulated and 310 downregulated. The gene functions and pathways for endocrine disorders, glucose metabolism, antioxidant defense, and immune responses were identified. In conclusion, although low-temperature acclimation inhibited the basal metabolism and immune ability of crayfish, it also increased the antioxidant substance content and anti-stress-gene expression to protect the organism from low-temperature damage. This study provided molecular insights into the study of low-temperature responses of low-temperature-tolerant crustacean species.

Keywords: antioxidant; *Cherax destructor*; heat shock proteins; low-temperature stress; transcriptomic responses

Citation: Yang, Y.; Xu, W.; Jiang, Q.; Ye, Y.; Tian, J.; Huang, Y.; Du, X.; Li, Y.; Zhao, Y.; Liu, Z. Effects of Low Temperature on Antioxidant and Heat Shock Protein Expression Profiles and Transcriptomic Responses in Crayfish (*Cherax destructor*). *Antioxidants* **2022**, *11*, 1779. https://doi.org/10.3390/antiox11091779

Academic Editors: Marcelo Hermes-Lima, Daniel Carneiro Moreira and Tania Zenteno-Savín

Received: 8 August 2022
Accepted: 6 September 2022
Published: 9 September 2022

1. Introduction

Water temperature is an inevitable factor causing aquatic animal stress, affecting almost all physiological and biochemical processes of poikilothermic animals. It can inhibit individual growth or even lead to death at low temperatures [1,2]. When the water temperature changes beyond the tolerance temperature range of poikilothermic animals, oxygen free radicals increase, antioxidant-enzyme activity decreases, oxidative damage is aggravated, and immunity becomes suppressed in the organism [3,4]. To adapt to the drastic changes in the surrounding environment, individuals have evolved a series of temperature stress response mechanisms, such as unfolded-protein response and apoptosis [5], activation of desaturases [6,7], and increased lipid metabolism [8]. In addition, as a class of molecular chaperones, heat shock proteins can prevent protein denaturation and play a critical role in an organism's resistance to environmental stress [9,10]. Heat shock proteins are associated with temperature tolerance and organism upregulation of *HSP* expression to protect cells from damage under temperature stress [10–12]. When an organism is subjected to environmental stress, *HSP70* can enhance cell viability by protecting cells from oxidative or nitrative stress, while *HSP90* can defend against pathogenic infection [13], thus improving the environmental adaptation of aquatic animals [14,15]. However, most of these studies are concentrated in species that cannot tolerate low temperature, such as *Litopenaeus vannamei* [3,16], *Marsupenaeus japonicus* [17], and *Cherax quadricarinatus* [18,19], whereas less research has been done on low-temperature-tolerant crustaceans.

Cherax destructor is one of the Australian freshwater crayfish with a large size, fast growth, high survival rate, and wide distribution [20]. Recently, *C. destructor* has been commercialized gradually as a potential economic species [21]. Different from other economically important Australian freshwater crayfish such as *Cherax tenuimanus* and *C. quadricarinatus*, *C. destructor* has strong temperature adaptability, and it can tolerate low temperatures [20]. At present, there are many studies on the tolerance of *C. destructor* to environmental factors such as salinity [22], dissolved oxygen [23], and pH [24], but there are fewer studies on low-temperature tolerance.

Investigating changes in gene expression patterns and metabolic pathways is the focus of understanding the molecular-response mechanisms of crustaceans under low-temperature stress. Currently, transcriptomic approaches based on high-throughput RNA sequencing (RNA-seq) have provided new analytical avenues for revealing potential genes and related pathways of organism-specific physiological processes under different environmental conditions [19,25]. RNA-seq is now widely used to study the defense responses of organisms under stress conditions. It provides comprehensive information for identifying novel immune genes and reveals potential regulatory and adaptive mechanisms [26–28]. With the help of RNA-seq, previous studies have identified some functional genes and physiological pathways of cold-tolerance mechanisms [29,30]. Therefore, using RNA-seq to analyze *C. destructor* at the molecular level under low temperatures is expected to help us better understand the cold-response mechanism of crayfish.

In this study, the weight growth rate and molting rate, the activity of antioxidant enzymes and glutathione content, as well as the gene expression of *heat shock proteins* (*HSPs*) and *cold shock protein* (*CPS*) in the hepatopancreas of *C. destructor* at different temperatures were examined. Transcriptome sequencing and bioinformatics analysis were applied to identify the essential genes and the major pathways in response to cold stress. The results provide a theoretical basis and valuable insights for further exploring the regulatory mechanisms of crustaceans under low-temperature stress.

2. Materials and Methods

2.1. Experimental Organisms

Healthy *C. destructor* were obtained from the Shanghai Academy of Agricultural Sciences (Shanghai, China). The temperature acclimation experiment on crayfish was carried out in culture tanks for 7 days. The tanks contained aerated water and were maintained at 20 ± 1 °C. The crayfish were fed daily with commercial feed during acclimatization, and excrement and food residue were removed.

2.2. Experimental Procedures

After acclimatization, the crayfish (n = 150) were randomly assigned to tanks (66 × 45 × 36 cm) for each temperature treatment at a density of 10 crayfish per tank. Three tanks were used for each treatment group (30, 25, 20, 15, and 10 °C). For the temperature treatment, 20 °C was maintained until the end of sampling. For the other four temperature treatments, the water temperature was increased/decreased from 20 °C to 1 °C per day until the set temperature was reached. Subsequently, the temperatures were maintained at the set temperature for 8 weeks before sampling.

During cultivation, the initial body weight and initial body length were recorded, and the number of molting crayfish was recorded every day. After cultivation, final body weight and final body length were measured, and the weight gain, length gain, and molting rates were calculated. Hepatopancreas samples of *C. destructor* were rapidly removed, flash frozen in liquid nitrogen, and stored at −80 °C for transcriptomic analysis and other biochemical assays. Only the hepatopancreas from the 25 °C and 10 °C groups were used for transcriptomic analysis.

The crayfish were fed commercial feed, and the food residues and animal feces were cleaned daily. Water temperatures were checked, and the water was changed every day. The amount of water exchanged was one-third of the total, and the tank was continuously aerated.

2.3. Antioxidant Index Detection

The hepatopancreas tissues from the different temperature groups (30, 25, 20, 15, and 10 °C) were combined with physiological saline. The hepatopancreas weight (g)/physiological saline (mL) ratio was 1:9. Homogenate was prepared with a homogenizer on ice. The contents of total glutathione (T-GSH), reduced glutathione (GSH), and oxidized glutathione (GSSG), as well as the enzyme activity of glutathione reductase (GR) and glutathione-S-transferase (GST), were detected using commercial kits purchased from Nanjing Jiancheng Co., Ltd. (Nanjing, China). The experimental operation was carried out in strict accordance with the instructions.

2.4. Detection of Anti-Stress-Gene Expression

Total RNA of hepatopancreas tissues was extracted by TRIzol reagent (Aidlab, Beijing, China), following the manufacturer's instructions. The concentration and quality of RNA were detected by a NanoDrop-2000C (Thermo Scientific, Wilmington, NC, USA) and 1% agarose gel electrophoresis, respectively. Total RNA first-strand cDNA of each sample was generated using the PrimeScript™ RT Master Mix Real Time Kit (Takara, Japan). The transcribed cDNA was stored at −20 °C for subsequent experiments. The fluorescent quantification dye was TransStart Top Green qPCR SuperMix (TransGen, Beijing, China). Primer sequences for anti-stress genes are shown in Table S1. *18S rRNA* was used as an internal reference gene, and the $2^{-\Delta\Delta ct}$ method was used for calculation [31].

2.5. Transcriptomic Analysis

2.5.1. Transcriptome Sequencing

Total RNA from hepatopancreas tissues of *C. destructor* in the 25 °C and 10 °C groups was extracted using the mirVanaTM miRNA Isolation Kit (Ambion, Austin, TX, USA). An Agilent 2100 Bioanalyzer (Agilent Technologies, Santa Clara, CA, USA) and a NanoDrop 2000 spectrophotometer (Thermo Scientific) were used to analyze the RNA integrity and quality, respectively. The TruSeq Stranded mRNA LT Sample Prep Kit (Illumina, San Diego, CA, USA) was used to construct cDNA libraries, following standardized kit procedures. The libraries were sequenced on the Illumine sequencing platform (Illumina HiSeq X Ten) at Shanghai OE Biotech Co., Ltd. (Shanghai, China), and 150 bp paired-end reads were generated.

2.5.2. RNA-Seq Read Processing and Mapping

The raw image data obtained by high-throughput sequencing were converted into raw sequence data by base calling analysis. They contained the sequence information of the raw sequence data and the sequencing quality information. Trimmomatic software [32], which controls the quality of raw data, including removal of adapters, low-quality reads, and low-quality bases, was utilized as it also obtains high-quality clean reads. Clean reads were mapped to the reference genome of *C. destructor* using HISAT2, and every sample was independently mapped. The software parameters were default values.

2.5.3. Differentially Expressed Genes (DEGs) and Functional Enrichment

Htseq-count software was used to obtain the read counts of genes in each sample. Cufflinks software was used to calculate the FPMK (fragments per kb per million reads) value of each gene [33,34]. The estimates SizeFactors function of the DESeq R package was used to normalize the data, and the nbinom Test function was used to calculate the *p*-value and fold change value. Genes with $p < 0.05$ and fold change ≥ 2 were considered DEGs. M-versus-A (MA) and volcano plots were created to visualize the overall distribution of DEGs. Gene Ontology (GO) enrichment and Kyoto Encyclopedia of Genes and Genomes (KEGG) enrichment analyses were performed on DEGs to determine the biological functions and pathways that differential genes mainly affect. The GO enrichment analysis was divided into biological processes, cellular components, and molecular functions.

2.6. Transcriptome Data Validation

To ensure the reliability of transcriptome data, fourteen genes were randomly selected from DEGs and measured using qRT-PCR. The gene expressions were detected using the qRT-PCR method, described in Section 2.4. Primer sequences for transcriptome data validation are shown in Table S2. *18S rRNA* was used as an internal reference gene, and the $2^{-\Delta\Delta ct}$ method was used for calculation [31].

2.7. Statistical Analysis

All data were presented as the mean ± standard deviation (SD). One-way analysis of variance and Tukey's test were used to determine the differences among different temperature groups. Statistical analyses were conducted with SPSS 19.0 (IBM, Chicago, IL, USA), and graphs were constructed using Graph Pad Prism 5 (Graph Pad Software, La Jolla, CA, USA). Significant differences were indicated when $p < 0.05$.

3. Results

3.1. Effects of Temperature on the Growth Indicators of C. destructor

As shown in Table 1, the weight gain rate and length gain rate were significantly decreased in the 15 °C and 10 °C groups compared to the 30, 25, and 20 °C groups ($p < 0.05$). There was no significant difference among the 30, 25, and 20 °C groups, as well as between the 15 °C and 10 °C groups ($p > 0.05$).

Table 1. The growth indicators of *C. destructor* at different temperatures (30, 25, 20, 15, and 10 °C) for 8 weeks.

	30	25	20	15	10
Initial crayfish wet weight (g)	3.40 ± 0.25	3.61 ± 0.32	3.47 ± 0.25	3.41 ± 0.41	3.39 ± 0.43
Initial crayfish body length (cm)	3.31 ± 0.45	3.29 ± 0.26	3.06 ± 0.40	3.32 ± 0.48	3.22 ± 0.35
Final crayfish wet weight (g)	10.86 ± 1.39	10.44 ± 0.74	9.95 ± 0.95	4.64 ± 0.26	4.46 ± 0.23
Final crayfish body length (cm)	7.95 ± 0.29	7.67 ± 0.37	7.62 ± 0.36	4.67 ± 0.53	4.63 ± 0.29
Weight gain rate (WG, %)	68.67% ^{b}	65.41% ^{b}	65.14% ^{b}	26.40% ^{a}	24.05% ^{a}
Length gain rate (LG, %)	58.37% ^{b}	57.08% ^{b}	59.86% ^{b}	28.82% ^{a}	30.55% ^{a}

Values are presented as the mean ± SD ($n = 10$); significant differences are indicated with different letters in the same row ($p < 0.05$).

3.2. Effects of Temperature on the Molting Rate of C. destructor

In the 30 °C and 25 °C groups, the molting rate in the second and eighth weeks was higher than in the fourth and sixth weeks. In the 20 °C group, the molting rate in the second and sixth weeks was lower than in the fourth and eighth weeks. In the 15 °C and 10 °C groups, the molting rate was enhanced as the time increased (Figure 1).

3.3. Effects of Temperature on Antioxidant Indicators of C. destructor

To evaluate the antioxidant properties of *C. destructor* at different temperatures, the content of glutathione and the activities of antioxidant enzymes were detected. T-GSH had maximum values in the 30 °C group, except in the 30 °C group, T-GSH increased as the temperature decreased, and it significantly increased in the 15 °C and 10 °C groups compared with the 25 °C group ($p < 0.05$, Figure 2A). GSSG and GSH peaked in the 30 °C group and the 10 °C group, respectively, and there were no significant differences among the other temperature groups ($p > 0.05$, Figure 2B,C). GSH/GSSG had an increasing trend with the temperature decrease, and it significantly increased in the 10 °C group compared with the other groups ($p < 0.05$, Figure 2D). Except for the GST of the 10 °C group, the activity of both GR and GST first decreased and then increased as the temperature decreased. Compared with the 25 °C group, GST activity was significantly decreased in the 10 °C group ($p < 0.05$), while GR activity had no significant (Figure 2E,F).

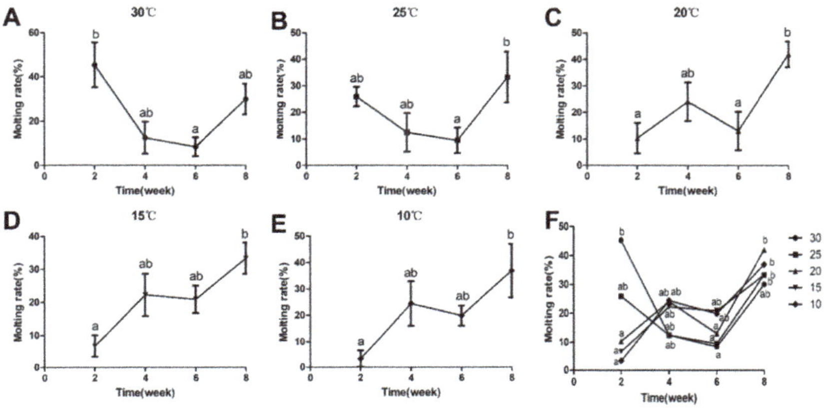

Figure 1. (**A–F**) The molting rate (%) of *C. destructor* at different temperatures (30, 25, 20, 15, and 10 °C) for 2, 4, 6, and 8 weeks. Data are presented as the mean ± SD. Means with different letters are significantly different ($p < 0.05$).

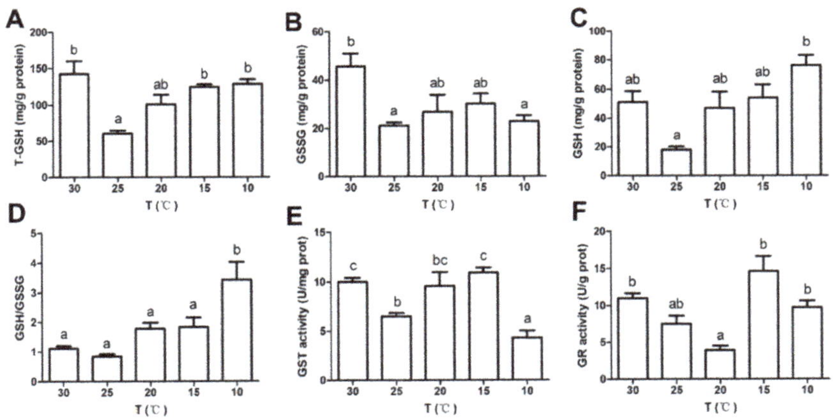

Figure 2. (**A–D**) The contents of T-GSH, GSSG, GSH, and GSH/GSSG; and (**E**,**F**) activities of GST and GR in the hepatopancreas of *C. destructor* at different temperatures (30, 25, 20, 15, and 10 °C) for 8 weeks. Data are presented as the mean ± SD of three independent experiments. Means with different letters are significantly different ($p < 0.05$).

3.4. Effects of Temperature on the Expression of Anti-Stress Genes in C. destructor

The results of the 25 °C group are set as 1. Compared with the 25 °C group, the expression of *HSP20* significantly decreased as the temperature decreased ($p < 0.05$, Figure 3A). The expressions of *HSP21* and *HSP90* were not significantly different in the low-temperature groups compared with the 25 °C group ($p > 0.05$), while they significantly increased in the 30 °C group ($p < 0.05$, Figure 3B,E). As the temperature decreased, the expressions of *HSP60*, *HSP70*, and *CSP* significantly increased compared with the 25 °C group ($p < 0.05$), and the expression of *CSP* significantly increased in the 30 °C group ($p < 0.05$, Figure 3C,D,F).

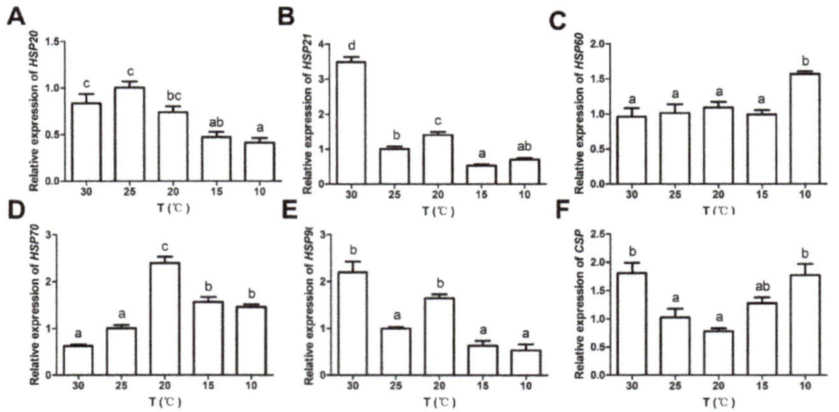

Figure 3. (**A–F**) The levels of *HSP20, HSP21, HSP60, HSP70, HSP90,* and *CSP* expression in the hepatopancreas of *C. destructor* at different temperatures (30, 25, 20, 15, and 10 °C) for 8 weeks. Data are presented as the mean ± SD of three independent experiments. Means with different letters are significantly different at $p < 0.05$.

3.5. GO and KEGG Pathway Enrichment Analysis

The DEGs were visualized on the MA and volcano plots (Figure 4). A total of 589 DEGs were identified, of which 279 were upregulated and 310 were downregulated.

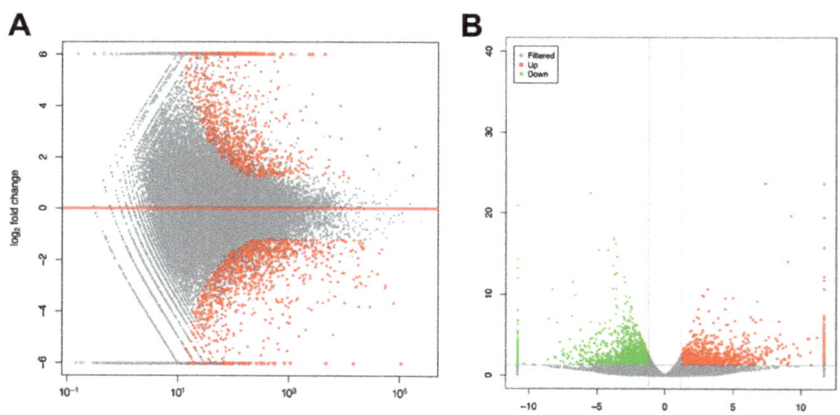

Figure 4. Distribution of DEGs in *C. destructor* among the low-temperature and control groups. (**A**) MA plot for DEG visualization. Each dot represents one gene, and the red dots represent DEGs. (**B**) Volcano plot map for visualization. Red and green dots represent upregulated and downregulated DEGs, respectively.

To identify the key functions in crayfish affected by cold temperature, 589 DEGs were mapped into three GO items. The results showed that the top three categories of biological processes enriched by DEGs were regulation of pupariation (GO: 0106023), negative regulation of ecdysone receptor-mediated signaling pathway (GO: 0120143), and cellulose catabolic process (GO: 0030245). The top three cellular components were extracellular space (GO: 0005615), anchored component of the external side of the plasma membrane (GO: 0005615), and mitochondrial envelope (GO: 0005740). The top three molecular functions were oxidoreductase activity (GO: 0016491), cellulase activity (GO: 0008810), and cellulose binding (GO: 0030248) (Figure 5).

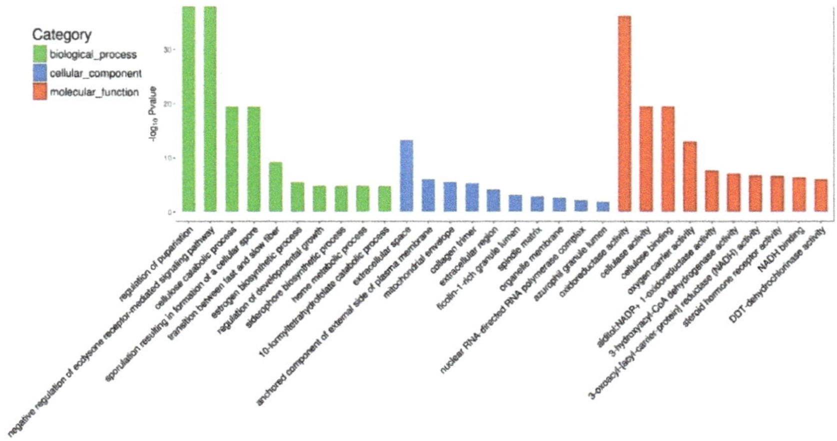

Figure 5. GO functional classification of DEGs in *C. destructor*. Three items are included: biological processes; cellular components; molecular functions.

As shown in Figure 6, KEGG enrichment was performed to further understand which pathways were significantly affected in the cold temperature group. The top 20 KEGG pathways that were significantly enriched were mainly involved in metabolic regulation (carbohydrate metabolism), innate immune response (glutathione metabolism, cytochrome P450 metabolism, neomycin, kanamycin, gentamicin biosynthesis, and retinol metabolism), endocrine system (steroid hormone biosynthesis and thyroid hormone synthesis), and low-temperature protective response (longevity-regulating pathway).

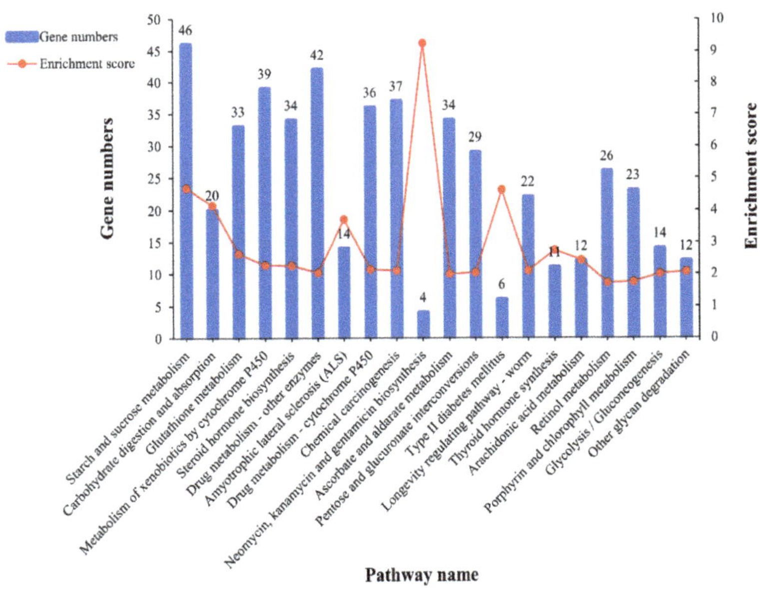

Figure 6. KEGG enrichment analysis of DEGs in *C. destructor*. The *x*-axis represents the pathway name. The *y*-axis represents the gene numbers enriched in the pathway (left) and enrichment score (right), corresponding to the blue bars and red lines, respectively.

3.6. Data Validation

To verify the reliability of transcriptome sequencing results, we detected 14 randomly selected genes using qRT-PCR, as shown in Figure 7. A significant positive correlation (R = 0.810) between the RNA-seq results and qRT-PCR results indicates that the transcriptome results were validated.

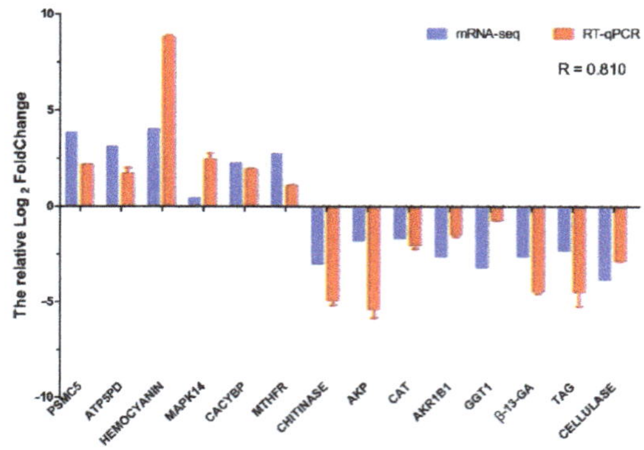

Figure 7. Validation of RNA-Seq results by qRT-PCR.

4. Discussion

Changes in water temperature profoundly impact the growth and metabolism of aquatic ectothermic animals. Previous studies have shown that growth was significantly inhibited in many species at low temperatures, such as *Macrobtachium nipponense* and *C. quadricarinatus* [35,36]. Similar results were found in our experiments, where the weight growth rate and length growth rate of *C. destructor* significantly decreased in the low-temperature groups (15 °C and 10 °C) compared with other temperature groups (30, 25, and 20 °C). The results indicated that when the temperature was lower than 15 °C, the growth of *C. destructor* was inhibited. Crustacean growth is accomplished through a series of molts [37]. In the present study, as the cultivation time increased, the molting rate of *C. destructor* showed different trends in the different temperature groups. The molting rate of crayfish was lower in the 15 °C and 10 °C groups than in the other groups after two weeks of culture. The results showed that low temperature inhibited the molting rate of *C. destructor*, thus likely leading to prolonged molting intervals in crustaceans [38,39]. With the increase in cultivation time, the molting rate of crayfish in the low-temperature groups showed an upward trend, while the other groups showed a downward trend. We speculate this may be related to the prolonged molt period in low temperatures. That is, after 4–6 weeks of culture, the crayfish in the low-temperature group may gradually enter the molting period, while the other groups were in the transition between two molts. Further studies are still required to understand molting fully. These results suggested that low temperature may inhibit the growth of *C. destructor* by reducing the molting rate and prolonging the molting cycle.

To explore the physiological processes and metabolic pathways of *C. destructor* in response to low temperature, we evaluated antioxidant and anti-stress indexes at different temperatures and combined RNA-seq technology to analyze the hepatopancreas of crayfish at 25 °C and 10 °C. GO enrichment results showed DEGs were mainly enriched in biological processes related to growth and development, and molecular functions related to oxidoreductase. These results were in agreement with the growth parameters and antioxidant indicators results.

The endocrine system regulates various physiological functions of organisms to cope with environmental changes. In *Macrobrachium rosenbergii* and *L. vannamei*, changes in endocrine hormone levels were found during cold stress [40,41]. These findings confirm that the endocrine system affects individuals' cold tolerance. In the present study, low temperature significantly altered the steroid hormone biosynthesis, including ecdysone and estrogen. Ecdysone and estrogen control molting, embryogenesis, gonadal development, and reproductive ability in aquatic animals [42–44]. Changes in the levels of these hormones suggest that they are involved in suppressing molting and sexual maturation at low temperatures. Responses of molting-related hormones and sexual-maturation-related hormones to temperature changes were also found in *L. vannamei* and *M. nipponense* [45,46]. Except for the steroid hormone, thyroid hormone biosynthesis was also affected in the low-temperature group. Thyroid hormone regulates the individual basal metabolic rate and is known for regulating body temperature in homothermic animals, as well as growth and development in poikilotherms [46–48]. It is affected by temperature in various aquatic animals [49–51]. Iodide is an important micronutrient in thyroid hormone biosynthesis, and iodide homeostasis within the thyroid gland is critical for thyroid hormone synthesis [52]. Iodotyrosine deiodinase is necessary to keep the balance between iodide and thyroid hormones [53]. Transcriptome results showed that the gene expression of *iodotyrosine deiodinase* was significantly down-regulated in the low-temperature group, suggesting that iodotyrosine deiodinase can be an important indicator of low temperature inhibiting thyroid hormone synthesis in crayfish. In addition, the thyroid hormone regulates thermal acclimation in the zebrafish during temperature changes [54]. Little is known about the regulatory control of thyroid hormone on thermogenesis in aquatic animals. Therefore, the thermal regulation of thyroid hormone on crayfish at low temperatures needs further exploration. The above results suggest that low temperature may lead to the synthesis and secretion of hormone disorders and then cause endocrine disorders, which finally affect individual growth and sexual maturity.

Previous studies have shown that amino acid metabolism and lipid metabolism are the main energy sources for crustaceans to cope with cold stress, and acute cold stress can lead to their dysregulation [17,55]. In the present study, among the top 20 KEGG pathways significantly enriched by DEGs under long-term low-temperature acclimation, five glucose-metabolism-related pathways (starch and sucrose metabolism, carbohydrate digestion and absorption, pentose and glucuronate interconversion, glycolysis/gluconeogenesis, and other glycan degradation) were affected. In addition, the qRT-PCR and transcriptome results showed a low-temperature-induced significant decrease in the expression of glycogenolysis-related genes, such as *β-1,3-GA*, *CELLULASE*, *amylase*, and *endoglucanase*. But the rate-limiting enzyme of glucose decomposition, *hexokinase*, was significantly upregulated. These results showed that long-term low-temperature stress could lead to abnormal glucose metabolism in individuals. Glucose is an important energy source that can be broken down by the hepatopancreas as a rapid energy source under cold stress to meet the needs of individual energy metabolism. Changes in glucose metabolism are an adaption mechanism in individuals to low temperatures; for example, *M. rosenbergii* and *L. vannamei* have increased blood glucose levels and decreased glycogen content during the adaptation process [40,55]. This result indicates the importance of a sugar source for crayfish in a long-term low-temperature environment.

The balance between the production and elimination of cellular reactive oxygen species (ROS) is disrupted when an organism is stressed. The activation of antioxidant enzymes and antioxidants is induced to prevent oxidative damage caused by excess ROS. Catalase is an antioxidant enzyme that detoxifies ROS [56]. In the transcriptome and qRT-PCR results, the expression of *CAT* in crayfish was decreased in the low-temperature group. A decrease in antioxidant enzyme activity was also observed in *Portunus trituberculatus* and *C. quadricarinatus* after low-temperature acclimation [10,18]. GSH is an antioxidant that acts on ROS [57], maintains the balance of oxidative stress and the antioxidant system, and converts GSSG to GSH via GR [58]. In *L. vannamei*, GSH has a regulatory role in temperature stress [4]. In the present study, DEGs were significantly enriched in glutathione metabolism, and the GSH content and GR activity were increased in the low-temperature group. In addition, the down-regulation of glutathione hydrolase and the up-regulation of isocitrate dehydrogenase prevent GSH breakdown and provide energy, respectively, for the GSH synthesis process [59]. They are both important enzymes that ensure GSH content. Our results showed that glutathione metabolism is involved in long-term cold acclimation. Low temperature regulates the activity of enzymes related to glutathione metabolism to increase the content of GSH, which is very important in the low-temperature adaption mechanism in *C. destructor*. Ascorbate is another antioxidant in organisms that scavenges H_2O_2, O_2^-, $HO^\bullet$ and lipid hydroperoxides [60,61]. Previous studies found that ascorbate is associated with resistance to environmental stress in aquatic animals [62,63]. The ascorbate-related pathway is significantly altered in *C. destructor* in the low-temperature group. These results suggest that long-term low temperature inhibits the activity of antioxidant enzymes but activates antioxidant-related metabolism, implying the important protective role of antioxidants during oxidative stress caused by long-term low temperature. This may be a physiological compensation mechanism of crayfish under long-term low-temperature acclimation [64].

Low temperature suppresses the immune performance of crayfish, such as antibacterial and anti-inflammatory activity. Neomycin, kanamycin, and gentamicin are from the aminoglycoside family of antibiotics [65–67]. They act against most gram-negative organisms and exert their antibacterial effect by blocking bacterial protein synthesis [65,68]. Retinol is involved in innate immunity and downregulates the expression of pro-inflammatory factors to exert anti-inflammatory effects [69,70]. In humans and rats, retinol has been found to have a regulatory effect at low temperatures [71,72]. In the present study, antibiotic biosynthesis and retinol metabolism were significantly changed in the low-temperature group. Mitogen-activated protein kinases (MAPKs) are involved in various physiological processes and respond to various extracellular stimuli. In *Meretrix petechialis*, *MAPK14* was activated after being challenged by *Vibrio* and elicited a series of immune responses [73]. In *Lateolabrax maculatus*, the expression of *MAPK14* was upregulated under hypoxia and salinity stress [74]. In the present study, the expression of *MAPK14* was increased in the low-temperature group, while the expression of immunity genes such as *AKP* [75], *GGT* [76], and *AKR1B1* [77] was significantly decreased. These results further confirmed that low temperature inhibited the innate immunity of crayfish. The above results indicate that the antibacterial and anti-inflammatory abilities of individuals are reduced at low temperatures and that the individual may be more susceptible to pathogen infection at this time.

Low temperature can induce a disorder of free-radical metabolism, destroy the physiological functions of cells and tissues, and cause apoptosis [19]. *HSPs* maintain homeostasis by protecting the structure and function of cells and tissues from various stressors, including temperature stress [78]. According to the molecular weight, *HSPs* can be divided into high-molecular-weight *HSPs* (i.e., *HSP100, HSP90, HSP70,* and *HSP60*) and small-molecule *HSPs* (i.e., *HSP40, HSP21,* and *HSP10*) [79]. In *M. rosenbergii* and *P. trituberculatus*, *HSP70* and *HSP90* were protective against temperature stress [10,13]. The expression of *HSP40* and *HSP21* was induced under temperature stress in *L. vannamei* and *C. quadricarinatus*, respectively [18,80]. In the present study, the detection of expressions for three high-molecular-weight *HSPs* (*HSP90, HSP70,* and *HSP60*) and two small-molecular-weight *HSPs* (*HSP21* and *HSP20*) suggest that *HSPs* with different molecular weights have different expression patterns during low-temperature adaptation. The different expression profiles were observed in the same *HSP* genes under different environmental stressors in *L. vannamei* [81], implying the necessity of different *HSP* expression patterns for individual survival. The expression of *HSP60* and *HSP70* was significantly upregulated, while other *HSPs* decreased or had no significant change at low temperatures in crayfish. The results indicated that *HSP60* and *HSP70* might have stronger protective effects at low-temperature acclimation in *C. destructor*. This may be related to the ATP dependence of high-molecular-weight *HSPs* [82].

ATP synthase is a protein that catalyzes the synthesis of ATP [83], and it provides energy for individual protective mechanisms, including *HSP*. The upregulation of ATP synthase was observed in both transcriptome and qRT-PCR results. In addition, it is well studied that *HSPs* are also involved in immune responses such as antibacterial, anti-inflammatory, and disease control [84,85]. *HSP70* is also a negative regulator of apoptosis, interfering with the occurrence of apoptosis [86–88]. This is related to the fact that the pathways involved in life regulation in crayfish are affected in the low-temperature group. It emphasizes the protective role of *HSPs* in crayfish at low-temperature acclimation. In transcriptome data, the longevity regulating pathway was significantly affected in the low-temperature group. Previous studies have pointed out that low temperature can improve the health and longevity of *Drosophila melanogaster* [89], *Brachionus horeanus* [90], *Caenorhabditis elegans*, and other species. The effect of temperature on the lifespan of *C. elegans* is related to *daf-16* (a key gene regulating longevity) [91]. In the present study, the expressions of *HSP60*, *SOD* (*superoxide dismutase*), *SMK-1*, and *TCERG1* (*transcription elongation regulator1*) in *C. destructor* were upregulated. SOD is an antioxidant enzyme that prevents oxidative damage. SMK-1 is required for innate immune and oxidative stress and modulates *daf-16* transcriptional specificity that regulates longevity [92]. Moreover, the overexpression of *TCERG1* extends the lifespan of *C. elegans* [93]. The increase of these genes showed low temperature could directly activate key genes that regulate lifespan in *C. destructor*. The results provide important information for the survival of cold-tolerant species at low temperatures.

CSPs are a class of RNA/DNA-binding proteins with one or more cold shock domains. *CSPs* are a critical factor required for cellular adaptation to low temperatures and overcoming the deleterious effects of cold stress [94,95]. Currently, increasing the expression level of *CSP* at low temperatures has been found in crustaceans such as *M. nipponense*, *C. quadricarinatus*, and *Alvinocaris longirostris* [96–98]. Similar results were also found in our experiments. The expression of *CSP* significantly increased in *C. destructor* in the low-temperature group, indicating *CSP* is indispensable for the survival of crayfish during cold acclimation. The specific protective mechanism of *CSP* in crustaceans is less studied and needs further exploration.

In general, stress responses are classified as primary stress responses (short term), secondary stress responses, and tertiary stress responses (long term) [99]. Primary stress responses are for the neuroendocrine system, and secondary and tertiary stress responses are for cellular responses and individual or population changes, respectively [100]. In crayfish, low-temperature acclimation affected the release of steroid hormone and thyroid hormone, and it inhibited the individual basic metabolism level and developmental process. Subsequently, a series of cellular immune responses such as antioxidant, antibacterial, anti-inflammatory, and disease resistance of crayfish were suppressed. However, during the process, anti-stress factors such as GSH, *HSP60*, *HSP70*, and *CSP* were activated to resist the deleterious effects of low temperatures. Thus, the survival of crayfish is ensured when the individual metabolism is slowed down, the molting is inhibited, and the growth is restricted.

5. Conclusions

This study evaluated the growth, detected the antioxidant enzyme activity and antioxidant substance content, measured the gene expression of *HSPs* and *CSP*, and used transcriptomic to explore the response mechanism at the molecular level in *C. destructor* at low temperature. The results showed that low temperature induces endocrine disorders, affects basal metabolism and glucose metabolism, inhibits antioxidant enzyme activity (GST) and immune gene expression (*AKP*, *CAT*, *GGT*, and *AKR1B1*), and slows individual growth. However, low temperatures activate the synthesis of antioxidants (GSH, antibiotics, and retinol), the expression of anti-stress genes (*HSP60*, *HSP70*, *CSP*, and *MAPK14*), and lifespan-related genes (*SMK-1* and *TCERG1*), which have important protective effects on the survival of individuals at low temperatures. Our results provide valuable information on the response metabolism of crustaceans at low temperatures.

Supplementary Materials: The following supporting information can be downloaded at: https://www.mdpi.com/article/10.3390/antiox11091779/s1, Table S1: Primer sequences of anti-stress genes; Table S2: Primer sequences of transcriptome validation.

Author Contributions: The work was carried out in collaboration among all authors. Conceptualization, Y.Y. (Ying Yang), W.X., Q.J., Y.Y. (Yucong Ye), J.T., Y.H., X.D., Z.L. and Y.Z.; methodology, Y.Y. (Ying Yang), W.X., Q.J., Y.Y. (Yucong Ye), J.T., Y.H., X.D., Z.L. and Y.Z.; software, Y.Y. (Ying Yang), W.X. and J.T.; validation, Y.Y. (Ying Yang), W.X., Q.J., Y.H. and X.D.; formal analysis, Y.Y. (Ying Yang) and W.X.; investigation, Y.Y. (Ying Yang), W.X., Q.J. and Y.Y. (Yucong Ye); resources, Y.Y. (Ying Yang), J.T., Z.L. and Y.Z.; data curation, Y.Y. (Ying Yang), W.X., J.T. and Y.H.; writing—original draft preparation, Y.Y. (Ying Yang); writing—review and editing, Y.Y. (Ying Yang), W.X., Q.J., Y.Y. (Yucong Ye), J.T., Y.H., X.D., Y.L., Z.L. and Y.Z.; visualization, Y.Y. (Ying Yang), W.X., Z.L. and Y.Z.; supervision, Z.L. and Y.Z.; project administration, Z.L. and Y.Z.; funding acquisition, Z.L. and Y.Z. All authors have read and agreed to the published version of the manuscript.

Funding: The research was funded by the Project of Shanghai Municipal Commission of Science and Technology, grant number 21DZ23S0300.

Institutional Review Board Statement: Not applicable.

Informed Consent Statement: Not applicable.

Data Availability Statement: The authors declare that all data supporting the conclusions of this study are available within the article and its supplementary material.

Conflicts of Interest: The authors declare no conflict of interest.

References

1. Xu, D.; Wu, J.; Sun, L.; Qin, X.; Fan, X.; Zheng, X. Combined stress of acute cold exposure and waterless duration at low temperature induces mortality of shrimp *Litopenaeus vannamei* through injuring antioxidative and immunological response in hepatopancreas tissue. *J. Therm. Biol.* **2021**, *100*, 103080. [CrossRef] [PubMed]
2. Ren, X.; Wang, Q.; Shao, H.; Xu, Y.; Liu, P.; Li, J. Effects of Low Temperature on Shrimp and Crab Physiology, Behavior, and Growth: A Review. *Front. Mar. Sci.* **2021**, *8*, 746177. [CrossRef]
3. Qiu, J.; Wang, W.-N.; Wang, L.-j.; Liu, Y.-F.; Wang, A.-L. Oxidative stress, DNA damage and osmolality in the Pacific white shrimp, *Litopenaeus vannamei* exposed to acute low temperature stress. *Comp. Biochem. Physiol. Part C Toxicol. Pharmacol.* **2011**, *154*, 36–41. [CrossRef] [PubMed]
4. Estrada-Cárdenas, P.; Cruz-Moreno, D.G.; González-Ruiz, R.; Peregrino-Uriarte, A.B.; Leyva-Carrillo, L.; Camacho-Jiménez, L.; Quintero-Reyes, I.; Yepiz-Plascencia, G. Combined hypoxia and high temperature affect differentially the response of antioxidant enzymes, glutathione and hydrogen peroxide in the white shrimp *Litopenaeus vannamei*. *Comp. Biochem. Physiol. Part A Mol. Integr. Physiol.* **2021**, *254*, 110909. [CrossRef]
5. Wang, Z.; Qu, Y.; Yan, M.; Li, J.; Zou, J.; Fan, L. Physiological responses of Pacific white shrimp *Litopenaeus vannamei* to temperature fluctuation in low-salinity water. *Front. Physiol.* **2019**, *10*, 1025. [CrossRef]
6. Wu, D.-L.; Huang, Y.-H.; Liu, Z.-Q.; Yu, P.; Gu, P.-H.; Fan, B.; Zhao, Y.-L. Molecular cloning, tissue expression and regulation of nutrition and temperature on Δ6 fatty acyl desaturase-like gene in the red claw crayfish (*Cherax quadricarinatus*). *Comp. Biochem. Physiol. Part B Biochem. Mol. Biol.* **2018**, *225*, 58–66. [CrossRef]
7. Wu, D.-L.; Rao, Q.-X.; Cheng, L.; Lv, W.-W.; Zhao, Y.-L.; Song, W.-G. Cloning and characterisation of a Δ9 fatty acyl desaturase-like gene from the red claw crayfish (*Cherax quadricarinatus*) and its expression analysis under cold stress. *J. Therm. Biol.* **2021**, *102*, 103122. [CrossRef]
8. Sun, Z.; Tan, X.; Liu, Q.; Ye, H.; Zou, C.; Xu, M.; Zhang, Y.; Ye, C. Physiological, immune responses and liver lipid metabolism of orange-spotted grouper (*Epinephelus coioides*) under cold stress. *Aquaculture* **2019**, *498*, 545–555. [CrossRef]
9. Chen, L.; Gómez, R.; Weiss, L.C. Distinct gene expression patterns of two heat shock protein 70 members during development, diapause, and temperature stress in the freshwater crustacean *Daphnia magna*. *Front. Cell Dev. Biol.* **2021**, *9*, 1531. [CrossRef]
10. Meng, X.-l.; Liu, P.; Li, J.; Gao, B.-Q.; Chen, P. Physiological responses of swimming crab *Portunus trituberculatus* under cold acclimation: Antioxidant defense and heat shock proteins. *Aquaculture* **2014**, *434*, 11–17. [CrossRef]
11. Kim, J.H.; Park, H.J.; Kim, D.H.; Oh, C.W.; Lee, J.S.; Kang, J.C. Changes in hematological parameters and heat shock proteins in juvenile sablefish depending on water temperature stress. *J. Aquat. Anim. Health* **2019**, *31*, 147–153. [CrossRef] [PubMed]
12. Zhou, J.; Wang, L.; Xin, Y.; Wang, W.-N.; He, W.-Y.; Wang, A.-L.; Liu, Y. Effect of temperature on antioxidant enzyme gene expression and stress protein response in white shrimp, *Litopenaeus vannamei*. *J. Therm. Biol.* **2010**, *35*, 284–289. [CrossRef]
13. Ju-Ngam, T.; McMillan, N.; Yoshimizu, M.; Kasai, H.; Wongpanya, R.; Srisapoome, P. Functional and Stress Response Analysis of Heat Shock Proteins 40 and 90 of Giant River Prawn (*Macrobrachium rosenbergii*) under Temperature and Pathogenic Bacterial Exposure Stimuli. *Biomolecules* **2021**, *11*, 1034. [CrossRef] [PubMed]

14. Padmini, E.; Rani, M.U. Impact of seasonal variation on HSP70 expression quantitated in stressed fish hepatocytes. *Comp. Biochem. Physiol. Part B Biochem. Mol. Biol.* **2008**, *151*, 278–285. [CrossRef] [PubMed]
15. Pelham, H.R. Speculations on the functions of the major heat shock and glucose-regulated proteins. *Cell* **1986**, *46*, 959–961. [CrossRef]
16. de Souza, D.M.; Borges, V.D.; Furtado, P.; Romano, L.A.; Wasielesky, W., Jr.; Monserrat, J.M.; de Oliveira Garcia, L. Antioxidant enzyme activities and immunological system analysis of *Litopenaeus vannamei* reared in biofloc technology (BFT) at different water temperatures. *Aquaculture* **2016**, *451*, 436–443. [CrossRef]
17. Ren, X.; Yu, Z.; Xu, Y.; Zhang, Y.; Mu, C.; Liu, P.; Li, J. Integrated transcriptomic and metabolomic responses in the hepatopancreas of kuruma shrimp (*Marsupenaeus japonicus*) under cold stress. *Ecotoxicol. Environ. Saf.* **2020**, *206*, 111360. [CrossRef]
18. Wu, D.-L.; Liu, Z.-Q.; Huang, Y.-H.; Lv, W.-W.; Chen, M.-H.; Li, Y.-M.; Zhao, Y.-L. Effects of cold acclimation on the survival, feeding rate, and non-specific immune responses of the freshwater red claw crayfish (*Cherax quadricarinatus*). *Aquac. Int.* **2018**, *26*, 557–567. [CrossRef]
19. Wu, D.; Huang, Y.; Chen, Q.; Jiang, Q.; Li, Y.; Zhao, Y. Effects and transcriptional responses in the hepatopancreas of red claw crayfish *Cherax quadricarinatus* under cold stress. *J. Therm. Biol.* **2019**, *85*, 102404. [CrossRef]
20. McCormack, R.B. New records and review of the translocation of the yabby Cherax destructor into eastern drainages of New South Wales, Australia. *Aust. Zool.* **2014**, *37*, 85–94. [CrossRef]
21. Mauro, M.; Arizza, V.; Arculeo, M.; Attanzio, A.; Pinto, P.; Chirco, P.; Badalamenti, G.; Tesoriere, L.; Vazzana, M. Haemolymphatic Parameters in Two Aquaculture Crustacean Species Cherax destructor (Clark, 1836) and Cherax quadricarinatus (Von Martens, 1868). *Animals* **2022**, *12*, 543. [CrossRef]
22. Mills, B.; Geddes, M. Salinty tolerance and osmoregulation of the Australian freshwater crayfish Cherax destructor Clark (Decapoda: Parastacidae). *Mar. Freshw. Res.* **1980**, *31*, 667–676. [CrossRef]
23. Morris, S.; Callaghan, J. The emersion response of the Australian Yabby Cherax destructor to environmental hypoxia and the respiratory and metabolic responses to consequent air-breathing. *J. Comp. Physiol. B* **1998**, *168*, 389–398. [CrossRef]
24. Ellis, B.; Morris, S. Effects of extreme pH on the physiology of the Australian'yabby'Cherax destructor: Acute and chronic changes in haemolymph oxygen levels, oxygen consumption and metabolic levels. *J. Exp. Biol.* **1995**, *198*, 409–418. [CrossRef] [PubMed]
25. Chen, K.; Li, E.; Li, T.; Xu, C.; Wang, X.; Lin, H.; Qin, J.G.; Chen, L. Transcriptome and molecular pathway analysis of the hepatopancreas in the Pacific white shrimp *Litopenaeus vannamei* under chronic low-salinity stress. *PLoS ONE* **2015**, *10*, e0131503. [CrossRef] [PubMed]
26. Li, Y.; Zhou, F.; Huang, J.; Yang, L.; Jiang, S.; Yang, Q.; He, J.; Jiang, S. Transcriptome reveals involvement of immune defense, oxidative imbalance, and apoptosis in ammonia-stress response of the black tiger shrimp (*Penaeus monodon*). *Fish Shellfish Immunol.* **2018**, *83*, 162–170. [CrossRef]
27. Zhong, S.; Mao, Y.; Wang, J.; Liu, M.; Zhang, M.; Su, Y. Transcriptome analysis of Kuruma shrimp (*Marsupenaeus japonicus*) hepatopancreas in response to white spot syndrome virus (WSSV) under experimental infection. *Fish Shellfish Immunol.* **2017**, *70*, 710–719. [CrossRef]
28. Huang, W.; Li, H.; Cheng, C.; Ren, C.; Chen, T.; Jiang, X.; Cheng, K.; Luo, P.; Hu, C. Analysis of the transcriptome data in *Litopenaeus vannamei* reveals the immune basis and predicts the hub regulation-genes in response to high-pH stress. *PLoS ONE* **2018**, *13*, e0207771. [CrossRef]
29. Lou, F.; Gao, T.; Han, Z. Transcriptome analyses reveal alterations in muscle metabolism, immune responses and reproductive behavior of Japanese mantis shrimp (*Oratosquilla oratoria*) at different cold temperature. *Comp. Biochem. Physiol. D-Genom. Proteom.* **2019**, *32*, 100615. [CrossRef]
30. Li, Y.; Wang, J.; Jin, Y.; Ji, G.; Zhang, X. Significant genes in response to low temperature in *Penaeus chinensis* screened from multiple groups of transcriptome comparison. *J. Therm. Biol.* **2022**, *107*, 103198. [CrossRef]
31. Livak, K.J.; Schmittgen, T.D. Analysis of relative gene expression data using real-time quantitative PCR and the $2-\Delta\Delta$CT method. *Methods* **2001**, *25*, 402–408. [CrossRef] [PubMed]
32. Bolger, A.M.; Lohse, M.; Usadel, B. Trimmomatic: A flexible trimmer for Illumina sequence data. *Bioinformatics* **2014**, *30*, 2114–2120. [CrossRef] [PubMed]
33. Roberts, A.; Trapnell, C.; Donaghey, J.; Rinn, J.L.; Pachter, L. Improving RNA-Seq expression estimates by correcting for fragment bias. *Genome Biol.* **2011**, *12*, R22. [CrossRef] [PubMed]
34. Trapnell, C.; Williams, B.A.; Pertea, G.; Mortazavi, A.; Kwan, G.; Van Baren, M.J.; Salzberg, S.L.; Wold, B.J.; Pachter, L. Transcript assembly and quantification by RNA-Seq reveals unannotated transcripts and isoform switching during cell differentiation. *Nat. Biotechnol.* **2010**, *28*, 511–515. [CrossRef] [PubMed]
35. Wang, W.-N.; Wang, A.-L.; Liu, Y.; Xiu, J.; Liu, Z.-B.; Sun, R.-Y. Effects of temperature on growth, adenosine phosphates, ATPase and cellular defense response of juvenile shrimp *Macrobrachium nipponense*. *Aquaculture* **2006**, *256*, 624–630. [CrossRef]
36. García-Guerrero, M.; Hernández-Sandoval, P.; Orduña-Rojas, J.; Cortés-Jacinto, E. Effect of temperature on weight increase, survival, and thermal preference of juvenile redclaw crayfish *Cherax quadricarinatus*. *Hidrobiológica* **2013**, *23*, 73–81.
37. Nagasawa, H. The crustacean cuticle: Structure, composition and mineralization. *Front. Biosci.* **2012**, *4*, 711–720. [CrossRef]
38. Gong, J.; Yu, K.; Shu, L.; Ye, H.; Li, S.; Zeng, C. Evaluating the effects of temperature, salinity, starvation and autotomy on molting success, molting interval and expression of ecdysone receptor in early juvenile mud crabs, *Scylla paramamosain*. *J. Exp. Mar. Biol. Ecol.* **2015**, *464*, 11–17. [CrossRef]

39. Travis, D.F. The molting cycle of the spiny lobster, *Panulirus argus* Latreille. I. Molting and growth in laboratory-maintained individuals. *Biol. Bull.* **1954**, *107*, 433–450. [CrossRef]
40. Hsieh, S.; Chen, S.; Yang, Y.; Kuo, C. Involvement of norepinephrine in the hyperglycemic responses of the freshwater giant prawn, *Macrobrachium rosenbergii*, under cold shock. *Comp. Biochem. Physiol. Part A Mol. Integr. Physiol.* **2006**, *143*, 254–263. [CrossRef]
41. Pan, L.-Q.; Hu, F.-W.; Jing, F.-T.; Liu, H.-J. The effect of different acclimation temperatures on the prophenoloxidase system and other defence parameters in *Litopenaeus vannamei*. *Fish Shellfish Immunol.* **2008**, *25*, 137–142. [CrossRef] [PubMed]
42. Subramoniam, T. *Crustacean ecdysteriods* in reproduction and embryogenesis. *Comp. Biochem. Physiol. Part C Pharmacol. Toxicol. Endocrinol.* **2000**, *125*, 135–156. [CrossRef]
43. Nelson, E.R.; Habibi, H.R. Estrogen receptor function and regulation in fish and other vertebrates. *Gen. Comp. Endocrinol.* **2013**, *192*, 15–24. [CrossRef] [PubMed]
44. Ye, H.; Huang, H.; Li, S.; Wang, G. Immunorecognition of estrogen and androgen receptors in the brain and thoracic ganglion mass of mud crab, *Scylla paramamosain*. *Prog. Nat. Sci.* **2008**, *18*, 691–695. [CrossRef]
45. Lago-Lestón, A.; Ponce, E.; Muñoz, M.E. Cloning and expression of hyperglycemic (CHH) and molt-inhibiting (MIH) hormones mRNAs from the eyestalk of shrimps of *Litopenaeus vannamei* grown in different temperature and salinity conditions. *Aquaculture* **2007**, *270*, 343–357. [CrossRef]
46. Qiao, H.; Xiong, Y.; Zhang, W.; Fu, H.; Jiang, S.; Sun, S.; Bai, H.; Jin, S.; Gong, Y. Characterization, expression, and function analysis of gonad-inhibiting hormone in Oriental River prawn, *Macrobrachium nipponense* and its induced expression by temperature. *Comp. Biochem. Physiol. Part A Mol. Integr. Physiol.* **2015**, *185*, 1–8. [CrossRef] [PubMed]
47. Mullur, R.; Liu, Y.-Y.; Brent, G.A. Thyroid hormone regulation of metabolism. *Physiol. Rev.* **2014**, *94*, 355–382. [CrossRef]
48. Little, A.G.; Seebacher, F. The evolution of endothermy is explained by thyroid hormone-mediated responses to cold in early vertebrates. *J. Exp. Biol.* **2014**, *217*, 1642–1648. [CrossRef]
49. Moriya, T. The effect of temperature on the action of thyroid hormone and prolactin in larvae of the salamander *Hynobius retardatus*. *Gen. Comp. Endocrinol.* **1983**, *49*, 1–7. [CrossRef]
50. Politis, S.N.; Servili, A.; Mazurais, D.; Zambonino-Infante, J.-L.; Miest, J.J.; Tomkiewicz, J.; Butts, I. Temperature induced variation in gene expression of thyroid hormone receptors and deiodinases of European eel (*Anguilla anguilla*) larvae. *Gen. Comp. Endocrinol.* **2018**, *259*, 54–65. [CrossRef]
51. Hammond, S.A.; Veldhoen, N.; Helbing, C.C. Influence of temperature on thyroid hormone signaling and endocrine disruptor action in Rana (Lithobates) catesbeiana tadpoles. *Gen. Comp. Endocrinol.* **2015**, *219*, 6–15. [CrossRef] [PubMed]
52. Bizhanova, A.; Kopp, P. The sodium-iodide symporter NIS and pendrin in iodide homeostasis of the thyroid. *Endocrinology* **2009**, *150*, 1084–1090. [CrossRef] [PubMed]
53. Moreno, J.C.; Klootwijk, W.; van Toor, H.; Pinto, G.; D'Alessandro, M.; Lèger, A.; Goudie, D.; Polak, M.; Grüters, A.; Visser, T.J. Mutations in the iodotyrosine deiodinase gene and hypothyroidism. *N. Engl. J. Med.* **2008**, *358*, 1811–1818. [CrossRef]
54. Little, A.G.; Kunisue, T.; Kannan, K.; Seebacher, F. Thyroid hormone actions are temperature-specific and regulate thermal acclimation in zebrafish (*Danio rerio*). *BMC Biol.* **2013**, *11*, 26. [CrossRef]
55. Zhou, M.; Wang, A.-L.; Xian, J.-A. Variation of free amino acid and carbohydrate concentrations in white shrimp, *Litopenaeus vannamei*: Effects of continuous cold stress. *Aquaculture* **2011**, *317*, 182–186. [CrossRef]
56. Ighodaro, O.; Akinloye, O. First line defence antioxidants-superoxide dismutase (SOD), catalase (CAT) and glutathione peroxidase (GPX): Their fundamental role in the entire antioxidant defence grid. *Alex. J. Med.* **2018**, *54*, 287–293. [CrossRef]
57. Hayes, J.D.; McLellan, L.I. Glutathione and glutathione-dependent enzymes represent a co-ordinately regulated defence against oxidative stress. *Free Radic. Res.* **1999**, *31*, 273–300. [CrossRef] [PubMed]
58. Schafer, F.Q.; Buettner, G.R. Redox environment of the cell as viewed through the redox state of the glutathione disulfide/glutathione couple. *Free Radic. Biol. Med.* **2001**, *30*, 1191–1212. [CrossRef]
59. Gálvez, S.; Gadal, P. On the function of the NADP-dependent isocitrate dehydrogenase isoenzymes in living organisms. *Plant Sci.* **1995**, *105*, 1–14. [CrossRef]
60. Frei, B.; England, L.; Ames, B.N. Ascorbate is an outstanding antioxidant in human blood plasma. *Proc. Natl. Acad. Sci. USA* **1989**, *86*, 6377–6381. [CrossRef]
61. Lesser, M.P. Oxidative stress in marine environments: Biochemistry and physiological ecology. *Annu. Rev. Physiol.* **2006**, *68*, 253–278. [CrossRef] [PubMed]
62. Guerriero, G.; Di Finizio, A.; Ciarcia, G. Stress-induced changes of plasma antioxidants in aquacultured sea bass, *Dicentrarchus labrax*. *Comp. Biochem. Physiol. Part A Mol. Integr. Physiol.* **2002**, *132*, 205–211. [CrossRef]
63. Han, B.; Kaur, V.I.; Baruah, K.; Nguyen, V.D.; Bossier, P. High doses of sodium ascorbate act as a prooxidant and protect gnotobiotic brine shrimp larvae (*Artemia franciscana*) against Vibrio harveyi infection coinciding with heat shock protein 70 activation. *Dev. Comp. Immunol.* **2019**, *92*, 69–76. [CrossRef] [PubMed]
64. Bullock, T.H. Compensation for temperature in the metabolism and activity of poikilotherms. *Biol. Rev.* **1955**, *30*, 311–342. [CrossRef]
65. Sasseville, D. Neomycin. *Dermatitis* **2010**, *21*, 3–7. [CrossRef]
66. Umezawa, H.; Ueda, M.; Maeda, K.; Yagishita, K.; Kondō, S.; Okami, Y.; Utahara, R.; Ōsato, Y.; Nitta, K.; Takeuchi, T. Production and isolation of a new antibiotic, kanamycin. *J. Antibiot. Ser. A* **1957**, *10*, 181–188.

67. Yoshizawa, S.; Fourmy, D.; Puglisi, J.D. Structural origins of gentamicin antibiotic action. *EMBO J.* **1998**, *17*, 6437–6448. [CrossRef]
68. Beard, E.L., Jr. The American Society of Health System Pharmacists. *JONA'S Healthc. Law Ethics Regul.* **2001**, *3*, 78–79. [CrossRef]
69. Emanuele, E.; Bertona, M.; Altabas, K.; Altabas, V.; Alessandrini, G. Anti-inflammatory effects of a topical preparation containing nicotinamide, retinol, and 7-dehydrocholesterol in patients with acne: A gene expression study. *Clin. Cosmet. Investig. Dermatol.* **2012**, *5*, 33. [CrossRef]
70. Ruamrak, C.; Lourith, N.; Natakankitkul, S. Comparison of clinical efficacies of sodium ascorbyl phosphate, retinol and their combination in acne treatment. *Int. J. Cosmet. Sci.* **2009**, *31*, 41–46. [CrossRef]
71. Fenzl, A.; Kulterer, O.C.; Spirk, K.; Mitulović, G.; Marculescu, R.; Bilban, M.; Baumgartner-Parzer, S.; Kautzky-Willer, A.; Kenner, L.; Plutzky, J. Cold-mediated regulation of systemic retinol transport controls adipose tissue browning. *bioRxiv* **2020**. [CrossRef]
72. Sundaresan, P.; Winters, V.G.; Therriault, D.G. Effect of low environmental temperature on the metabolism of vitamin A (retinol) in the rat. *J. Nutr.* **1967**, *92*, 474–478. [CrossRef]
73. Zhang, S.; Yu, J.; Wang, H.; Liu, B.; Yue, X. p38 MAPK is involved in the immune response to pathogenic Vibrio in the clam *Meretrix petechialis*. *Fish Shellfish Immunol.* **2019**, *95*, 456–463. [CrossRef] [PubMed]
74. Tian, Y.; Wen, H.; Qi, X.; Zhang, X.; Li, Y. Identification of mapk gene family in *Lateolabrax maculatus* and their expression profiles in response to hypoxia and salinity challenges. *Gene* **2019**, *684*, 20–29. [CrossRef] [PubMed]
75. Lallès, J.-P. Biology, environmental and nutritional modulation of skin mucus alkaline phosphatase in fish: A review. *Fish Shellfish Immunol.* **2019**, *89*, 179–186. [CrossRef] [PubMed]
76. van Beek, J.H.; de Moor, M.H.; de Geus, E.J.; Lubke, G.H.; Vink, J.M.; Willemsen, G.; Boomsma, D.I. The genetic architecture of liver enzyme levels: GGT, ALT and AST. *Behav. Genet.* **2013**, *43*, 329–339. [CrossRef]
77. Penning, T.M. The aldo-keto reductases (AKRs): Overview. *Chem.-Biol. Interact.* **2015**, *234*, 236–246. [CrossRef]
78. Morimoto, R.I.; Santoro, M.G. Stress–inducible responses and heat shock proteins: New pharmacologic targets for cytoprotection. *Nat. Biotechnol.* **1998**, *16*, 833–838. [CrossRef]
79. Ahn, Y.-J.; Im, E. Heterologous expression of heat shock proteins confers stress tolerance in *Escherichia coli*, an industrial cell factory: A short review. *Biocatal. Agric. Biotechnol.* **2020**, *29*, 101833. [CrossRef]
80. Chen, T.; Lin, T.; Li, H.; Lu, T.; Li, J.; Huang, W.; Sun, H.; Jiang, X.; Zhang, J.; Yan, A. Heat shock protein 40 (HSP40) in pacific white shrimp (*Litopenaeus vannamei*): Molecular cloning, tissue distribution and ontogeny, response to temperature, acidity/alkalinity and salinity stresses, and potential role in ovarian development. *Front. Physiol.* **2018**, *9*, 1784. [CrossRef]
81. Qian, Z.; Liu, X.; Wang, L.; Wang, X.; Li, Y.; Xiang, J.; Wang, P. Gene expression profiles of four heat shock proteins in response to different acute stresses in shrimp, Litopenaeus vannamei. *Comp. Biochem. Physiol. Part C Toxicol. Pharmacol.* **2012**, *156*, 211–220. [CrossRef] [PubMed]
82. Mallouk, Y.; Vayssier-Taussat, M.; Bonventre, J.V.; Polla, B.S. Heat shock protein 70 and ATP as partners in cell homeostasis. *Int. J. Mol. Med.* **1999**, *4*, 463–537. [CrossRef] [PubMed]
83. Walker, J.E. The ATP synthase: The understood, the uncertain and the unknown. *Biochem. Soc. Transact.* **2013**, *41*, 1–16. [CrossRef] [PubMed]
84. Srivastava, P. Roles of heat-shock proteins in innate and adaptive immunity. *Nat. Rev. Immunol.* **2002**, *2*, 185–194. [CrossRef]
85. Pockley, A.G.; Henderson, B. Extracellular cell stress (heat shock) proteins—Immune responses and disease: An overview. *Philos. Trans. R. Soc. B Biol. Sci.* **2018**, *373*, 20160522. [CrossRef]
86. Beere, H.M.; Green, D.R. Stress management–heat shock protein-70 and the regulation of apoptosis. *Trends Cell Biol.* **2001**, *11*, 6–10. [CrossRef]
87. Mosser, D.D.; Caron, A.W.; Bourget, L.; Denis-Larose, C.; Massie, B. Role of the human heat shock protein hsp70 in protection against stress-induced apoptosis. *Mol. Cell. Biol.* **1997**, *17*, 5317–5327. [CrossRef]
88. Ravagnan, L.; Gurbuxani, S.; Susin, S.A.; Maisse, C.; Daugas, E.; Zamzami, N.; Mak, T.; Jäättelä, M.; Penninger, J.M.; Garrido, C. Heat-shock protein 70 antagonizes apoptosis-inducing factor. *Nat. Cell Biol.* **2001**, *3*, 839–843. [CrossRef]
89. Luckinbill, L. Selection for longevity confers resistance to low-temperature stress in *Drosophila melanogaster*. *J. Gerontol. Ser. A Biol. Sci. Med. Sci.* **1998**, *53*, B147–B153. [CrossRef]
90. Lee, M.-C.; Yoon, D.-S.; Lee, Y.; Choi, H.; Shin, K.-H.; Park, H.G.; Lee, J.-S. Effects of low temperature on longevity and lipid metabolism in the marine rotifer *Brachionus koreanus*. *Comp. Biochem. Physiol. Part A Mol. Integr. Physiol.* **2020**, *250*, 110803. [CrossRef]
91. Zhang, B.; Xiao, R.; Ronan, E.A.; He, Y.; Hsu, A.-L.; Liu, J.; Xu, X.S. Environmental temperature differentially modulates *C. elegans* longevity through a thermosensitive TRP channel. *Cell Rep.* **2015**, *11*, 1414–1424. [CrossRef] [PubMed]
92. Wolff, S.; Ma, H.; Burch, D.; Maciel, G.A.; Hunter, T.; Dillin, A. SMK-1, an essential regulator of DAF-16-mediated longevity. *Cell* **2006**, *124*, 1039–1053. [CrossRef] [PubMed]
93. Ghazi, A.; Henis-Korenblit, S.; Kenyon, C. A transcription elongation factor that links signals from the reproductive system to lifespan extension in *Caenorhabditis elegans*. *PLoS Genet.* **2009**, *5*, e1000639. [CrossRef] [PubMed]
94. Lindquist, J.A.; Mertens, P.R. Cold shock proteins: From cellular mechanisms to pathophysiology and disease. *Cell Commun. Signal.* **2018**, *16*, 63. [CrossRef] [PubMed]
95. Phadtare, S.; Alsina, J.; Inouye, M. Cold-shock response and cold-shock proteins. *Curr. Opin. Microbiol.* **1999**, 2, 175–180. [CrossRef]

96. Sun, S.; Fu, H.; Ge, X.; Zhu, J.; Qiao, H.; Jin, S.; Zhang, W. Molecular cloning and expression analysis of cold shock protein Y-box gene from oriental river pawn (*Macrobrachium nipponense*). *J. Fish. China* **2017**, *41*, 1345–1354.
97. Meng, Q.; Chen, J.; Huang, Y.; Jin, M.; Wei, G.; Wang, W. Molecular cloning and expression analysis of the cold shock protein Y-box coding gene of red claw crayfish, *Cherax quadricarinatus*. *Freshw. Fish.* **2012**, *42*, 14–20.
98. Hui, M.; Cheng, J.; Sha, Z. Adaptation to the deep-sea hydrothermal vents and cold seeps: Insights from the transcriptomes of *Alvinocaris longirostris* in both environments. *Deep Sea Res. Part I Oceanogr. Res. Pap.* **2018**, *135*, 23–33. [CrossRef]
99. Wendelaar Bonga, S.E. The stress response in fish. *Physiol. Rev.* **1997**, *77*, 591–625. [CrossRef]
100. Manfrin, C.; Pallavicini, A.; Battistella, S.; Lorenzon, S.; Giulianini, P.G. Crustacean immunity: The modulation of stress responses. In *Lessons in Immunity*; Elsevier: Amsterdam, The Netherlands, 2016; pp. 107–116.

antioxidants

Article

Mercury Induced Tissue Damage, Redox Metabolism, Ion Transport, Apoptosis, and Intestinal Microbiota Change in Red Swamp Crayfish (*Procambarus clarkii*): Application of Multi-Omics Analysis in Risk Assessment of Hg

Lang Zhang [1], Yuntao Zhou [1], Ziwei Song [2], Hongwei Liang [1], Shan Zhong [2,3], Yali Yu [1], Ting Liu [1], Hang Sha [1], Li He [1,4] and Jinhua Gan [1,4,*]

1 Yangtze River Fisheries Research Institute, Chinese Academy of Fishery Sciences, Wuhan 430223, China
2 Department of Genetics, Wuhan University, Wuhan 430071, China
3 Hubei Province Key Laboratory of Allergy and Immunology, Wuhan 430071, China
4 Key Laboratory of Control of Quality and Safety for Aquatic Products, Ministry of Agriculture and Rural Affairs, Beijing 100141, China
* Correspondence: gjh@yfi.ac.cn

Abstract: As one of the most toxic elements, mercury (Hg) is a widespread toxicant in aquatic environments. Crayfish are considered suitable for indicating the impact of heavy metals on aquatic crustaceans. Nevertheless, Hg toxicity on *Procambarus clarkii* is largely unknown. In this research, the acute Hg-induced alterations of biochemical responses, histopathology, hepatopancreatic transcriptome, and intestinal microbiome of *Procambarus clarkii* were studied. Firstly, Hg induced significant changes in reactive oxygen species (ROS) and malonaldehyde (MDA) content as well as antioxidant enzyme activity. Secondly, Hg exposure caused structural damage to the hepatopancreas (e.g., vacuolization of the epithelium and dilatation of the lumen) as well as to the intestines (e.g., dysregulation of lamina epithelialises and extension of lamina proprias). Thirdly, after treatment with three different concentrations of Hg, RNA-seq assays of the hepatopancreas revealed a large number of differentially expressed genes (DEGs) linked to a specific function. Among the DEGs, a lot of redox metabolism- (e.g., ACOX3, SMOX, GPX3, GLO1, and P4HA1), ion transport- (e.g., MICU3, MCTP, PYX, STEAP3, and SLC30A2), drug metabolism- (e.g., HSP70, HSP90A, CYP2L1, and CYP9E2), immune response- (e.g., SMAD4, HDAC1, and DUOX), and apoptosis-related genes (e.g., CTSL, CASP7, and BIRC2) were identified, which suggests that Hg exposure may perturb the redox equilibrium, disrupt the ion homeostasis, weaken immune response and ability, and cause apoptosis. Fourthly, bacterial 16S rRNA gene sequencing showed that Hg exposure decreased bacterial diversity and dysregulated intestinal microbiome composition. At the phylum level, there was a marked decrease in *Proteobacteria* and an increase in *Firmicutes* after exposure to high levels of Hg. With regards to genus, abundances of *Bacteroides*, *Dysgonomonas*, and *Arcobacter* were markedly dysregulated after Hg exposures. Our findings elucidate the mechanisms involved in Hg-mediated toxicity in aquatic crustaceans at the tissue, cellular, molecular as well as microbial levels.

Keywords: *Procambarus clarkii*; mercury; histopathology; intestinal microbiota; hepatopancreatic transcriptome

Citation: Zhang, L.; Zhou, Y.; Song, Z.; Liang, H.; Zhong, S.; Yu, Y.; Liu, T.; Sha, H.; He, L.; Gan, J. Mercury Induced Tissue Damage, Redox Metabolism, Ion Transport, Apoptosis, and Intestinal Microbiota Change in Red Swamp Crayfish (*Procambarus clarkii*): Application of Multi-Omics Analysis in Risk Assessment of Hg. *Antioxidants* **2022**, *11*, 1944. https://doi.org/10.3390/antiox11101944

Academic Editors: Marcelo Hermes-Lima, Daniel Carneiro Moreira and Tania Zenteno-Savín

Received: 8 August 2022
Accepted: 24 September 2022
Published: 29 September 2022

1. Introduction

In recent years, there has been increasing concern about aquatic heavy metal pollution [1–3]. Mercury, a toxic element, reaches aquatic environments mainly via anthropogenic activities. Globally, it is the third most common environmental contaminant [4,5]. Mercury is transported and biomagnified in aquatic ecosystems via aquatic animal food webs, such as algae, sediments, carnivorous fish, and benthic crustaceans.

Methyl mercury (MeHg), a Hg compound, is obtained from Hg^{2+} via microbial activities [6]. It bio-accumulates in higher trophic consumers, particularly long-living, slow-growing species [7–9]. In lakes as well as estuaries, sediments are common sites for MeHg [10–13]. However, anthropogenic mercury input is highly associated with elevated Hg^{2+} levels. Thus, the accumulation of Hg^{2+} in benthic crustaceans found at the sediment and water interface is a focus of research [14,15]. Unfortunately, in crustaceans, acute toxic responses to inorganic Hg have not been fully established.

Adverse outcomes of Hg on organisms have been documented. Toxic effects of mercury on animal models (e.g., fish and mouse), depending on their exposure duration and dose, could cause hepatotoxicity [16], neurotoxicity [17], as well as endocrine [18] and reproductive disruption [19]. The toxicity of Hg on crustaceans has also been documented; however, most of them were focused on the larval stage and the median lethal concentration (LC50). For instance, 96 h Hg LC50 values in various species were: 20 μg/L in *Penaeus monodon* postlarvae [20], 1.2 μg/L in *Penaeus japonicus* postlarvae [21], 15 μg/L in *Neomysis awatschensis* [22], 18 μg/L in *Penaeus japonicus* embryos [23], and 40 μg/L in *Scylla serrata* juveniles [24]. Despite these studies, more research should be undertaken to determine the impact of Hg on antioxidant enzymes, histopathology, hepatopancreatic transcriptome, and intestinal microbiome structures of crustaceans.

It has been found that metals induced the production of ROS, which causes oxidative stress, resulting in several detrimental effects on cells [25]. The overproduction of ROS leads to the formation of malonaldehyde-like species in lipids [26,27]. Consequently, ROS and MDA levels reflect the degree of oxidative damage. In order to protect themselves from ROS, organisms (such as crustaceans) have mechanisms of non-enzymatic and enzymatic antioxidants [28]. Non-enzymatic antioxidants contain tocopherols, ascorbic acid, and glutathione (GSH). GSH is the most abundant cellular thiol and well-studied antioxidant compound in organisms [29]. In terms of enzymatic defenses, superoxide dismutase (SOD), catalase (CAT), and glutathione S-transferase (GST) are antioxidant enzymes responsible for maintaining cellular redox status [30]. Higher enzyme activity means higher detoxification capacity, which is important for counteracting ROS-induced cellular damage [31,32]. However, not much is known concerning the effects of Hg on oxidative stress damage and the antioxidant system in crustaceans.

In crustaceans, the hepatopancreas is involved in metabolism, immune functions, nutrient absorption, and xenobiotic detoxification [33,34]. In addition, the hepatopancreas is one of the key organs affected by environmental stressors [35–37]. In crustaceans, transcriptome analysis is an effective method to provide information about the global expression profiles of genes and related mechanisms involved in the toxicity of heavy metals [38–40]. The intestinal microbiome is important in the sustenance of health and in the regulation of many vital physiological host functions [41–43]. Studies on gut microbial communities suggest that diseased and healthy shrimp have different intestinal bacterial communities [43]. In addition, gut microbiota can be used to measure shrimp health [44]. Research on crustaceans has revealed the importance of diets [45], developmental stage [46], health status [43], and risk factors [47,48] on the gut microbiota. Furthermore, the relationship between heavy metal toxicity and intestinal microbiota alteration in crustaceans has been studied. The concentration of 0.5 mg L^{-1} Cu^{2+} or more increases the abundance of intestinal pathogens in *Litopenaeus vannamei* [49]. Cd exposure could alter the richness, diversity, and composition of intestinal microbiota in *Procambarus clarkii* (*P. clarkii*) [47]. However, the toxicity of inorganic Hg to crustaceans has not been fully established. Deep sequencing data of hepatopancreatic transcriptome and intestinal microbiota will reveal abundant genetic and bacterial signatures of Hg toxicity.

Procambarus clarkii, the freshwater crayfish is an important commercial species [50,51]. Considering its long lifecycle, wide distribution, and simple anatomy, *P. clarkii* is often utilized as a typical bioindicator of toxic pollutants in studies on aquatic environments [38,52–54]. Additionally, *P. clarkii* is considered a model organism for research in aquatic crustaceans [55]. In the present research, adult *P. clarkii* was exposed to acute Hg. On the basis of these previous studies, we investigated the bioaccumulation together with antioxidant enzymes, histological variations, the hepatopancreatic transcriptome, and intestinal microbiota changes to further reveal how *P. clarkii* responds to inorganic Hg^{2+} at the biochemical, physiological, molecular, and intestinal microbiota levels. Another crucial question in our study is how Hg exposure damages crustaceans' tissues. The results of this study could serve as a physiological reference for mitigating the negative effects of Hg stress on crustacean aquaculture.

2. Materials and Methods

2.1. Experimental Animals

To avoid gender-related differences and the influence of female oviposition, only male adult freshwater crayfish with similar sizes (weight: 19.46 ± 2.84 g, length: 9.8 ± 1.4 cm) were acquired from a commercial crayfish farm in Lixian (Changde, China). They were kept in several glass aquaria (60 cm length, 40 cm width, and 35 cm depth) containing 30 L of dechlorinated tap water (temperature 24.0 ± 0.8 °C, $CaCO_3$ hardness 44.21 ± 0.28 mg/L, pH 7.10 ± 0.07, and DO 6.76 ± 0.21 mg/L) for 7 days for acclimatization. Before experimental procedures, all crayfish were fed once daily on red worms (*Limnodrilus*), but the crayfish were not fed during exposure experiment. In addition, the photoperiod was 12 h/12 h dark/light.

2.2. Toxicity Test

Crayfish were randomized into 4 groups (three replicates per group); in addition, twelve specimens per replicate. There was no replacement of water during the static exposure experiment. $HgCl_2$ (Sinopharm Chemical Reagent Company, Shanghai, China) at analytical grades was used. Previous study has indicated that 96 h LC50 concentration of Hg^{2+} for freshwater crayfish at 24 °C is 0.35 mg/L [56], in order to better evaluate the relationship between Hg concentration and toxicity; therefore, an appropriate Hg^{2+} exposure time of 96 h and temperature of 24 °C were chosen for this toxicity test. The Hg exposure concentration drew on research conducted with another heavy metal cadmium exposure on the crayfish [38], exposures of *P. clarkii* to Hg were respectively conducted at doses of 0 (Control group, Ctrl, 0 μg/L Hg^{2+}), 1/40 LC50 (Low concentration group, Low, 8.75 μg/L Hg^{2+}), 1/16 LC50 (Medium concentration group, Med, 21.875 μg/L Hg^{2+}), and 1/8 LC50 (High concentration group, High, 43.75 μg/L Hg^{2+}). Figure 1 illustrates the experimental procedure schematically. The doses of Hg^{2+} solutions were established by dissolving the desired amount of Hg^{2+} stock solution in dechlorinated tap water. The other exposure conditions were similar to those for the acclimatization described above. There was no crayfish death during the exposure experiments.

2.3. Sampling

In the Hg^{2+} exposure experiment, all samples from each group were randomly selected at 96 h and anesthetized with eugenol bath (1:10,000). To examine histopathology, crayfish hepatopancreas and gut tissue samples were fixed in 4% paraformaldehyde. To avoid inter-individual variation, tissues of three specimens were pooled together per biological replicate. Three replicates of each Hg^{2+} treatment were collected for the analysis of total mercury content, enzyme activities, hepatopancreas transcriptome, as well as intestinal microbiota. To assess transcriptomic changes caused by Hg^{2+}, 300 mg of the hepatopancreas were acquired, instantly frozen in liquid nitrogen, and then kept at −80 °C for RNA extractions. Subsequently, intestines were flushed thrice using PBS and dissected. Then, intestinal contents were cautiously obtained in a 1.5 mL sterile centrifuge tube, immediately frozen in liquid nitrogen, and kept at −80 °C until required. During sampling, all operations were conducted on super clean workbench.

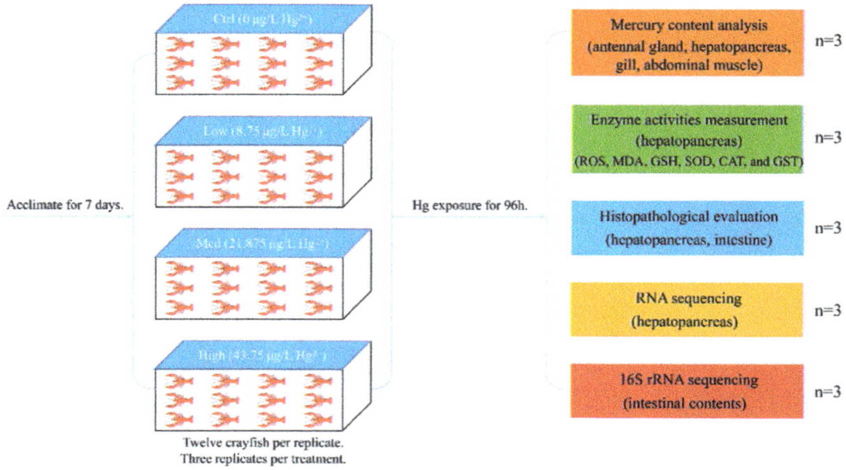

Figure 1. Schematic representation of the experimental procedure.

2.4. Total Mercury Content Analysis

At 0 h and 96 h, water samples (10 mL) from each aquarium were taken and acidified with HNO_3 for later use. Samples of crayfish were digested according to a previous publication [57]. In order to digest the tissue samples, a microwave digestion system (MARS, CEM) was utilized. A total of 300 mg of each sample and 10 mL mixed liquid of $HClO_4$ (perchloric acid 70%) and HNO_3 (Nitric acid 65% Suprapur®) were added to the digestion vessel. Following is the digested program: 5 min to temperature 120 °C, 5 min at temperature 120 °C, 5 min to temperature 150 °C, 10 min at temperature 150 °C, 5 min to temperature 190 °C, and 20 min at temperature 190 °C. Once the cooling process is complete, 2% HNO3 was added to the digestion solution to dilute it to 50 mL for later use. The concentration of Hg was determined by atomic fluorescence spectrometer (AFS, Wuhan, China).

2.5. Measurement of Enzyme Activities

The protein concentration, content of ROS (Cat. No. E004-1-1), MDA (Cat. No. A003-1-1) as well as GSH (Cat. No. A006-2-1), and enzyme activities of SOD (Cat. No. A001-3-2), CAT (Cat. No. A007-1), as well as GST (Cat. No. A004-1-1), were examined using Testing Kit (Nanjing Jiancheng Bioengineering, Nanjing, China). DCFH oxidation method by Keston and Brandt was used to determine the ROS levels in hepatopancreas [58]. SOD was quantified based on the method of Marklund and Marklund [59]. MDA, CAT, and GST were measured based on the methods of Satoh [60], Sinha [61], and Habig et al. [62], respectively. Concentrations of GSH were estimated by the method of Moron et al. [63]. The levels of ROS, MDA, GSH, SOD, CAT, and GST were normalized with the corresponding protein content.

2.6. Histopathological Evaluation

Dehydration of the fixed hepatopancreas as well as the gut was performed through a graded-ethanol serial, made transparent by soaking in xylene, and then paraffin-embedded. After that, sections (4 μm thick) were prepared utilizing a rotary microtome followed by hematoxylin and eosin staining. At last, stained sections were analyzed by a microscope (Olympus IX73).

2.7. RNA Isolation, Preparation of the RNA-Seq Library, and Sequencing

Isolation of total RNA from the hepatopancreas was performed using the Tri Reagent, as instructed by the manufacturer. With the Agilent Bioanalyzer 2100 system (Agilent

Technologies, Shanghai, China), RNA Nano 6000 Assay Kit was used for the evaluation of RNA integrity and quantification. In accordance with the manufacturer's instructions, sequencing libraries were prepared using NEBNext® Ultra™ RNA Library Prep Kit for Illumina® (NEB, Ipswich, MA, USA). After DNase I treatment, poly-T oligo (dT) magnetic beads were used to purify mRNA from RNA. After purification and fragmentation, the mRNA was utilized to synthesize cDNA. The downstream experiments were conducted on samples with an RNA integrity number (RIN) > 7.0. SMARTer PCR cDNA synthesis kit was used to prepare libraries for RNA sequencing. Twelve cDNA libraries were constructed from the Ctrl, Low, Med, and High groups, each group with three biological replications. To generate paired-end reads from the libraries, the Illumina HiSeq 2500 sequencing platform was used.

2.8. Transcriptome Assembly and Annotation

The in-house Perl scripts were initially used to process raw reads in fastq format. To obtain clean reads, adaptor sequences, low-quality sequences, as well as poly-N were eliminated from raw reads. Subsequently, Q20, Q30, and GC levels of the clean reads were evaluated. The Raw RNA-seq data were deposited in the NCBI Sequence Read Archive (SRA) under BioProject PRJNA788175 (www.ncbi.nlm.nih.gov/bioproject/PRJNA788175, accessed on 12 December 2021). Additionally, de-novo assembly of *P. clarkii* transcriptome was accomplished utilizing Trinity software [64]. Gene function annotation was based on GO (gene ontology), KO (KEGG Ortholog), NR, Pfam (protein family), STRING, SWISSPROT, and KOG databases.

2.9. Identification of DEGs

DEGs were identified by "DESeq" in R [65]. Adjustments of *p* values were made to control the FDR (false discovery rate) [66]. Genes with $\log_2$ |Fold Change| > 1 and adjusted *p*-value (FDR) < 0.05 were assigned as DEGs. The R function prcomp was used to perform the principal component analyses (PCA) for all genes. Hierarchical DEG clustering was conducted using the R 'heatmap' package. Then, WEGO as well as Blast2GO v2.5 programs were used to map the DEGs to the GO database for functional annotations. Furthermore, all DEGs were mapped to terms in the KEGG database to establish the markedly enriched KEGG terms.

2.10. Quantitative RT-PCR (qPCR) Assay

Primers of peroxisome-related genes were designed utilizing Primer Premier 6 software. 18S rRNA gene was utilized as the reference "housekeeping" gene of *P. clarkii* according to previous publications [38,51]. The primer sequences, amplicon size, and amplification efficiency were displayed in Table S1. According to the documentation, qPCR was performed using SYBR Green [67,68]. Briefly, qPCR was performed in 20 μL reactions comprised of 2 μL of cDNA, 0.4 μL of 10 μM both the reverse as well as forward primers, 10 μL of 2 × SybrGreen qPCR Master Mix, as well as 7.2 μL of RNase-free H_2O. The PCR thermal cycle program consisted of 95 °C for 2 min, 45 cycles of 95 °C for 3 s, and 60 °C for 30 s. Analysis of relative expressions was conducted using the $2^{-\triangle\triangle Ct}$ method [69].

2.11. Extraction of DNA and PCR-Amplifications

Total intestinal DNA was extracted using the E.Z.N.A.™ Mag-Bind Soil DNA Kit (Omega Bio-Tek, Norcross, GA, USA). PCR amplification of the 16S rDNA hypervariable V3-V4 regions was performed using primers 338F 5′-ACTCCTACGGGAGGCAGCA-3′ and 806R 5′-GGACTACHVGGGTWTCTAAT-3′. All PCR reactions were conducted in triplicate using a total volume of 20 μL reaction system containing 2 μL 2.5 mM dNTPs, 4 μL 5 × FastPfu Buffer, 0.4 μL FastPfu Polymerase, 0.8 μL each primer (5 mM), and 10 ng template DNA. According to Zhang et al., PCR amplification was conducted [37].

2.12. Illumina Miseq and Sequencing

The AxyPrep DNA Gel Extraction Kit (Axygen Biosciences, Union City, CA, USA) was used to extract amplicons from 2% agarose gels after which they were purified. Quantification of the PCR product was performed using the QuantiFluor^{TM}-ST fluorescence system (Promega, Madison, WI, USA). After that, the purified amplicons were pooled in equimolar concentrations, then sequenced (2 × 300) on an Illumina MiSeq platform. Demultiplexed 16S rRNA data were quality-filtered using the Quantitative Insights Into Microbial Ecology (QIIME 1.8.0) software package. Firstly, low-quality reads with scores < 20 or with a read length < 200 bp were filtered out. Secondly, barcodes were matched, while ambiguous bases and unmatched barcodes were removed. Thirdly, unassembled reads were discarded. Overlapping sequences longer than 10 bp were assembled based on their overlap sequences. UPARSE 7.1 (http://drive5.com/uparse/ accessed on 12 December 2021) was used to cluster operational taxonomic units (OTUs). UCHIME was used to identify and remove chimeric sequences. Using a confidence threshold of 70%, RDP Classifier (http://rdp.cme.msu.edu/ accessed on 12 December 2021) was used to analyze the taxonomy of individual 16S rRNA gene sequences against Silva (SSU115). Sequences obtained by 16S rRNA sequencing were deposited in NCBI SRA under BioProject PRJNA788294 (www.ncbi.nlm.nih.gov/bioproject/PRJNA788294, accessed on 12 December 2021).

2.13. Biodiversity Analysis

Alpha diversity assessments, such as Community coverage index (Coverage), diversity parameters (Simpson, Shannon), and richness parameters (Ace, Chao) were performed using the MOTHUR software (v.1.30.1) [70]. QIIME (version 1.8.0) was used to construct the rarefaction curves to evaluate sequencing depth. As for measurements of beta diversity, unweighted Unifrac was utilized for Principal Coordinate Analysis (PCoA). Unweighted pair-group method with arithmetic means (UPGMA) hierarchical clustering was performed using QIIME and displayed using R. Venn diagrams were performed using R to display shared, unique OTUs (operational taxonomic units) [71]. Diagrams of intestine microbial community composition were plotted utilizing Origin 8.0 software. Differences in communal composition differences between the Hg^{2+} exposure and control groups were evaluated by using one-way ANOVA, with $p \leq 0.05$ signifying statistical significance.

3. Results

3.1. The Bioaccumulation of Hg in the Tissues of P. clarkii

As shown in Table S2, a slight decrease in Hg concentration in water was observed over time due to accumulation by crayfish and adsorption on aquarium walls. After exposure for 96 h, the measured Hg concentrations (mean ± SD) were 0, 8.56 ± 0.16, 21.39 ± 0.42, and 43.72 ± 0.84 μg/L in Control, Low, Med, and High group, respectively.

Bioaccumulation of Hg in tissues during Hg^{2+} exposure is shown in Table S3. The crayfish in control group showed very low Hg accumulation levels. In control tissues, Hg accumulation levels were relative to each other as follows: gill > hepatopancreas > abdominal muscle > antennal gland. With increasing Hg concentrations, the accumulation level of Hg in all examined tissues significantly increased ($p < 0.05$) after 96 h of exposure. In all tissues with Hg^{2+}-treated for 96 h, Hg accumulation levels were relative to each other as follows: gill > antennal gland > hepatopancreas > abdominal muscle. These results indicated that Hg bioaccumulation in tissues (gill, antennal gland, hepatopancreas, abdominal muscle) was dose-dependent.

3.2. Oxidative Stress and Antioxidant Parameters

As a result of Hg exposure, enzymes involved in the response to oxidative stress were affected. As shown in Table S4, the ROS and MDA content of hepatopancreas in all Hg-treated crayfish increased significantly ($p < 0.05$). After Hg exposure for 96 h, ROS levels in Low, Med, and High groups increased by 37.21%, 41.24%, and 57.20%, respectively. MDA levels increased by 23.77%, 29.60%, and 37.22%, respectively. With increasing Hg concentrations, the enzymatic activities of hepatopancreatic SOD and CAT decreased gradually. The effects of Hg exposure on glutathione-mediated antioxidant enzyme activities were also significant. Concentrations of GSH in crayfish hepatopancreas were 17.45%, 23.73%, and 26.86% less, respectively, relative to that of control. Activity of GST in crayfish hepatopancreas was 14.48%, 19.26%, and 26.46% greater, respectively, relative to that of control.

3.3. Histopathology

Histological sections from hepatopancreas and intestines of Ctrl and Hg^{2+}-treated crayfish are shown in Figure 2. As shown in Figure 2A, the hepatopancreas cells from Ctrl group displayed well-organized structures and asterisk-like shapes in the tubule lumens. As shown in Figure 2B–D, compared with Ctrl group, the hepatopancreas cells of crayfish treated with Hg exhibited histological changes. All Hg-treated (Low, Med, and High concentrations of Hg^{2+}) hepatopancreas cells exhibited tubule lumen dilatations and apparent epithelium vacuolizations. In addition, this research demonstrated that crayfish intestines treated with Hg^{2+} displayed signs of damage. The intestine cells from Ctrl group showed normal palisade arrangements and a regular nucleus (Figure 2E). Compared with Ctrl group, the intestine in Low group was nearly identical (Figure 2F). By contrast, the crayfish intestines treated with Med and High concentrations of Hg displayed histological differences from intestines of Ctrl group, including apparent vacuoles in the microvilli, extended lamina proprias, and disordered lamina epithelialises (Figure 2G–H). As shown in Figure S1, increased Hg concentration significantly increased the proportions of hepatopancreatic tubule lumen dilatation and intestine microvilli vacuolization. After Hg exposure for 96 h, the mean proportion of hepatopancreatic tubule lumen dilatation in Control, Low, Med, and High groups were 5.6%, 32.9%, 63.8%, and 84.7%, respectively. The mean proportion of intestine microvilli vacuolization in Control, Low, Med, and High groups were 5.4%, 9.9%, 33.8%, and 44.7%, respectively. The above results indicate that the degree of hepatopancreas and intestine tissue damage showed a dose-dependent relationship with the concentration of Hg.

3.4. Transcriptome Sequencing and Assembly

In this study, we constructed cDNA libraries of mRNAs using the hepatopancreas RNA isolated from the Ctrl and Hg-treated groups. RNA-seq generated 42,795,604 to 61,860,116 raw reads (Table S5). Subsequent to the initial quality control, 42,694,670 to 61,735,916 clean reads (99.69–99.82% of the raw data) were generated. From the 12 constructed libraries, the clean reads were utilized for sequence assembly. Additionally, a total of 192,666 unigenes were generated by de novo assembly of RNA sequencing data using Trinity software. The length distribution of the unigenes was as presented in Figure S2. The average, largest, and smallest lengths for all unigenes were 711.6 bp, 36,069 bp, and 197 bp, respectively.

3.5. Functional Annotations and Classification

In our study, after transcriptome assembly, A total of 192,666 unigenes were identified from the hepatopancreas of *P. clarkii*. By using BLASTx and BLASTn, 192,666 unigenes were further classified based on their functional predictions. BLAST results showed that 6006, 5091, 11564, 5357, 7694, 6068, and 6409 unigenes matched with the annotated sequences in GO, KO, NR, PFAM, STRING, SWISSPROT, and KOG databases, respectively (Table S6).

GO classification is a standardized system to categorize genes. According to Figure S3A–C, 31, 20, and 15 of these subcategories were grouped into biological processes, cellular components, and molecular functions, respectively. Among the biological processes, cellular processes were the most abundant with 5478 unigenes, metabolic processes with 4828 unigenes, and biological regulation with 4293 unigenes. Among the sequences categorized as molecular function, 4924 unigenes were included in binding and 3245 unigenes were predicted to possess catalytic activity. Among the sequences categorized as component categories, cell was the most abundant with 5535 unigenes, cell part with 5530 unigenes, and organelle with 5178 unigenes, which are involved in the basic functional and structural unit of organisms. These results indicated that most of the annotated unigenes were related to various types of biological processes. KOG database classified 6409 unigenes into 25 functional categories (Figure S3D). The predominant category contains signal transduction mechanisms, general function prediction only, cytoskeleton, function unknown, ribosomal structure and biogenesis, transcription, and translation. In the KEGG pathway database, based on the KO database, a total of 5091 unigenes were grouped into 364 pathways. The majority of the unigenes were divided into the categories of signal transduction (813), transport and catabolism (502), endocrine system (466), immune system (404), and translation (394). The top 34 of these KEGG biological pathway classifications are shown in Figure S3E.

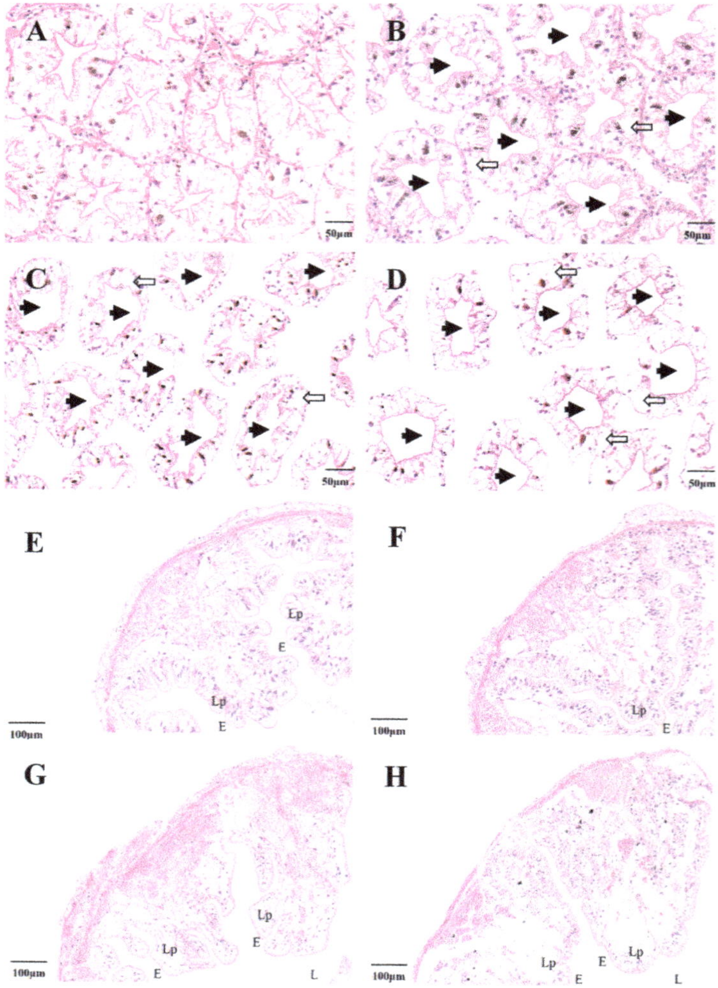

Figure 2. Hg exposure caused hepatopancreas and intestines injury. Hepatopancreas histology from Ctrl (**A**), Low (**B**), Med (**C**), and High (**D**) group in 96 h. Intestines histology from Ctrl (**E**), Low (**F**), Med (**G**), and High (**H**) group in 96 h. Black solid arrow: dilatation of tubule lumen; Black hollow arrow: vacuolization; Lp: lamina propria; E: epithelium; L: lumen. H&E stain (100×). Ctrl: 0 μg/L Hg^{2+}; Low: 8.75 μg/L Hg^{2+}; Med: 21.875 μg/L Hg^{2+}; High: 43.75 μg/L Hg^{2+}.

3.6. The DEGs

Biological repetition is particularly important for biological experiments. To do so, principal component analysis was used, as shown in Figure S4; the 3 parallels in each group are very similar, indicating general comparability between the four groups. To evaluate Hg toxicity, DEGs in Low, Med, and High groups relative to Ctrl group hepatopancreas were determined. Compared with Ctrl group, 1670 genes were up-regulated, and 1534 genes were down-regulated in Low group; 2025 genes were up-regulated, and 1948 genes were down-regulated in Med group; 2249 genes were up-regulated, and 2043 genes were down-regulated in High group (Figure 3A–C). The above results suggest that the number of DEGs showed a dose-dependent relationship with the concentration of Hg. As shown in Figure 3D, the Venn diagram depicts overlapping and non-overlapping numbers of DEGs among three comparisons. The Venn diagram demonstrates that 641 genes were all significantly changed in the three exposure groups (Low, Med, and High), indicating that these genes might play important role in Hg exposure. The expressions of the 641 genes were significantly changed under their exposure to Hg in different concentrations (Figure 3E).

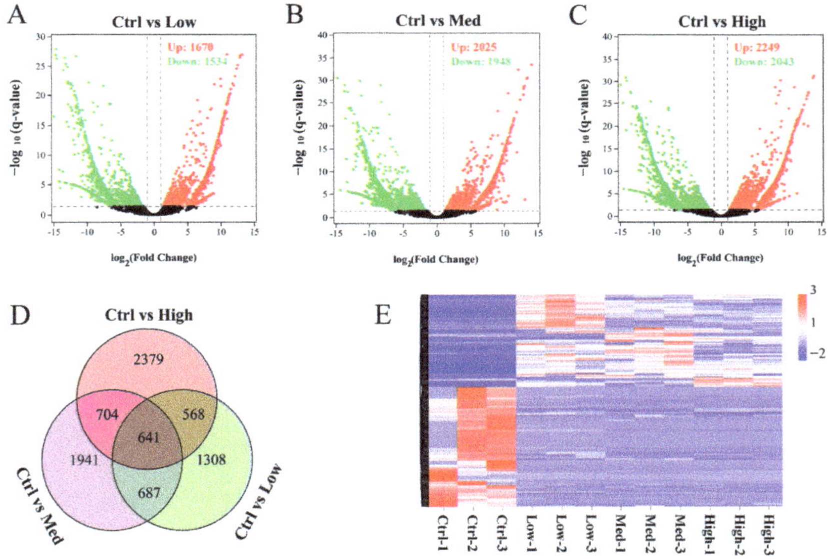

Figure 3. DEGs were identified after treatment with different concentrations of Hg. Volcano plots for DEGs in the 3 comparisons: (**A**) Ctrl vs. Low; (**B**) Ctrl vs. Med; (**C**) Ctrl vs. High. The red dots indicate genes that are up-regulated, while the green dots indicate genes that are down-regulated. (**D**) Venn diagram of DEGs in the 3 comparisons. (**E**) Heatmap based on fragments per kilobase of transcript per million mapped reads (FPKM) values showing the variations in expressions of overlapping DEGs. Genes whose expressions were greater than the mean are colored red while those below the mean are colored blue. Ctrl: 0 μg/L Hg^{2+}; Low: 8.75 μg/L Hg^{2+}; Med: 21.875 μg/L Hg^{2+}; High: 43.75 μg/L Hg^{2+}.

3.7. GO and KEGG Analyses of DEGs

GO analysis (Table S7) showed that the hormone metabolic process, anion transport, response to drug, drug transmembrane transport, and drug transport were markedly enriched (corrected $p < 0.05$) after exposure to low concentrations of Hg. Moreover, organic acid metabolic processes, carboxylic acid metabolic processes, and small molecule metabolic processes were significantly enriched after exposure to Med concentrations of Hg. Long-chain fatty acid metabolic processes, small molecule metabolic processes, and fatty acid metabolic processes were significantly enriched after exposure to high concentrations of Hg.

Analysis of KEGG was conducted to confirm significantly enriched pathways in crayfish after Hg treatment. Each of the ten enriched KEGG pathways in the three comparisons is presented in Figure 4. KEGG enrichment analysis showed the DEGs were enriched in phagosome, peroxisome, apoptosis-multiple species, and biosynthesis of antibiotics signaling pathways in comparisons (Ctrl vs. Low, Ctrl vs. Med, and Ctrl vs. High). Drug metabolism—cytochrome P450, and arachidonic acid metabolism were significantly enriched in comparisons (Ctrl vs. Low, Ctrl vs. Med). Ascorbate and aldarate metabolism and ABC transporters were significantly enriched in comparisons (Ctrl vs. Med, Ctrl vs. High).

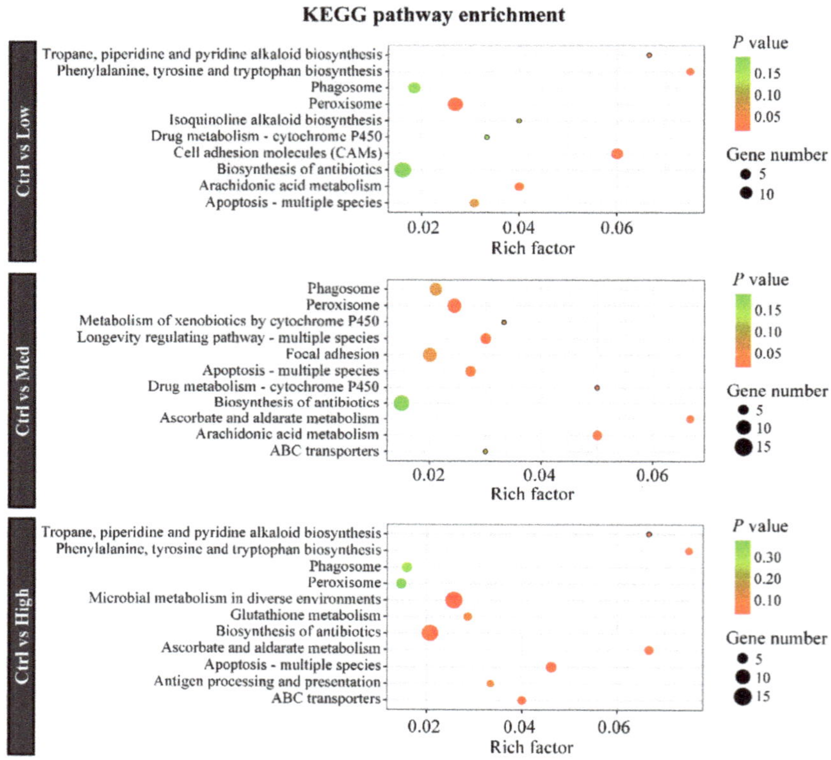

Figure 4. KEGG enrichments of the DEGs after treatment with different concentrations of Hg. The horizontal axis denotes the rich factor while the vertical axis denotes pathways. Color shades represent different *p*-values while dot sizes denote the number of DEGs. A larger dot indicates more DEGs. Ctrl: 0 μg/L Hg^{2+}; Low: 8.75 μg/L Hg^{2+}; Med: 21.875 μg/L Hg^{2+}; High: 43.75 μg/L Hg^{2+}.

In particular, the DEGs linked to redox metabolism (oxidoreductase activity and oxidation-reduction process) (Figure 5A,B), ion transport (calcium ion transport, sodium ion transport, potassium ion transport, phosphate ion transport, iron ion transport, zinc ion transport, copper ion transport, and magnesium ion transport) (Figure S5A–I), drug metabolic process (Figure 6A), immune response (Figure 6B), as well as apoptosis (Figure 6C), were analyzed.

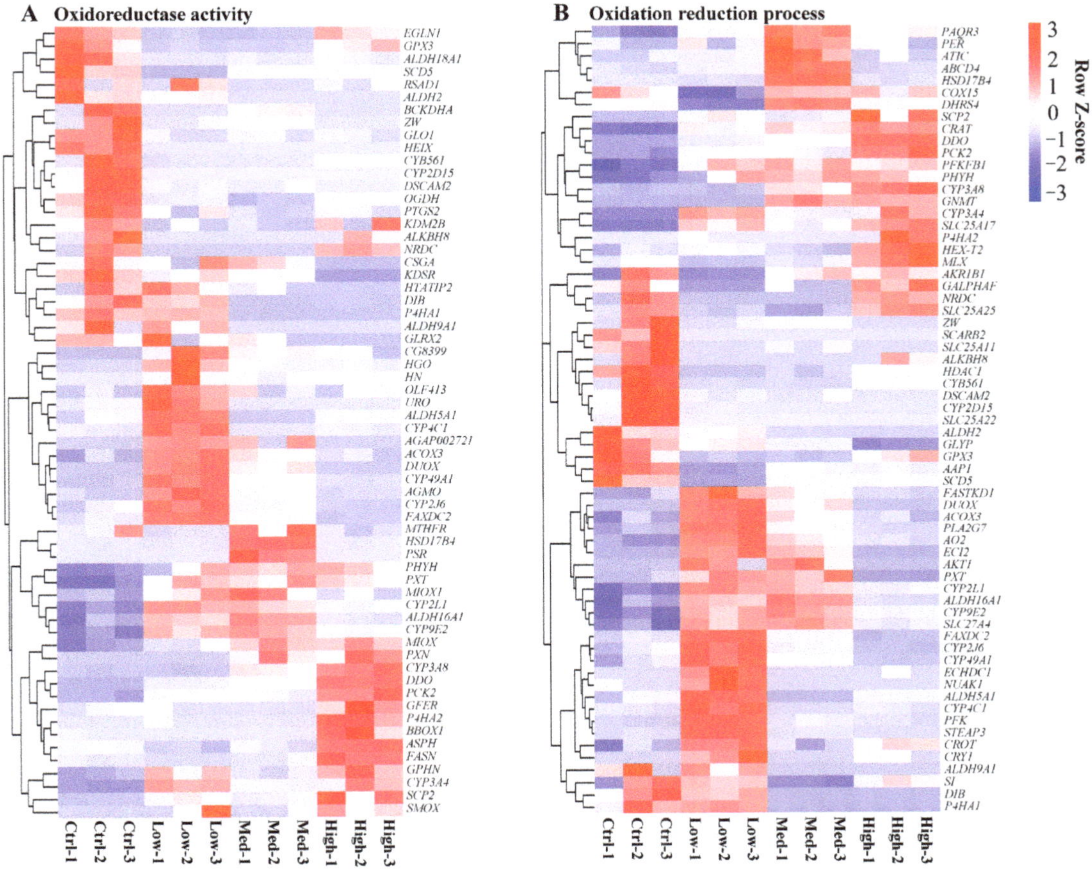

Figure 5. The effects of treatment with different concentrations of Hg on redox metabolism-related gene expression. Terms linked to redox metabolism, such as oxidoreductase activity (**A**) and oxidation-reduction process (**B**). Each gene was assessed based on its average FPKM value. The genes with higher expression levels are colored red and those with lower expression levels are colored blue. Ctrl: 0 μg/L Hg^{2+}; Low: 8.75 μg/L Hg^{2+}; Med: 21.875 μg/L Hg^{2+}; High: 43.75 μg/L Hg^{2+}.

3.8. qPCR Analysis for Verification of Transcriptome Data

A total of 641 DEGs were shared under exposure to Hg in different concentrations (Figure 3D). In order to validate the RAN-seq results, six of these shared DEGs, along with being enriched into the peroxisome pathway (Figure S6), including PECI, CRAT, XDH, DDO, ACOX1, and SCP2, were selected for qPCR analyses. As shown in Figure 7, even though there were some variations in fold changes by different computing as well as measuring methods, expression profiles of the chosen DEGs analyzed by qPCR were consistent with the transcriptome data, indicating that the transcriptome data were reliable and accurate.

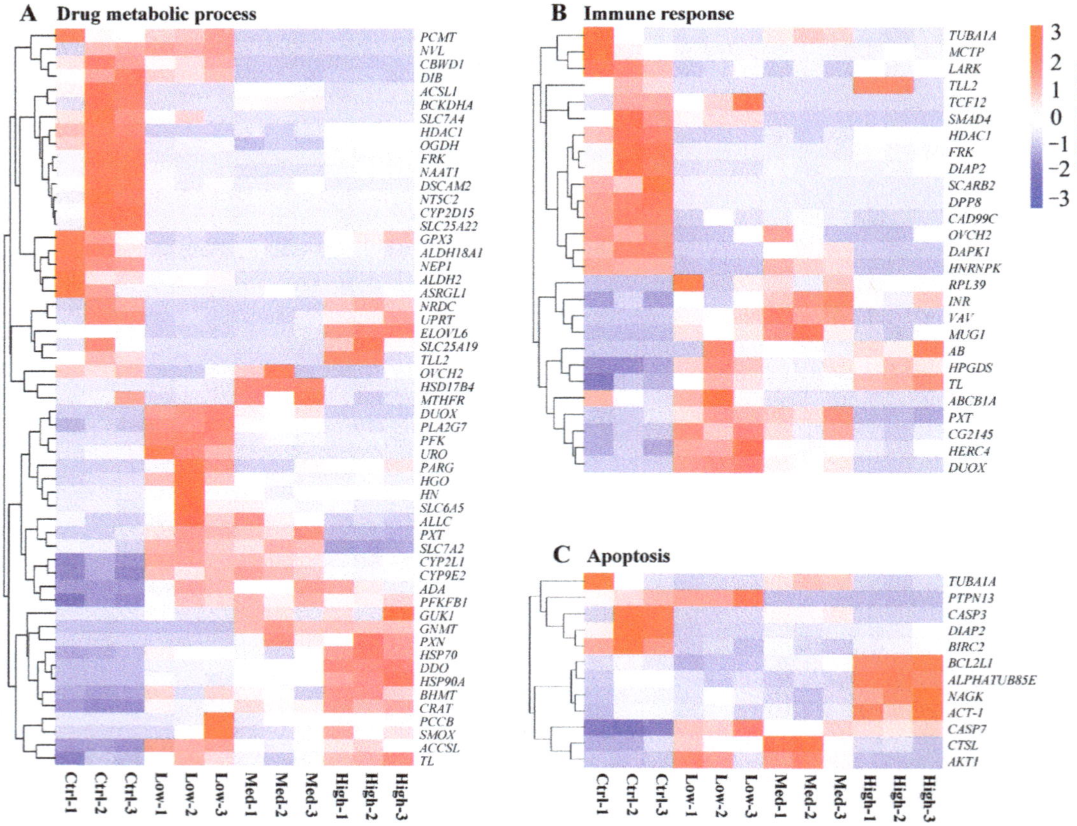

Figure 6. The effects of treatment with different concentrations of Hg on drug metabolism, immune response, and apoptosis-related gene expression. The heat map illustrates variations in drug metabolism (**A**), immune response (**B**), and apoptosis-related gene expression (**C**). Each gene was assessed based on its average FPKM value. The genes with higher expression levels are colored red and those with lower expression levels are colored blue. Ctrl: 0 μg/L Hg^{2+}; Low: 8.75 μg/L Hg^{2+}; Med: 21.875 μg/L Hg^{2+}; High: 43.75 μg/L Hg^{2+}.

3.9. Sequencing

After screening for quality, 1,008,924 valid reads were obtained from all crayfish intestinal samples. These valid sequences ranged from 65,704 to 94,362 (Table S8). As displayed in Figure S7A, there was a fast ascension of the species accumulation curve, which reached saturation as the sample number increased, implying that there were enough samples in this assay to fully reflect microbial richness. Moreover, rarefaction curves (Figure S7B) revealed high sampling coverage (>99%) in every sample with sufficient sequencing depth.

3.10. Alpha-Diversity, Beta-Diversity, and OTU Distribution

Alpha-diversity indices for the intestinal microbiome in the Ctrl and Hg-treated crayfish are shown in Figure 8A,B and Table S8. The Chao index of *P. clarkii* was decreased significantly in the Hg treatment groups, relative to Ctrl group. However, there was no significant difference in the Shannon index of Hg-treated groups compared with Ctrl group. Based on unweighted UniFrac distance PCoA analyses (Figure 8C), samples from Ctrl and three Hg-treated were divided into four groups. It was found that Hg exerted potential effects on the intestinal microbiota structures of crayfish.

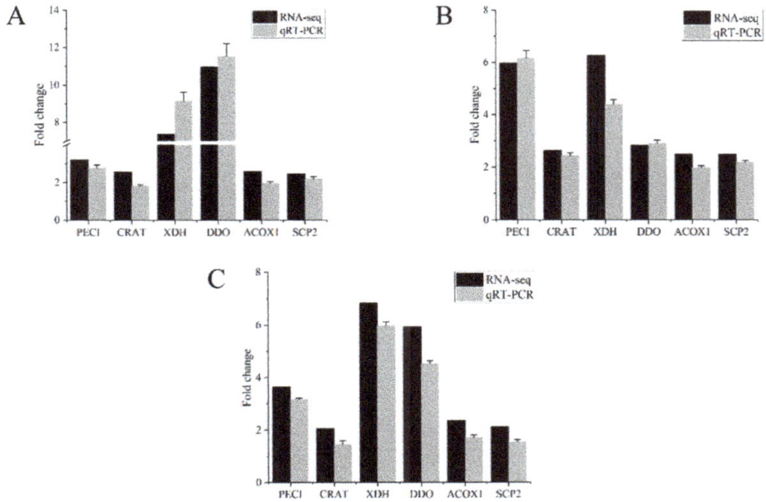

Figure 7. Validation of RNA-seq gene expressions in *P. clarkii* hepatopancreas by qRT-PCR: (**A**) Low concentration of Hg (8.75 μg/L Hg^{2+}). (**B**) Med concentration of Hg (21.875 μg/L Hg^{2+}). (**C**) High concentration of Hg (43.75 μg/L Hg^{2+}).

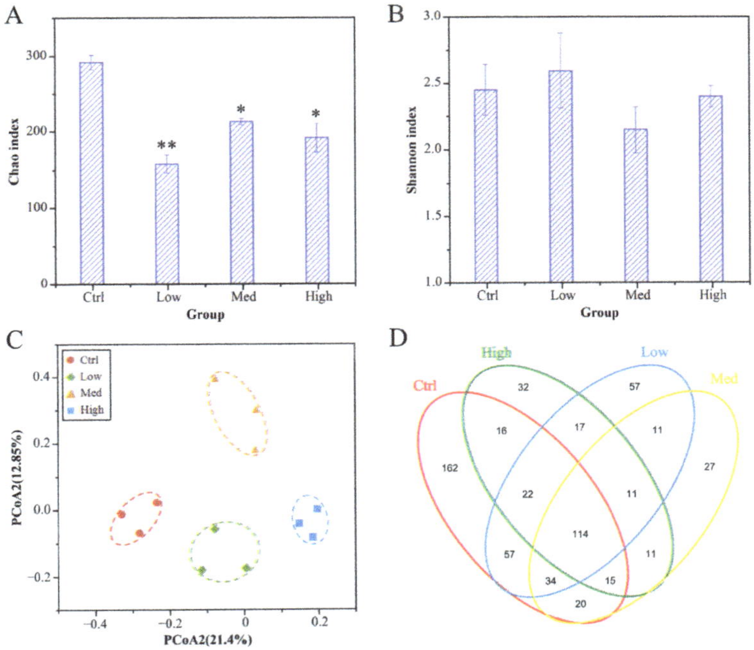

Figure 8. Alpha- and Beta-diversities of intestinal microbiome of *P. clarkii* in the Hg-treated groups: (**A**) Chao indices of bacterial community. (**B**) Shannon indices of bacterial community. (**C**) PCoA of the microbiome. (**D**) Venn diagram showing the number of shared and unshared OTUs among the different Hg-treated groups. * $p \leq 0.05$, ** $p \leq 0.01$. Data are shown as mean ± SD. Ctrl: 0 μg/L Hg^{2+}; Low: 8.75 μg/L Hg^{2+}; Med: 21.875 μg/L Hg^{2+}; High: 43.75 μg/L Hg^{2+}.

The Venn diagram in Figure 8D shows that 114 OTUs were common in all four groups, suggesting that the OTUs could be insensitive to Hg. In addition, 162 OTUs were limited to the control group, implying that they were sensitive to Hg. In comparison, 57, 27, and 32 OTUs were, respectively, exclusive in Low, Med, and High groups. Thus, we hypothesized that Hg altered the gut microbiome of crayfish.

3.11. Intestinal Microbiome Composition

Figure 9A showed that the most abundant bacterial phyla were *Proteobacteria* (43.97–61.83%), *Firmicutes* (12.90–23.40%), *Bacteroidetes* (10.05–19.43%), and *Tenericutes* (1.23–3.84%) in the intestine of *P. clarkii*. At the genus level (Figure 9B), the top four genera were *Shewanella* (16.08–28.60%), *Vibrio* (5.98–19.45%), *Bacteroides* (6.45–13.28%), and *Aeromonas* (2.50–6.00%). Relative to Ctrl group, the high concentration of Hg markedly decreased *Proteobacteria* abundance, whereas it induced a significant increase in *Firmicutes* (Table S9). The relative abundance of the genera exhibited marked variations in at least one Hg-treated group (Table S10). *Arcobacter* showed a significant decrease in abundance, while *Bacteroides* and *Dysgonomonas* showed a significant increase in abundance. Notably, these changes were primarily significant in the High concentration treatment group.

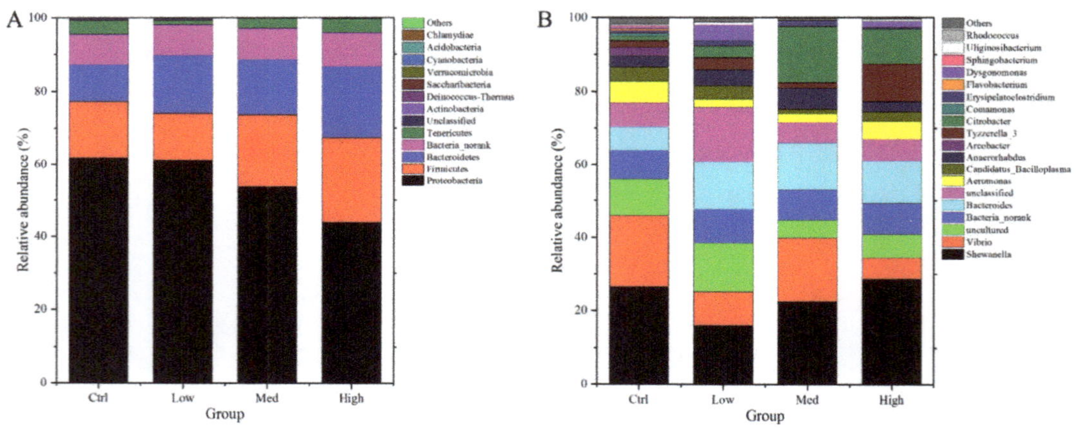

Figure 9. The influences of Hg exposure on intestinal microbiome composition at phyla and genera level: (**A**) Relative abundances of the dominant bacteria at phyla level among the four Hg exposure groups. (**B**) Relative abundances of the dominant bacteria at genera level among the four Hg exposure groups. Ctrl: 0 μg/L Hg^{2+}; Low: 8.75 μg/L Hg^{2+}; Med: 21.875 μg/L Hg^{2+}; High: 43.75 μg/L Hg^{2+}.

4. Discussion

Hg has previously been evaluated for its toxic effects on crustaceans, and most of these studies were focused on the median lethal concentration (LC50). In these articles, the effect of temperature on the LC50 of Hg, the effect of Hg exposure on ovarian maturation, and a health risk assessment on humans of Hg accumulation in *P. clarkii* were researched [56,72,73]; however, Hg toxicity to crayfish has not been fully established. Therefore, we evaluated histological variations, the hepatopancreatic transcriptome, and intestinal microbiota to establish how *P. clarkii* responds to Hg at biochemical, physiological, tissue, molecular, as well as gut flora levels.

4.1. Influence of Hg on Biochemical and Physiological Variations in P. clarkii

In this research, we examined the concentration-dependent uptake of Hg into tissues (gill, antennal gland, hepatopancreas, and abdominal muscle) of *P. clarkii* after aqueous exposure. As a primary organ of detoxification, the hepatopancreas is a major reservoir of xenobiotics and is vulnerable to metal accumulation [74,75]. In our study, with increasing

Hg concentrations, the accumulation level of Hg in hepatopancreas significantly increased ($p < 0.01$) after 96 h of exposure. These results are consistent with the above literature that the hepatopancreas of *P. clarkii* could accumulate Hg effectively.

Apoptosis and the death of cells are linked to excessive ROS [76]. MDA is considered a biomarker of oxidative damage and the main mechanism by which ROS induces tissue injury [77]. In this research, increasing Hg concentrations significantly increased ROS and MDA levels, suggesting that Hg exposure can induce oxidative stress and result in membrane polyunsaturated fatty acid peroxidation. Antioxidants play a significant role in regulating ROS levels in cells and preventing oxidative damage to the organism [78]. Biological systems establish a first-line defense against ROS by controlling SOD and CAT activity [77]. CAT is an antioxidant that can protect cells and organs from H_2O_2-induced damage [79]. In the present study, a decrease in SOD and CAT activity observed is consistent with ROS being generated in response to Hg. The results are in accordance with the literature that exposure to Hg reduced the SOD and CAT activity in *Scylla serrata* [80]. It has been reported that the superoxide anion radical inhibits CAT activity, which may also explain the lesser activity of CAT [81]. As a tripeptide, GSH stimulates strong antioxidant activity by reversing ROS-induced damage [82]. In our study, as a result of Hg exposure for 96 h, GSH levels in hepatopancreas decreased significantly in a concentration-dependent manner, indicating the rapid and sustained responses of GSH in Hg detoxification. Lower concentrations of GSH are consistent with the greater utilization of GSH in reducing H_2O_2 to H_2O. As a detoxifying enzyme, GST suppresses ROS production [83]. Hg exposure results in higher GST activity, which indicates detoxification mechanisms are activated. There has been evidence that Hg causes oxidative stress in several species [80,84]; as a result, it could be hypothesized that GST's greater activity in crayfish exposed to Hg is an adaptation to oxidative stress. A dose-dependent increase in ROS levels was observed after MeHg exposure in *Brachionus koreanus* and *Paracyclopina nana* [85]. The GSH levels in *Brachionus koreanus* were also significantly decreased after exposure to 500 and 1000 ng/L MeHg [86]. The above results of MeHg toxicity on ROS and GSH levers coincide with the influence of inorganic Hg^{2+} on *Procambarus clarkii* in our study.

4.2. Influence of Hg on Histological Variations in P. clarkii

In this study, Hg toxicity was assessed by examining the histological alternations of tissues. Hg was associated with various injuries to the hepatopancreas, including epithelium vacuolization and lumen dilatation. Other studies showed that Hg caused severe alterations in the hepatopancreas of prawn *Macrobrachium malcolmsonii* [87] and tropical *Macrobrachium rosenbergii* [88]. In the intestine, after treatment with med and high concentrations of Hg, there was apparent damage to epithelial cells. There were irregular-shaped, abnormal palisade-arranged epithelial cells, implying that Hg damaged the intestinal microvilli of *P. clarkii*. Intestinal histological damage or alterations were also reported in common carp after exposure to Hg [89]. Thus, intestinal abnormalities imply that intestinal structures of *P. clarkii* were damaged by Hg, which might, in turn, lead to variations in the diversity, richness, and composition of the intestinal microbiome.

4.3. Influence of Hg on the Hepatopancreatic Transcriptome of P. clarkii

Using hepatopancreatic transcriptome analysis, the influences of Hg on redox metabolism, ion transport, drug metabolism, immune response, as well as apoptosis-associated genes and signaling pathways in *P. clarkii* were revealed. The following are the details of the classified discussion.

4.3.1. Influence of Hg on Redox Metabolism in P. clarkii

Redox homeostasis is essential to sustain metabolism and growth [90]. As a result of redox metabolism, excess levels of ROS are removed and cellular redox balance is reestablished [91,92]. The related genes involved in the redox metabolism of *P. clarkii* under exposure to Hg are still little known. In this study, a large number of genes related to oxidoreductase activity (Figure 5A) and the oxidation-reduction process (Figure 5B) were identified to be involved in the molecular regulatory network of *P. clarkii* under the exposure of Hg.

During the peroxisomal beta-oxidation of fatty acids, acyl-CoA oxidase 3 (ACOX3) produces hydrogen peroxide (H_2O_2) [93]. Oxidative stress can be induced by H_2O_2, a highly reactive oxidant [94]. Spermine oxidase (SMOX) can produce reactive oxygen species (ROS) while degrading polyamines [95], further inducing oxidative stress. During xenobiotic chemical metabolism, cytochrome P450 (CYP) produces oxygen free radicals, which can lead to oxidative stress. In the early stages of exposure to chemicals, P450 is normally expressed [96]. In this study, ACOX3, SMOX, CYP2J6, CYP3A4, CYP3A8, CYP4C1, CYP9E2, and CYP49A1 were found to be significantly up-regulated under single or multiple concentrations of Hg exposure, which indicate that Hg exposure increases the production of ROS and H_2O_2. These results provide evidence supporting our findings that Hg exposure significantly increases the production of ROS, further causing oxidative stress in *P. clarkii*.

Molecular antioxidants can prevent oxidative damage to target molecules [97]. As part of the major oxidative pathway that involves alcohol metabolism, aldehyde dehydrogenase (ALDH) plays a significant role [98]. As a result of the activation of ALDH isozymes, reactive aldehydes within cells can be detoxified and oxidative insults from ROS could be prevented [99]. Through the reduction in peroxides, glutathione peroxidase 3 (GPX3) protects cells from oxidative damage [100]. Knockdown of glyoxalase 1 (GLO1) in nondiabetic mice can induce oxidative stress [101]. Additionally, the mammalian GLO1 gene contains an antioxidant-responsive element, which helps GLO1 participate in the major cytoprotective antioxidant system [102]. Through the HIF-1 pathway, Prolyl 4-hydroxylase subunit alpha 1 (P4HA1) can inhibit oxidative phosphorylation and ROS generation [103]. In our study, these antioxidant-related genes include ALDH2, ALDH9A1, ALDH18A1, GPX3, GLO1, as well as P4HA1, which were found to be significantly down-regulated under single or multiple concentrations of Hg exposure. These results indicate that Hg exposure decreases the antioxidant production related to these genes, further perturbing the redox equilibrium of *P. clarkii*.

4.3.2. Influence of Hg on Ion Transport in *P. clarkii*

A number of crucial physiological parameters are controlled by ion transport, including ion balance and membrane potential, which are prerequisites for regulating the vital functions in life processes [104,105]. In this study, abundant genes related to ion transport (Figure S5) were identified to be involved in the molecular regulatory network of *P. clarkii* under the exposure of Hg.

Aspartyl/asparaginyl beta-hydroxylase (ASPH) regulates calcium (Ca) cycling in cardiomyocytes, knockout ASPH in mice exhibits impaired fertility, morphological defects, and abnormal heart function [106,107]. Calcium uptake protein 3 (MICU3) can increase mitochondrial Ca^{2+} uptake dramatically. Multiple C2 and transmembrane domain-containing protein (MCTP) stabilize baseline calcium release by acting downstream of calcium influx [108]. In our study, ASPH was significantly changed, MICU3 was significantly upregulated in Med and High groups, and MCTP was downregulated in all Hg-treated groups; these above results indicated that Hg exposure showed a stronger effect on the calcium cycling, mitochondrial Ca^{2+} uptake, and calcium influx.

In the liver and intestine, solute carrier family 10 member 2 (SLC10A2) encodes a sodium-dependent bile acid transporter [109]. As a result of pyrexia (PYX) activation, potassium will be effluxed from the cell [110]. The STEAP3 metalloreductase (STEAP3) converts copper and iron from trivalent to divalent cationic forms and maintains homeostasis [111]. Tumor Suppressor Candidate 3 (TUSC3) plays a role in embryonic development, protein glycosylation, and cellular magnesium uptake [112]. Zinc transporter 1 is encoded by the soluble carrier family 39 member 1 gene (SLC39A1). Studies have shown that SLC39A1 is overexpressed in prostate cancer, causing depletion of zinc in the glands [113,114]. Zinc transporter ZnT2 is also known as SLC30A2 (Solute carrier family 30 member 2) and is crucial for transporting zinc into the mammary epithelium [115]. The over-expression of SLC30A2 results in Zn vesicularization, reduced proliferation, and enhanced apoptosis [116]. In this study, these prominent changes in genes, including SLC10A2, PYX, STEAP3, TUSC3, SLC39A1, and SLC30A2, indicate that Hg exposure has a negative influence on the ion transport of sodium, potassium, iron, copper, magnesium, as well as zinc, and may well disrupt the ion homeostasis of *P. clarkii*. In addition, the significant up-regulation of SLC30A2 with rising Hg concentrations suggest that Hg exposure may cause apoptosis via Zn accumulation in hepatopancreas cell.

4.3.3. Influence of Hg on Drug Metabolism in *P. clarkii*

In response to Hg toxicity, a large number of drug metabolism-related DEGs were identified (Figure 6A). The heat shock proteins (HSPs), also known as "stress proteins", play an important role in protecting the body from environmental and cellular stress [117,118]. Moreover, HSPs can protect against oxidative stress [119]. In the current study, the up-regulation of HSP70 and HSP90A may protect hepatopancreas cells against the oxidative stress and protein damage induced by Hg exposure. In addition, CYPs are involved in exogenous substance detoxification and are a potential biomarker for evaluating pollutants in aquatic environments [120]. The influence of Hg on CYPs is less studied in crustaceans. In mouse, Hg was shown to up-regulate Cyp1a1, Cyp2b9, Cyp2b10, Cyp2b19, Cyp4a10, Cyp4a12, and Cyp4a14 genes [121]. We found up-regulations of CYP2L1 and CYP9E2 in *P. clarkii* after exposure to Hg, which suggests that they play an important role in the detoxification of Hg.

4.3.4. Influence of Hg on Immune Response in *P. clarkii*

In crustaceans, the hepatopancreas is a critical immune organ responsible for regulating metabolic processes and immune responses [122]. In our study, several genes related to immune response (Figure 6B) were identified as belonging to the hepatopancreas in *P. clarkii* under Hg exposure. Loss or deficiency of SMAD family member 4 (SMAD4) in T cells often causes immune diseases [123]. By binding to the NF-kB co-repressor, histone deacetylase 1 (HDAC1) has a wide range of effects on the immune system in mammals [124]. DUOX (dual oxidase) suppression promotes effective immune responses, leading to lower infection loads [125]. After Hg exposure, the SMAD4 and HDAC1 were down-regulated and DUOX was up-regulated in our study; these results suggest that Hg exposure could weaken the immune response and ability, as well as induce immune diseases.

4.3.5. Influence of Hg on Apoptosis in *P. clarkii*

Apoptosis is important in homeostasis maintenance and in aquatic organisms; it is initiated by several environmental stressors [126]. Hg was shown to initiate apoptosis in various cell types, such as grass carp cell ZC7901 and marine teleost fish SAF-1 cell lines [127,128]. In our study, expressions of various apoptosis-associated genes, such as cathepsin L (CTSL), baculoviral IAP repeat containing 2 (BIRC2), and caspase 7 (CASP7) were markedly altered in the *P. clarkii* hepatopancreas after Hg exposure. As a cysteine protease, CTSL is highly involved in various apoptosis-associated pathways [129]. Elevated CTSL induces apoptosis by the release of cytochrome c from the mitochondria and the activation of mitochondrial apoptosis [130,131]. BIRC2 belongs to the inhibitor of the apoptosis family of antiapoptotic proteins. Apoptosis inhibitors suppress apoptosis by downregulating procaspase activation, as well as by directly suppressing activated caspases [132]. CASP7 has been shown to be an important executioner protein of apoptosis [133]. In our study, the expression of CTSL and CASP7 was overall increased and the expression of BIRC2 was decreased under Hg exposure. These results indicate that apoptosis occurred in *P. clarkii* hepatopancreas after acute exposure to Hg. Furthermore, the obvious tissue damage in the hepatopancreas caused by Hg exposure may be mediated by the activation of apoptosis.

Integrating these above findings, we speculate a potential mechanism of how Hg exposure damages crustaceans' tissues. Our results lead to the following deduction, exposure to inorganic Hg^{2+} induces apoptosis via oxidative stress and the dysregulation of ion homeostasis, further causing tissue damage in crustaceans. On the one hand, exposure of *P. clarkii* to Hg^{2+} significantly increases ROS and MDA levels and inhibits the activity of SOD and CAT, further inducing oxidative stress and apoptosis. On the other hand, exposure to Hg^{2+} downregulates MCTP expression and enhances Ca^{2+} influx, upregulating SLC30A2 and causing Zn vesicularization, further triggering apoptosis. However, this deduction needs to be confirmed by further study.

4.4. Effect of Hg on Intestinal Microbiota in P. clarkii

The intestine is a complex micro-ecosystem, in which inhabits a huge and varied microbial community consisting of archaea, viruses, bacteria, and various unknown eukaryotes. Intestinal histopathology, morphology, and microbiota balance are important indicators of intestinal health [134]. In our study, Hg exposure damaged the intestine resulting in microvilli vacuolization and extended lamina proprias, which might lead to variations in the diversity, richness, and composition of the intestinal microbiota.

The physiological status of the host, especially the immune system, is closely related to the intestinal microbiota [135,136]. In this research, Hg exposure dysregulated the intestinal microbiota in *P. clarkii*. The decreased ACE and Chao indices in Hg-treated groups showed that intestinal microbiota exposed to Hg had a significantly low abundance ($p < 0.05$). Based on PCoA plots, samples were allocated into four parts, implying that Hg dysregulated the intestinal microbiome.

In this study, *Firmicutes*, *Proteobacteria*, and *Bacteroidetes* were established to be the dominant phyla in the intestines of *P. clarkii*. These findings are in tandem with those found in other aquatic crustaceans [137–139]. Thus, we concluded that these three phyla are predominant in the gut of aquatic crustaceans. *Proteobacteria* plays various roles in several biogeochemical processes, such as carbon, sulfur, and nitrogen cycling [140]. A high abundance of *Firmicutes* enhances fatty acid absorption [141]. *Bacteroidetes* are involved in the cycling of protein-rich substances as well as complex carbon [142]. In this study, *Proteobacteria, Firmicutes*, as well as *Bacteroidetes* accounted for over 86.80% of the total phyla in the gut of *P. clarkii*, implying that they are of significance in digestion as well as absorption. Hence, the considerable decrease in *Proteobacteria* and increase in *Firmicutes* suggest that carbon, sulfur, and nitrogen cycling were weakened, while fatty acid absorption was promoted in the gut of *P. clarkii* after exposure to high concentrations of mercury.

Bacteroides secretes various capsular polysaccharides to change its surface antigenicity in the human colon [143]. It can regulate the milieu via interactions with host immune systems to control the proliferation of other bacteria [144]. *Bacteroides* has an abundance of enzymes involved in carbohydrate transport as well as protein metabolism. Moreover, it has vitamins, glycans, and co-factor enzymes, which are important in digestion [145]. Thus, a high intestinal abundance of this genus implied an adaptation of *P. clarkii* to Hg. *Dysgonomonas macrotermitis* is the only *Dysgonomonas* species in invertebrates, and its major role in hind-intestines of *Macrotermes barneyi* is to decompose lignocellulose and provide nutrition to its host [146]. The high *Dysgonomonas* abundance indicates that Hg exposure enhances the digestion of plant raw materials by *P. clarkii*. Meanwhile, there were also some pathogenic bacteria, such as *Arcobacter* in the gut of Hg-exposed *P. clarkii*. Some *Arcobacter* species induce infections in animals and humans, and they are also considered zoonotic and enteropathogenic [147]. Taken together, Hg exposure damages the intestine tissue, reduces the abundance of intestinal microbiota, and dysregulates the intestinal microbiota composition, which suggests that Hg exposure can further affect the overall health of *P. clarkii*.

5. Conclusions

In this study, Hg accumulated in the hepatopancreas in a dose-dependent manner. Oxidative stress and antioxidant parameters, including ROS, MDA, SOD, CAT, GSH, and GST, showed significant variations after being challenged with Hg in the hepatopancreas. Hence, these parameters might serve as effective, rapid, and sensitive biomonitoring parameters of mercury. Additionally, Hg exposure caused tissue damage to the hepatopancreas and intestine, and the degree of tissue damage was shown to be dose-dependent. Using transcriptomes analysis, we identified a lot of redox metabolism, ion transport, drug metabolism, immune response, as well as apoptosis-associated DEGs under Hg exposure. Based on these genetic alterations, we clarified the potential mechanism involved in the enhanced production of ROS in Hg exposure and speculated that Hg exposure may perturb the redox equilibrium, disrupt the ion homeostasis, weaken immune response and ability, and cause apoptosis. Meanwhile, Hg exposure decreased microbiome richness, and markedly altered the microbial structure in the intestines of crayfish at the phylum as well as the genus level. As a result of this study, we gained valuable insight into the toxic mechanisms of heavy metals in crayfish. Moreover, the insights gained from this study will be useful for future research and the assessment of potential antioxidant and immunomodulatory effects in polluted environments and crustacean farming.

Supplementary Materials: The following supporting information can be downloaded at: https://www.mdpi.com/article/10.3390/antiox11101944/s1, Figure S1: Quantified results of the hepatopancreas tubule lumen dilatation (A) and intestine microvilli vacuolization (B) proportion in the Ctrl, Low, Med, and High groups (mean ± SD, n = 9). Figure S2: Length distribution of *P. clarkii* hepatopancreas unigenes. Figure S3: Classification of assembled unigenes. GO function classification (A–C), KOG classification histogram presentation (D), and KEGG classification (E) of assembled unigenes. Figure S4: Principal Component Analysis (PCA) of the gene expression of samples. Figure S5: The heat map illustrates variations in ion transport-related gene expression. Figure S6: Peroxisome pathway of *P. clarkii* induced by high concentration of Hg. Figure S7: (A) Species accumulation curves of *P. clarkii* intestinal microbial samples. (B) Rarefaction curves of all the samples based on Illumina MiSeq sequencing. Table S1: Primers used in the quantitative PCR analysis. Table S2: Measured and nominal Hg concentrations in water during the exposure experiment. Table S3: Bioaccumulation of Hg in tissues during exposure to different concentrations of Hg. Table S4: Effect of different concentrations of Hg on oxidative stress and antioxidant parameters in hepatopancreas of *P. clarkii*. Table S5: Statistics of hepatopancreas transcriptome sequencing. Table S6: Summary of the annotations. Table S7: GO enrichment analysis of DEGs. Table S8: Richness and diversity indices of bacterial communities for all intestinal samples. Table S9: The microbial composition (mean ± SE) of *P. clarkii* after exposure to different concentrations of Hg at the phylum level. Table S10: The microbial composition (mean ± SE) of *P. clarkii* after exposure to different concentrations of Hg at the genus level.

Author Contributions: Conceptualization, L.Z.; methodology, L.Z., Y.Z. and T.L.; software, L.Z., Y.Z., Z.S., H.L. and S.Z.; formal analysis, L.Z.; investigation, Z.S., Y.Y., H.S. and L.H.; validation, T.L.; resources, J.G.; data curation, Y.Z. and S.Z.; writing—original draft preparation, L.Z.; writing—review and editing, J.G.; supervision, J.G.; project administration, J.G.; funding acquisition, L.Z. and J.G. All authors have read and agreed to the published version of the manuscript.

Funding: This work was supported by Central Public-Interest Scientific Institution Basal Research Fund, CAFS (No. 2021JBF02, No. YFI202213, No. 2022XT01, and No. 2020TD74), and China Agriculture Research System of MOF and MARA (No. CARS-48).

Institutional Review Board Statement: The Animal Care and Use Committee of Yangtze River Fisheries Research Institute permitted this study (Permit No. 2020zhanglang003).

Informed Consent Statement: Not applicable.

Data Availability Statement: Data are contained within the article and Supplementary Materials.

Conflicts of Interest: The authors declare no conflict of interest. The funders had no role in the design of the study; in the collection, analyses, or interpretation of data; in the writing of the manuscript or in the decision to publish the results.

References

1. Joseph, L.; Jun, B.-M.; Flora, J.R.; Park, C.M.; Yoon, Y. Removal of heavy metals from water sources in the developing world using low-cost materials: A review. *Chemosphere* **2019**, *229*, 142–159. [CrossRef] [PubMed]
2. Goretti, E.; Pallottini, M.; Ricciarini, M.; Selvaggi, R.; Cappelletti, D. Heavy metals bioaccumulation in selected tissues of red swamp crayfish: An easy tool for monitoring environmental contamination levels. *Sci. Total Environ.* **2016**, *559*, 339–346. [CrossRef] [PubMed]
3. Zhang, Y.; Ding, Z.; Zhao, G.; Zhang, T.; Xu, Q.; Cui, B.; Liu, J.-X. Transcriptional responses and mechanisms of copper nanoparticle toxicology on zebrafish embryos. *J. Hazard. Mater.* **2018**, *344*, 1057–1068. [CrossRef] [PubMed]
4. Costa, S.; Viegas, I.; Pereira, E.; Duarte, A.C.; Palmeira, C.M.; Pardal, M.A. Differential sex, morphotype and tissue accumulation of mercury in the crab Carcinus maenas. *Water Air Soil Pollut.* **2011**, *222*, 65–75. [CrossRef]
5. Ares, A.; Aboal, J.; Carballeira, A.; Giordano, S.; Adamo, P.; Fernández, J. Moss bag biomonitoring: A methodological review. *Sci. Total Environ.* **2012**, *432*, 143–158. [CrossRef]
6. Lyons, K.; Carlisle, A.B.; Lowe, C.G. Influence of ontogeny and environmental exposure on mercury accumulation in muscle and liver of male round stingrays. *Mar. Environ. Res.* **2017**, *130*, 30–37. [CrossRef]
7. Morel, F.M.; Kraepiel, A.M.; Amyot, M. The chemical cycle and bioaccumulation of mercury. *Annu. Rev. Ecol. Syst.* **1998**, *29*, 543–566. [CrossRef]
8. Watras, C.; Back, R.; Halvorsen, S.; Hudson, R.; Morrison, K.; Wente, S. Bioaccumulation of mercury in pelagic freshwater food webs. *Sci. Total Environ.* **1998**, *219*, 183–208. [CrossRef]
9. La Colla, N.S.; Botté, S.E.; Marcovecchio, J.E. Mercury cycling and bioaccumulation in a changing coastal system: From water to aquatic organisms. *Mar. Pollut. Bull.* **2019**, *140*, 40–50. [CrossRef]
10. Benoit, J.; Gilmour, C.C.; Mason, R.; Riedel, G.S.; Riedel, G.F. Behavior of mercury in the Patuxent River estuary. *Biogeochemistry* **1998**, *40*, 249–265. [CrossRef]
11. Kolka, R.K.; Riggs, C.E.; Nater, E.A.; Wickman, T.R.; Witt, E.L.; Butcher, J.T. Temporal fluctuations in young-of-the-year yellow perch mercury bioaccumulation in lakes of northeastern Minnesota. *Sci. Total Environ.* **2019**, *656*, 475–481. [CrossRef] [PubMed]
12. Wang, Y.; Xie, Q.; Xu, Q.; Xue, J.; Zhang, C.; Wang, D. Mercury bioaccumulation in fish in an artificial lake used to carry out cage culture. *J. Environ. Sci.* **2019**, *78*, 352–359. [CrossRef] [PubMed]
13. Chen, C.-F.; Ju, Y.-R.; Lim, Y.C.; Chen, C.-W.; Wu, C.-H.; Lin, Y.-L.; Dong, C.-D. Dry and wet seasonal variation of total mercury, inorganic mercury, and methylmercury formation in estuary and harbor sediments. *J. Environ. Manag.* **2020**, *253*, 109683. [CrossRef] [PubMed]
14. Lawrence, A.L.; McAloon, K.M.; Mason, R.P.; Mayer, L.M. Intestinal solubilization of particle-associated organic and inorganic mercury as a measure of bioavailability to benthic invertebrates. *Environ. Sci. Technol.* **1999**, *33*, 1871–1876. [CrossRef]
15. Williams, J.J.; Dutton, J.; Chen, C.Y.; Fisher, N.S. Metal (As, Cd, Hg, and CH3Hg) bioaccumulation from water and food by the benthic amphipod Leptocheirus plumulosus. *Environ. Toxicol. Chem.* **2010**, *29*, 1755–1761. [CrossRef]
16. Ung, C.Y.; Lam, S.H.; Hlaing, M.M.; Winata, C.L.; Korzh, S.; Mathavan, S.; Gong, Z. Mercury-induced hepatotoxicity in zebrafish: In vivo mechanistic insights from transcriptome analysis, phenotype anchoring and targeted gene expression validation. *BMC Genom.* **2010**, *11*, 1–14. [CrossRef]
17. Ceccatelli, S.; Daré, E.; Moors, M. Methylmercury-induced neurotoxicity and apoptosis. *Chem.-Biol. Interact.* **2010**, *188*, 301–308. [CrossRef]
18. Tan, S.W.; Meiller, J.C.; Mahaffey, K.R. The endocrine effects of mercury in humans and wildlife. *Crit. Rev. Toxicol.* **2009**, *39*, 228–269. [CrossRef]
19. Birch, R.J.; Bigler, J.; Rogers, J.W.; Zhuang, Y.; Clickner, R.P. Trends in blood mercury concentrations and fish consumption among US women of reproductive age, NHANES, 1999–2010. *Environ. Res.* **2014**, *133*, 431–438. [CrossRef]
20. Zou, D.; Gao, S. Acute toxicity of Cu, Zn, Cd, Hg, Mn and Cr to postlarvae of Penaeus monodon. *Mar. Environ. Sci* **1994**, *13*, 13–18.
21. Gao, S. Acute toxicity of Hg, Cd, Zn and Mn to postlarvae of Penaeus japonicus Bate. *Mar. Sci. Bull.-Tianjin* **1999**, *18*, 93–96.
22. Li, J.; Yang, X.; Huang, R.; Wei, S. Acute toxicity test of five heavy metal ions to Neomysis awatschensis. *Mar. Environ. Sci.* **2006**, *25*, 51–53.
23. Jiang, H.; Liu, A.; Song, X.; Sun, G.; Liu, L. The toxic effects of heavy metals on the embryonic development of Penaeus japonicus. *Asian J. Ecotoxicol.* **2013**, *8*, 972–980. [CrossRef]
24. Monteclaro, H.M.; Babaran, R.P.; Sanares, R.C.; Quinitio, E.T. Physiological and avoidance responses of juvenile mud crab Scylla serrata to mercury. *Aquac. Aquar. Conserv. Legis.* **2014**, *7*, 441–448.
25. Sabatini, S.E.; Chaufan, G.; Juárez, A.B.; Coalova, I.; Bianchi, L.; Eppis, M.R.; Ríos de Molina Mdel, C. Dietary copper effects in the estuarine crab, Neohelice (Chasmagnathus) granulata, maintained at two different salinities. *Comp. Biochem. Physiol. Toxicol. Pharmacol. CBP* **2009**, *150*, 521–527. [CrossRef]
26. Livingstone, D.R. Contaminant-stimulated reactive oxygen species production and oxidative damage in aquatic organisms. *Mar. Pollut. Bull* **2001**, *42*, 656–666. [CrossRef]

27. Xu, Z.; Regenstein, J.M.; Xie, D.; Lu, W.; Ren, X.; Yuan, J.; Mao, L. The oxidative stress and antioxidant responses of Litopenaeus vannamei to low temperature and air exposure. *Fish Shellfish Immunol.* **2018**, *72*, 564–571. [CrossRef]
28. Frías-Espericueta, M.G.; Bautista-Covarrubias, J.C.; Osuna-Martínez, C.C.; Delgado-Alvarez, C.; Bojórquez, C.; Aguilar-Juárez, M.; Roos-Muñoz, S.; Osuna-López, I.; Páez-Osuna, F. Metals and oxidative stress in aquatic decapod crustaceans: A review with special reference to shrimp and crabs. *Aquat. Toxicol.* **2022**, *242*, 106024. [CrossRef]
29. Kristoff, G.; Verrengia Guerrero, N.R.; Cochón, A.C. Effects of azinphos-methyl exposure on enzymatic and non-enzymatic antioxidant defenses in Biomphalaria glabrata and Lumbriculus variegatus. *Chemosphere* **2008**, *72*, 1333–1339. [CrossRef]
30. Quintaneiro, C.; Ranville, J.; Nogueira, A.J.A. Effects of the essential metals copper and zinc in two freshwater detritivores species: Biochemical approach. *Ecotoxicol. Environ. Saf.* **2015**, *118*, 37–46. [CrossRef]
31. Capparelli, M.V.; Bordon, I.C.; Araujo, G.; Gusso-Choueri, P.K.; de Souza Abessa, D.M.; McNamara, J.C. Combined effects of temperature and copper on oxygen consumption and antioxidant responses in the mudflat fiddler crab Minuca rapax (Brachyura, Ocypodidae). *Comp. Biochem. Physiol. Toxicol. Pharmacol. CBP* **2019**, *223*, 35–41. [CrossRef] [PubMed]
32. Parrilla-Taylor, D.P.; Zenteno-Savín, T.; Magallón-Barajas, F.J. Antioxidant enzyme activity in pacific whiteleg shrimp (Litopenaeus vannamei) in response to infection with white spot syndrome virus. *Aquaculture* **2013**, *380*, 41–46. [CrossRef]
33. Rőszer, T. The invertebrate midintestinal gland ("hepatopancreas") is an evolutionary forerunner in the integration of immunity and metabolism. *Cell Tissue Res.* **2014**, *358*, 685–695. [CrossRef] [PubMed]
34. de Melo, M.S.; Dos Santos, T.P.G.; Jaramillo, M.; Nezzi, L.; Muller, Y.M.R.; Nazari, E.M. Histopathological and ultrastructural indices for the assessment of glyphosate-based herbicide cytotoxicity in decapod crustacean hepatopancreas. *Aquat. Toxicol.* **2019**, *210*, 207–214. [CrossRef] [PubMed]
35. Clark, K.F.; Acorn, A.R.; Greenwood, S.J. A transcriptomic analysis of American lobster (Homarus americanus) immune response during infection with the bumper car parasite Anophryoides haemophila. *Dev. Comp. Immunol.* **2013**, *40*, 112–122. [CrossRef]
36. Xie, L.; Hanyu, T.; Futatsugi, N.; Komatsu, M.; Steinman, A.D.; Park, H.-D. Inhibitory effect of naringin on microcystin-LR uptake in the freshwater snail Sinotaia histrica. *Environ. Toxicol. Pharmacol.* **2014**, *38*, 430–437. [CrossRef] [PubMed]
37. Sabatini, S.E.; Brena, B.M.; Pirez, M.; de Molina, M.d.C.R.; Luquet, C.M. Oxidative effects and toxin bioaccumulation after dietary microcystin intoxication in the hepatopancreas of the crab Neohelice (Chasmagnathus) granulata. *Ecotoxicol. Environ. Saf.* **2015**, *120*, 136–141. [CrossRef]
38. Zhang, Y.; Li, Z.; Kholodkevich, S.; Sharov, A.; Feng, Y.; Ren, N.; Sun, K. Cadmium-induced oxidative stress, histopathology, and transcriptome changes in the hepatopancreas of freshwater crayfish (Procambarus clarkii). *Sci. Total Environ.* **2019**, *666*, 944–955. [CrossRef]
39. Tang, D.; Shi, X.; Guo, H.; Bai, Y.; Shen, C.; Zhang, Y.; Wang, Z. Comparative transcriptome analysis of the gills of Procambarus clarkii provides novel insights into the immune-related mechanism of copper stress tolerance. *Fish Shellfish Immunol.* **2020**, *96*, 32–40. [CrossRef]
40. Sun, M.; Ting Li, Y.; Liu, Y.; Chin Lee, S.; Wang, L. Transcriptome assembly and expression profiling of molecular responses to cadmium toxicity in hepatopancreas of the freshwater crab Sinopotamon henanense. *Sci. Rep.* **2016**, *6*, 19405. [CrossRef]
41. Jin, Y.; Wu, S.; Zeng, Z.; Fu, Z. Effects of environmental pollutants on gut microbiota. *Environ. Pollut.* **2017**, *222*, 1–9. [CrossRef] [PubMed]
42. Tran, N.T.; Zhang, J.; Xiong, F.; Wang, G.-T.; Li, W.-X.; Wu, S.-G. Altered gut microbiota associated with intestinal disease in grass carp (Ctenopharyngodon idellus). *World J. Microbiol. Biotechnol.* **2018**, *34*, 1–9. [CrossRef] [PubMed]
43. Xiong, J.; Wang, K.; Wu, J.; Qiuqian, L.; Yang, K.; Qian, Y.; Zhang, D. Changes in intestinal bacterial communities are closely associated with shrimp disease severity. *Appl. Microbiol. Biotechnol.* **2015**, *99*, 6911–6919. [CrossRef] [PubMed]
44. Xiong, J.; Zhu, J.; Dai, W.; Dong, C.; Qiu, Q.; Li, C. Integrating gut microbiota immaturity and disease-discriminatory taxa to diagnose the initiation and severity of shrimp disease. *Environ. Microbiol.* **2017**, *19*, 1490–1501. [CrossRef]
45. Chai, P.-C.; Song, X.-L.; Chen, G.-F.; Xu, H.; Huang, J. Dietary supplementation of probiotic Bacillus PC465 isolated from the gut of Fenneropenaeus chinensis improves the health status and resistance of Litopenaeus vannamei against white spot syndrome virus. *Fish Shellfish Immunol.* **2016**, *54*, 602–611. [CrossRef]
46. Ooi, M.C.; Goulden, E.F.; Smith, G.G.; Nowak, B.F.; Bridle, A.R. Developmental and gut-related changes to microbiomes of the cultured juvenile spiny lobster Panulirus ornatus. *FEMS Microbiol. Ecol.* **2017**, *93*, fix159. [CrossRef]
47. Zhang, Y.; Li, Z.; Kholodkevich, S.; Sharov, A.; Chen, C.; Feng, Y.; Ren, N.; Sun, K. Effects of cadmium on intestinal histology and microbiota in freshwater crayfish (Procambarus clarkii). *Chemosphere* **2020**, *242*, 125105. [CrossRef]
48. Zhang, Y.; Li, Z.; Kholodkevich, S.; Sharov, A.; Feng, Y.; Ren, N.; Sun, K. Microcystin-LR-induced changes of hepatopancreatic transcriptome, intestinal microbiota, and histopathology of freshwater crayfish (Procambarus clarkii). *Sci. Total Environ.* **2020**, *711*, 134549. [CrossRef]
49. Qian, D.; Xu, C.; Chen, C.; Qin, J.G.; Chen, L.; Li, E. Toxic effect of chronic waterborne copper exposure on growth, immunity, anti-oxidative capacity and gut microbiota of Pacific white shrimp Litopenaeus vannamei. *Fish Shellfish Immunol.* **2020**, *100*, 445–455. [CrossRef]
50. Peng, Q.; Nunes, L.M.; Greenfield, B.K.; Dang, F.; Zhong, H. Are Chinese consumers at risk due to exposure to metals in crayfish? A bioaccessibility-adjusted probabilistic risk assessment. *Environ. Int.* **2016**, *88*, 261–268. [CrossRef]

51. Zhang, L.; Song, Z.; Zhong, S.; Gan, J.; Liang, H.; Yu, Y.; Wu, G.; He, L. Acute hypoxia and reoxygenation induces oxidative stress, glycometabolism, and oxygen transport change in red swamp crayfish (Procambarus clarkii): Application of transcriptome profiling in assessment of hypoxia. *Aquac. Rep.* **2022**, *23*, 101029. [CrossRef]
52. Vioque-Fernandez, A.; de Almeida, E.A.; López-Barea, J. Assessment of Doñana National Park contamination in Procambarus clarkii: Integration of conventional biomarkers and proteomic approaches. *Sci. Total Environ.* **2009**, *407*, 1784–1797. [CrossRef] [PubMed]
53. Osuna-Jiménez, I.; Abril, N.; Vioque-Fernández, A.; Gómez-Ariza, J.L.; Prieto-Álamo, M.-J.; Pueyo, C. The environmental quality of Doñana surrounding areas affects the immune transcriptional profile of inhabitant crayfish Procambarus clarkii. *Fish Shellfish Immunol.* **2014**, *40*, 136–145. [CrossRef] [PubMed]
54. Marçal, R.; Pacheco, M.; Guilherme, S. DNA of crayfish spermatozoa as a target of waterborne pesticides–An ex vivo approach as a tool to short-term spermiotoxicity screening. *J. Hazard. Mater.* **2020**, *400*, 123300. [CrossRef] [PubMed]
55. Shi, L.; Han, S.; Fei, J.; Zhang, L.; Ray, J.W.; Wang, W.; Li, Y. Molecular Characterization and Functional Study of Insulin-Like Androgenic Gland Hormone Gene in the Red Swamp Crayfish, Procambarus clarkii. *Genes* **2019**, *10*, 645. [CrossRef] [PubMed]
56. Del Ramo, J.; Diaz-Mayans, J.; Torreblanca, A.; Nunez, A. Effects of temperature on the acute toxicity of heavy metals (Cr, Cd, and Hg) to the freshwater crayfish, Procambarus clarkii (Girard). *Bull. Environ. Contam. Toxicol.* **1987**, *38*, 736–741. [CrossRef] [PubMed]
57. Mo, A.; Huang, Y.; Gu, Z.; Liu, C.; Wang, J.; Yuan, Y. Health risk assessment and bioaccumulation of heavy metals in Procambarus clarkii from six provinces of China. *Environ. Sci. Pollut. Res.* **2022**, *29*, 2539–2546. [CrossRef] [PubMed]
58. Keston, A.S.; Brandt, R. The fluorometric analysis of ultramicro quantities of hydrogen peroxide. *Anal. Biochem.* **1965**, *11*, 1–5. [CrossRef]
59. Marklund, S.; Marklund, G. Involvement of the superoxide anion radical in the autoxidation of pyrogallol and a convenient assay for superoxide dismutase. *Eur. J. Biochem.* **1974**, *47*, 469–474. [CrossRef]
60. Kei, S. Serum lipid peroxide in cerebrovascular disorders determined by a new colorimetric method. *Clin. Chim. Acta* **1978**, *90*, 37–43. [CrossRef]
61. Sinha, A.K. Colorimetric assay of catalase. *Anal. Biochem.* **1972**, *47*, 389–394. [CrossRef]
62. Habig, W.H.; Pabst, M.J.; Jakoby, W.B. Glutathione S-Transferases: The first enzymatic step in mercapturic acid formation. *J. Biol. Chem.* **1974**, *249*, 7130–7139. [CrossRef]
63. Moron, M.S.; Depierre, J.W.; Mannervik, B. Levels of glutathione, glutathione reductase and glutathione S-transferase activities in rat lung and liver. *Biochim. et Biophys. Acta (BBA)-Gen. Subj.* **1979**, *582*, 67–78. [CrossRef]
64. Grabherr, M.G.; Haas, B.J.; Yassour, M.; Levin, J.Z.; Thompson, D.A.; Amit, I.; Adiconis, X.; Fan, L.; Raychowdhury, R.; Zeng, Q. Full-length transcriptome assembly from RNA-Seq data without a reference genome. *Nat. Biotechnol.* **2011**, *29*, 644–652. [CrossRef] [PubMed]
65. Anders, S.; Huber, W. Differential expression analysis for sequence count data. *Nat. Preced.* **2010**, *11*, r1062010. [CrossRef]
66. Benjamini, Y.; Yekutieli, D. The control of the false discovery rate in multiple testing under dependency. *Ann. Stat.* **2001**, *29*, 1165–1188. [CrossRef]
67. Zhang, L.; Li, X.; Yu, Y.; Zhang, L.; Dong, L.; Gan, J.; Mao, T.; Liu, T.; Peng, J.; He, L. Comparative analyses of liver transcriptomes reveal the effect of exercise on growth-, glucose metabolism-, and oxygen transport-related genes and signaling pathways in grass carp (Ctenopharyngodon idella). *Comp. Biochem. Physiol. Part A Mol. Integr. Physiol.* **2021**, *262*, 111081. [CrossRef]
68. Zhang, L.; Yu, Y.; Dong, L.; Gan, J.; Mao, T.; Liu, T.; Li, X.; He, L. Effects of moderate exercise on hepatic amino acid and fatty acid composition, liver transcriptome, and intestinal microbiota in channel catfish (Ictalurus punctatus). *Comp. Biochem. Physiol. Part D Genom. Proteom.* **2021**, *40*, 100921. [CrossRef]
69. Livak, K.J.; Schmittgen, T.D. Analysis of relative gene expression data using real-time quantitative PCR and the $2^{-\Delta\Delta CT}$ method. *Methods* **2001**, *25*, 402–408. [CrossRef]
70. Schloss, P.D.; Gevers, D.; Westcott, S.L. Reducing the effects of PCR amplification and sequencing artifacts on 16S rRNA-based studies. *PLoS ONE* **2011**, *6*, e273102011. [CrossRef]
71. Fouts, D.E.; Szpakowski, S.; Purushe, J.; Torralba, M.; Waterman, R.C.; MacNeil, M.D.; Alexander, L.J.; Nelson, K.E. Next generation sequencing to define prokaryotic and fungal diversity in the bovine rumen. *PLoS ONE* **2012**, *7*, e48289. [CrossRef] [PubMed]
72. Reddy, P.S.; Tuberty, S.R.; Fingerman, M. Effects of cadmium and mercury on ovarian maturation in the red swamp crayfish, Procambarus clarkii. *Ecotoxicol. Environ. Saf.* **1997**, *37*, 62–65. [CrossRef] [PubMed]
73. Anandkumar, A.; Li, J.; Prabakaran, K.; Jia, Z.X.; Leng, Z.; Nagarajan, R.; Du, D. Accumulation of toxic elements in an invasive crayfish species (Procambarus clarkii) and its health risk assessment to humans. *J. Food Compos. Anal.* **2020**, *88*, 103449. [CrossRef]
74. Alcorlo, P.; Otero, M.; Crehuet, M.; Baltanás, A.; Montes, C. The use of the red swamp crayfish (Procambarus clarkii, Girard) as indicator of the bioavailability of heavy metals in environmental monitoring in the River Guadiamar (SW, Spain). *Sci. Total Environ.* **2006**, *366*, 380–390. [CrossRef]
75. Brittle, S.W.; Paluri, S.L.; Foose, D.P.; Ruis, M.T.; Amato, M.T.; Lam, N.H.; Buttigieg, B.; Gagnon, Z.E.; Sizemore, I.E. Freshwater Crayfish: A Potential Benthic-Zone Indicator of Nanosilver and Ionic Silver Pollution. *Environ. Sci. Technol.* **2016**, *50*, 7056–7065. [CrossRef] [PubMed]

76. Sakkas, D.; Mariethoz, E.; Manicardi, G.; Bizzaro, D.; Bianchi, P.G.; Bianchi, U. Origin of DNA damage in ejaculated human spermatozoa. *Rev. Reprod.* **1999**, *4*, 31–37. [CrossRef] [PubMed]
77. Wei, K.; Yang, J. Oxidative damage of hepatopancreas induced by pollution depresses humoral immunity response in the freshwater crayfish Procambarus clarkii. *Fish Shellfish Immunol.* **2015**, *43*, 510–519. [CrossRef]
78. Chapman, P.M.; Wang, F.; Caeiro, S.S. Assessing and managing sediment contamination in transitional waters. *Environ. Int.* **2013**, *55*, 71–91. [CrossRef]
79. Dong, H.; Lu, G.; Yan, Z.; Liu, J.; Nkoom, M.; Yang, H. Responses of antioxidant and biotransformation enzymes in Carassius carassius exposed to hexabromocyclododecane. *Environ. Toxicol Pharm.* **2018**, *62*, 46–53. [CrossRef]
80. Singaram, G.; Harikrishnan, T.; Chen, F.-Y.; Bo, J.; Giesy, J.P. Modulation of immune-associated parameters and antioxidant responses in the crab (Scylla serrata) exposed to mercury. *Chemosphere* **2013**, *90*, 917–928. [CrossRef]
81. Kono, Y.; Fridovich, I. Superoxide radical inhibits catalase. *J. Biol. Chem.* **1982**, *257*, 5751–5754. [CrossRef]
82. Yu, K.; Hai, X.; Yue, S.; Song, W.; Bi, S. Glutathione-activated DNA-Au nanomachine as targeted drug delivery platform for imaging-guided combinational cancer therapy. *Chem. Eng. J.* **2021**, *419*, 129535. [CrossRef]
83. Cavin, C.; Marin-Kuan, M.; Langouët, S.; Bezençon, C.; Guignard, G.; Verguet, C.; Piguet, D.; Holzhäuser, D.; Cornaz, R.; Schilter, B. Induction of Nrf2-mediated cellular defenses and alteration of phase I activities as mechanisms of chemoprotective effects of coffee in the liver. *Food Chem. Toxicol. Int. J. Publ. Br. Ind. Biol. Res. Assoc.* **2008**, *46*, 1239–1248. [CrossRef] [PubMed]
84. Zaman, K.; MacGill, R.S.; Johnson, J.E.; Ahmad, S.; Pardini, R.S. An insect model for assessing mercury toxicity: Effect of mercury on antioxidant enzyme activities of the housefly (Musca domestica) and the cabbage looper moth (Trichoplusia ni). *Arch Environ. Contam Toxicol* **1994**, *26*, 114–118. [CrossRef]
85. Lee, Y.H.; Kim, D.-H.; Kang, H.-M.; Wang, M.; Jeong, C.-B.; Lee, J.-S. Adverse effects of methylmercury (MeHg) on life parameters, antioxidant systems, and MAPK signaling pathways in the rotifer Brachionus koreanus and the copepod Paracyclopina nana. *Aquat. Toxicol.* **2017**, *190*, 181–189. [CrossRef] [PubMed]
86. Ballatori, N.; Clarkson, T.W. Developmental changes in the biliary excretion of methylmercury and glutathione. *Science* **1982**, *216*, 61–63. [CrossRef]
87. Yamuna, A.; Bhavan, P.S.; Geraldine, P. Ultrastructural observations in gills and hepatopancreas of prawn Macrobrachium malcolmsonii exposed to mercury. *J. Environ. Biol.* **2009**, *30*, 693. [CrossRef]
88. Kaoud, H.A.; Zaki, M.M.; Ismail, M.M. Effect of exposure to mercury on health in tropical Macrobrachium rosenbergii. *Life Sci. J.* **2011**, *8*, 154–163.
89. Shang, X.; Yu, P.; Yin, Y.; Zhang, Y.; Lu, Y.; Mao, Q.; Li, Y. Effect of selenium-rich Bacillus subtilis against mercury-induced intestinal damage repair and oxidative stress in common carp. *Comp. Biochem. Physiol. Part C Toxicol. Pharmacol.* **2021**, *239*, 108851. [CrossRef]
90. Cao, A.; Zheng, Y.; Yu, Y.; Wang, X.; Shao, D.; Sun, J.; Cui, B. Comparative Transcriptome Analysis of SE initial dedifferentiation in cotton of different SE capability. *Sci. Rep.* **2017**, *7*, 8583. [CrossRef]
91. Kocsy, G.; Laurie, R.; Szalai, G.; Szilágyi, V.; Simon-Sarkadi, L.; Galiba, G.; De Ronde, J.A. Genetic manipulation of proline levels affects antioxidants in soybean subjected to simultaneous drought and heat stresses. *Physiol. Plant.* **2005**, *124*, 227–235. [CrossRef]
92. Valliyodan, B.; Nguyen, H.T. Understanding regulatory networks and engineering for enhanced drought tolerance in plants. *Curr. Opin. Plant Biol.* **2006**, *9*, 189–195. [CrossRef] [PubMed]
93. Ghorbel, M.T.; Patel, N.N.; Sheikh, M.; Angelini, G.D.; Caputo, M.; Murphy, G.J. Changes in renal medulla gene expression in a pre-clinical model of post cardiopulmonary bypass acute kidney injury. *BMC Genom.* **2014**, *15*, 916. [CrossRef] [PubMed]
94. Park, J.H.; Choi, J.W.; Ju, E.J.; Pae, A.N.; Park, K.D. Antioxidant and Anti-Inflammatory Activities of a Natural Compound, Shizukahenriol, through Nrf2 Activation. *Molecules* **2015**, *20*, 15989–16003. [CrossRef]
95. Bae, D.H.; Lane, D.J.R.; Jansson, P.J.; Richardson, D.R. The old and new biochemistry of polyamines. *Biochim. Et Biophys. Acta. Gen. Subj.* **2018**, *1862*, 2053–2068. [CrossRef]
96. Lee, S.J.; Yum, Y.N.; Kim, S.C.; Kim, Y.; Lim, J.; Lee, W.J.; Koo, K.H.; Kim, J.H.; Kim, J.E.; Lee, W.S.; et al. Distinguishing between genotoxic and non-genotoxic hepatocarcinogens by gene expression profiling and bioinformatic pathway analysis. *Sci. Rep.* **2013**, *3*, 2783. [CrossRef]
97. Ali, S.A.; Awad, S.M.; Said, A.M.; Mahgoub, S.; Taha, H.; Ahmed, N.M. Design, synthesis, molecular modelling and biological evaluation of novel 3-(2-naphthyl)-1-phenyl-1H-pyrazole derivatives as potent antioxidants and 15-Lipoxygenase inhibitors. *J. Enzym. Inhib. Med. Chem.* **2020**, *35*, 847–863. [CrossRef]
98. Wang, X.; Dong, C.; Sun, L.; Zhu, L.; Sun, C.; Ma, R.; Ning, K.; Lu, B.; Zhang, J.; Xu, J. Quantitative proteomic analysis of age-related subventricular zone proteins associated with neurodegenerative disease. *Sci. Rep.* **2016**, *6*, 37443. [CrossRef]
99. Kim, J.; Chen, C.H.; Yang, J.; Mochly-Rosen, D. Aldehyde dehydrogenase 2*2 knock-in mice show increased reactive oxygen species production in response to cisplatin treatment. *J. Biomed. Sci.* **2017**, *24*, 33. [CrossRef]
100. Ishibashi, N.; Mirochnitchenko, O. Chemokine expression in transgenic mice overproducing human glutathione peroxidases. *Methods Enzymol.* **2002**, *353*, 460–476. [CrossRef]
101. Giacco, F.; Du, X.; D'Agati, V.D.; Milne, R.; Sui, G.; Geoffrion, M.; Brownlee, M. Knockdown of glyoxalase 1 mimics diabetic nephropathy in nondiabetic mice. *Diabetes* **2014**, *63*, 291–299. [CrossRef] [PubMed]

102. Gambelunghe, A.; Giovagnoli, S.; Di Michele, A.; Boncompagni, S.; Dell'Omo, M.; Leopold, K.; Iavicoli, I.; Talesa, V.N.; Antognelli, C. Redox-Sensitive Glyoxalase 1 Up-Regulation Is Crucial for Protecting Human Lung Cells from Gold Nanoparticles Toxicity. *Antioxidants* **2020**, *9*, 697. [CrossRef] [PubMed]
103. Xiong, G.; Stewart, R.L.; Chen, J.; Gao, T.; Scott, T.L.; Samayoa, L.M.; O'Connor, K.; Lane, A.N.; Xu, R. Collagen prolyl 4-hydroxylase 1 is essential for HIF-1α stabilization and TNBC chemoresistance. *Nat. Commun.* **2018**, *9*, 4456. [CrossRef] [PubMed]
104. Gouaux, E.; Mackinnon, R. Principles of selective ion transport in channels and pumps. *Science* **2005**, *310*, 1461–1465. [CrossRef] [PubMed]
105. Franco, R.; Bortner, C.D.; Cidlowski, J.A. Potential roles of electrogenic ion transport and plasma membrane depolarization in apoptosis. *J. Membr. Biol.* **2006**, *209*, 43–58. [CrossRef] [PubMed]
106. Dinchuk, J.E.; Henderson, N.L.; Burn, T.C.; Huber, R.; Ho, S.P.; Link, J.; O'Neil, K.T.; Focht, R.J.; Scully, M.S.; Hollis, J.M.; et al. Aspartyl beta -hydroxylase (Asph) and an evolutionarily conserved isoform of Asph missing the catalytic domain share exons with junctin. *J. Biol. Chem.* **2000**, *275*, 39543–39554. [CrossRef]
107. Yuan, Q.; Fan, G.C.; Dong, M.; Altschafl, B.; Diwan, A.; Ren, X.; Hahn, H.H.; Zhao, W.; Waggoner, J.R.; Jones, L.R.; et al. Sarcoplasmic reticulum calcium overloading in junctin deficiency enhances cardiac contractility but increases ventricular automaticity. *Circulation* **2007**, *115*, 300–309. [CrossRef]
108. Genç, Ö.; Dickman, D.K.; Ma, W.; Tong, A.; Fetter, R.D.; Davis, G.W. MCTP is an ER-resident calcium sensor that stabilizes synaptic transmission and homeostatic plasticity. *eLife* **2017**, *6*, e22904. [CrossRef]
109. Xiao, L.; Pan, G. An important intestinal transporter that regulates the enterohepatic circulation of bile acids and cholesterol homeostasis: The apical sodium-dependent bile acid transporter (SLC10A2/ASBT). *Clin. Res. Hepatol. Gastroenterol.* **2017**, *41*, 509–515. [CrossRef]
110. Roessingh, S.; Rosing, M.; Marunova, M.; Ogueta, M.; George, R.; Lamaze, A.; Stanewsky, R. Temperature synchronization of the Drosophila circadian clock protein PERIOD is controlled by the TRPA channel PYREXIA. *Commun. Biol.* **2019**, *2*, 246. [CrossRef]
111. Ohgami, R.S.; Campagna, D.R.; McDonald, A.; Fleming, M.D. The Steap proteins are metalloreductases. *Blood* **2006**, *108*, 1388–1394. [CrossRef] [PubMed]
112. Sakharkar, M.K.; Kashmir Singh, S.K.; Rajamanickam, K.; Mohamed Essa, M.; Yang, J.; Chidambaram, S.B. A systems biology approach towards the identification of candidate therapeutic genes and potential biomarkers for Parkinson's disease. *PLoS ONE* **2019**, *14*, e0220995. [CrossRef] [PubMed]
113. Franklin, R.B.; Feng, P.; Milon, B.; Desouki, M.M.; Singh, K.K.; Kajdacsy-Balla, A.; Bagasra, O.; Costello, L.C. hZIP1 zinc uptake transporter down regulation and zinc depletion in prostate cancer. *Mol. Cancer* **2005**, *4*, 32. [CrossRef] [PubMed]
114. Desouki, M.M.; Franklin, R.B.; Costello, L.C.; Fadare, O. Persistent low expression of hZip1 in mucinous carcinomas of the ovary, colon, stomach and lung. *J. Ovarian Res.* **2015**, *8*, 40. [CrossRef]
115. Lopez, V.; Kelleher, S.L. Zinc transporter-2 (ZnT2) variants are localized to distinct subcellular compartments and functionally transport zinc. *Biochem. J.* **2009**, *422*, 43–52. [CrossRef]
116. Chandler, P.; Kochupurakkal, B.S.; Alam, S.; Richardson, A.L.; Soybel, D.I.; Kelleher, S.L. Subtype-specific accumulation of intracellular zinc pools is associated with the malignant phenotype in breast cancer. *Mol. Cancer* **2016**, *15*, 2. [CrossRef]
117. Kaźmierczuk, A.; Kiliańska, Z.M. The pleiotropic activity of heat-shock proteins. *Postepy Hig. I Med. Dosw. (Online)* **2009**, *63*, 502–521.
118. Polanowska-Grabowska, R.; Gear, A.R. Heat-shock proteins and platelet function. *Platelets* **2000**, *11*, 6–22. [CrossRef]
119. Garrido, C.; Gurbuxani, S.; Ravagnan, L.; Kroemer, G. Heat shock proteins: Endogenous modulators of apoptotic cell death. *Biochem. Biophys. Res. Commun.* **2001**, *286*, 433–442. [CrossRef]
120. Uno, T.; Ishizuka, M.; Itakura, T. Cytochrome P450 (CYP) in fish. *Environ. Toxicol. Pharmacol.* **2012**, *34*, 1–13. [CrossRef]
121. Amara, I.E.; Elshenawy, O.H.; Abdelrady, M.; El-Kadi, A.O. Acute mercury toxicity modulates cytochrome P450, soluble epoxide hydrolase and their associated arachidonic acid metabolites in C57Bl/6 mouse heart. *Toxicol. Lett.* **2014**, *226*, 53–62. [CrossRef] [PubMed]
122. Dai, L.S.; Abbas, M.N.; Kausar, S.; Zhou, Y. Transcriptome analysis of hepatopancraes of Procambarus clarkii challenged with polyriboinosinic polyribocytidylic acid (poly I:C). *Fish Shellfish Immunol.* **2017**, *71*, 144–150. [CrossRef] [PubMed]
123. Wan, R.; Feng, J.; Tang, L. Consequences of Mutations and Abnormal Expression of SMAD4 in Tumors and T Cells. *OncoTargets Ther.* **2021**, *14*, 2531–2540. [CrossRef] [PubMed]
124. Licciardi, P.V.; Karagiannis, T.C. Regulation of immune responses by histone deacetylase inhibitors. *ISRN Hematol.* **2012**, *2012*, 690901. [CrossRef] [PubMed]
125. Kakani, P.; Kajla, M.; Choudhury, T.P.; Gupta, L.; Kumar, S. Anopheles stephensi Dual Oxidase Silencing Activates the Thioester-Containing Protein 1 Pathway to Suppress Plasmodium Development. *J. Innate Immun.* **2019**, *11*, 496–505. [CrossRef]
126. AnvariFar, H.; Amirkolaie, A.K.; Miandare, H.K.; Ouraji, H.; Jalali, M.A.; Üçüncü, S.İ. Apoptosis in fish: Environmental factors and programmed cell death. *Cell Tissue Res.* **2017**, *368*, 425–439. [CrossRef]
127. Morcillo, P.; Esteban, M.Á.; Cuesta, A. Heavy metals produce toxicity, oxidative stress and apoptosis in the marine teleost fish SAF-1 cell line. *Chemosphere* **2016**, *144*, 225–233. [CrossRef]
128. Lixin, X.; Jianzhong, S.; Zhen, M. Apoptosis induction in fish cells under stress of six heavy metal ions. *Sheng Wu Hua Xue Yu Sheng Wu Wu Li Jin Zhan* **2001**, *28*, 866–869. [CrossRef]

129. Hsu, K.-F.; Wu, C.-L.; Huang, S.-C.; Wu, C.-M.; Hsiao, J.-R.; Yo, Y.-T.; Chen, Y.-H.; Shiau, A.-L.; Chou, C.-Y. Cathepsin L mediates resveratrol-induced autophagy and apoptotic cell death in cervical cancer cells. *Autophagy* **2009**, *5*, 451–460. [CrossRef]
130. Zhou, S.; Dong, J.; Liu, Y.; Yang, Q.; Xu, N.; Yang, Y.; Ai, X. Effects of acute deltamethrin exposure on kidney transcriptome and intestinal microbiota in goldfish (Carassius auratus). *Ecotoxicol. Environ. Saf.* **2021**, *225*, 112716. [CrossRef]
131. Droga-Mazovec, G.; Bojic, L.; Petelin, A.; Ivanova, S.; Repnik, U.; Salvesen, G.S.; Stoka, V.; Turk, V.; Turk, B. Cysteine cathepsins trigger caspase-dependent cell death through cleavage of bid and antiapoptotic Bcl-2 homologues. *J. Biol. Chem.* **2008**, *283*, 19140–19150. [CrossRef] [PubMed]
132. LaCasse, E.C.; Baird, S.; Korneluk, R.G.; MacKenzie, A.E. The inhibitors of apoptosis (IAPs) and their emerging role in cancer. *Oncogene* **1998**, *17*, 3247–3259. [CrossRef] [PubMed]
133. Elmore, S. Apoptosis: A review of programmed cell death. *Toxicol. Pathol.* **2007**, *35*, 495–516. [CrossRef] [PubMed]
134. Zhang, B.; Li, G.; Shahid, M.S.; Gan, L.; Fan, H.; Lv, Z.; Yan, S.; Guo, Y. Dietary l-arginine supplementation ameliorates inflammatory response and alters gut microbiota composition in broiler chickens infected with Salmonella enterica serovar Typhimurium. *Poult. Sci.* **2020**, *99*, 1862–1874. [CrossRef] [PubMed]
135. Sovran, B.; Lu, P.; Loonen, L.M.; Hugenholtz, F.; Belzer, C.; Stolte, E.H.; Boekschoten, M.V.; van Baarlen, P.; Smidt, H.; Kleerebezem, M.; et al. Identification of Commensal Species Positively Correlated with Early Stress Responses to a Compromised Mucus Barrier. *Inflamm. Bowel Dis.* **2016**, *22*, 826–840. [CrossRef] [PubMed]
136. Malmuthuge, N.; Griebel, P.J.; Guan, L.L. The Gut Microbiome and Its Potential Role in the Development and Function of Newborn Calf Gastrointestinal Tract. *Front. Vet. Sci.* **2015**, *2*, 36. [CrossRef]
137. Liu, J.; Wang, K.; Wang, Y.; Chen, W.; Jin, Z.; Yao, Z.; Zhang, D. Strain-specific changes in the gut microbiota profiles of the white shrimp Litopenaeus vannamei in response to cold stress. *Aquaculture* **2019**, *503*, 357–366. [CrossRef]
138. Parlapani, F.F.; Michailidou, S.; Anagnostopoulos, D.A.; Koromilas, S.; Kios, K.; Pasentsis, K.; Psomopoulos, F.; Argiriou, A.; Haroutounian, S.A.; Boziaris, I.S. Bacterial communities and potential spoilage markers of whole blue crab (Callinectes sapidus) stored under commercial simulated conditions. *Food Microbiol.* **2019**, *82*, 325–333. [CrossRef]
139. Shi, C.; Xia, M.; Li, R.; Mu, C.; Zhang, L.; Liu, L.; Ye, Y.; Wang, C. Vibrio alginolyticus infection induces coupled changes of bacterial community and metabolic phenotype in the gut of swimming crab. *Aquaculture* **2019**, *499*, 251–259. [CrossRef]
140. Klase, G.; Lee, S.; Liang, S.; Kim, J.; Zo, Y.-G.; Lee, J. The microbiome and antibiotic resistance in integrated fishfarm water: Implications of environmental public health. *Sci. Total Environ.* **2019**, *649*, 1491–1501. [CrossRef]
141. Semova, I.; Carten, J.D.; Stombaugh, J.; Mackey, L.C.; Knight, R.; Farber, S.A.; Rawls, J.F. Microbiota regulate intestinal absorption and metabolism of fatty acids in the zebrafish. *Cell Host Microbe* **2012**, *12*, 277–288. [CrossRef] [PubMed]
142. Shi, X.; Ng, K.K.; Li, X.-R.; Ng, H.Y. Investigation of intertidal wetland sediment as a novel inoculation source for anaerobic saline wastewater treatment. *Environ. Sci. Technol.* **2015**, *49*, 6231–6239. [CrossRef] [PubMed]
143. Krinos, C.M.; Coyne, M.J.; Weinacht, K.G.; Tzianabos, A.O.; Kasper, D.L.; Comstock, L.E. Extensive surface diversity of a commensal microorganism by multiple DNA inversions. *Nature* **2001**, *414*, 555–558. [CrossRef]
144. Wexler, H.M. Bacteroides: The good, the bad, and the nitty-gritty. *Clin. Microbiol. Rev.* **2007**, *20*, 593–621. [CrossRef] [PubMed]
145. Karlsson, F.H.; Ussery, D.W.; Nielsen, J.; Nookaew, I. A closer look at bacteroides: Phylogenetic relationship and genomic implications of a life in the human gut. *Microb. Ecol.* **2011**, *61*, 473–485. [CrossRef] [PubMed]
146. Sun, X.; Yang, Y.; Zhang, N.; Shen, Y.; Ni, J. Draft genome sequence of Dysgonomonas macrotermitis strain JCM 19375T, isolated from the gut of a termite. *Genome Announc.* **2015**, *3*, e00963-00915. [CrossRef] [PubMed]
147. Ho, H.T.; Lipman, L.J.; Gaastra, W. Arcobacter, what is known and unknown about a potential foodborne zoonotic agent! *Vet. Microbiol.* **2006**, *115*, 1–13. [CrossRef]

antioxidants

MDPI

Article

The Evolution and Ecology of Oxidative and Antioxidant Status: A Comparative Approach in African Mole-Rats

Paul. J. Jacobs [1,*], Daniel W. Hart [1], Hana N. Merchant [2], Cornelia Voigt [3] and Nigel C. Bennett [3]

[1] Department of Zoology and Entomology, University of Pretoria, Pretoria 0002, South Africa; daniel.hart@zoology.up.ac.za
[2] Department of Biological Sciences, School of Life and Environmental Sciences, Royal Holloway University of London, Egham Hill, Egham, Surrey TW20 0EX, UK; hana.merchant.2020@live.rhul.ac.uk
[3] Mammal Research Institute, Department of Zoology and Entomology, University of Pretoria, Pretoria 0002, South Africa; cvoigt@zoology.up.ac.za (C.V.); ncbennett@zoology.up.ac.za (N.C.B.)
* Correspondence: u10533207@tuks.co.za

Abstract: The naked mole-rat of the family Bathyergidae has been the showpiece for ageing research as they contradict the traditional understanding of the oxidative stress theory of ageing. Some other bathyergids also possess increased lifespans, but there has been a remarkable lack of comparison between species within the family Bathyergidae. This study set out to investigate how plasma oxidative markers (total oxidant status (TOS), total antioxidant capacity (TAC), and the oxidative stress index (OSI)) differ between five species and three subspecies of bathyergids, differing in their maximum lifespan potential (MLSP), resting metabolic rate, aridity index (AI), and sociality. We also investigated how oxidative markers may differ between captive and wild-caught mole-rats. Our results reveal that increased TOS, TAC, and OSI are associated with increased MLSP. This pattern is more prevalent in the social-living species than the solitary-living species. We also found that oxidative variables decreased with an increasing AI and that wild-caught individuals typically have higher antioxidants. We speculate that the correlation between higher oxidative markers and MLSP is due to the hypoxia-tolerance of the mole-rats investigated. Hormesis (the biphasic response to oxidative stress promoting protection) is a likely mechanism behind the increased oxidative markers observed and promotes longevity in some members of the Bathyergidae family.

Keywords: oxidative stress; oxidative status; longevity; ageing; maximum lifespan potential; sociality; metabolism; aridity; antioxidants; reactive oxygen species

Citation: Jacobs, P.J.; Hart, D.W.; Merchant, H.N.; Voigt, C.; Bennett, N.C. The Evolution and Ecology of Oxidative and Antioxidant Status: A Comparative Approach in African Mole-Rats. *Antioxidants* **2023**, *12*, 1486. https://doi.org/10.3390/antiox12081486

Academic Editors: Marcelo Hermes-Lima, Daniel Carneiro Moreira and Tania Zenteno-Savín

Received: 14 June 2023
Revised: 18 July 2023
Accepted: 20 July 2023
Published: 25 July 2023

1. Introduction

The 'Great oxidation event' (some 2.45 billion years ago) saw the Earth's atmosphere and the shallow ocean experience a significant rise in the amount of oxygen, likely a result of the evolution of photosynthesis by Archean, anaerobic cyanobacteria [1–3]. The increase in the oxygen composition of the atmosphere acted as an intense selective pressure on anaerobic organisms, paving the way forward to a world dominated by aerobic organisms [1,4,5]. However, despite the increased concentration of oxygen providing a more energy-efficient means for metabolism, it also came with its own set of problems, the most significant being the formation of oxygen4-derived free radicals or non-radical reactive oxygen species, typically known as reactive oxygen species (ROS) [6–8]. ROS can harm biological systems [9–11] and likely mediates ageing [12,13]. Thus, the study of ROS and oxidative stress {the balance between oxidative damage and antioxidant (compounds and mechanisms that inhibit oxidation) measures} has been of vital interest for investigating life-history ecology and evolution [4,14–18].

Ageing has been described as gradual fitness loss due to detrimental changes at the cell and molecular level over time [19–22]. Initially, it was proposed that the resultant damage due to unscavenged ROS, which in turn can damage proteins, lipids, and DNA [15,18,23,24], may cause cellular attrition, mediating the ageing process through physiological decline, resulting in eventual death [12–14,22,25–28]. This hypothesis was named the oxidative stress theory of ageing (OSTA) or the free radical theory of ageing (FRTA) [12–14,22,25–28]. Despite the negative connotation to ROS, they are essential for normal physiological function and are involved in cellular signalling [29], inflammation response [30], altering glucose uptake and metabolism [31], and the immune response [11], allowing for the processes associated with the dealing of tolerating hypoxic stress [32], osmoprotective signalling [33], and the control of ventilation, nerve transmission, and immune regulatory processes [34]. However, it is crucial to consider that not all ROS can be scavenged and that some damage will always transpire [18,22,28,35,36]. In general, oxidative stress and the resulting damage can be due to a variety of factors, but it is primarily due to either the increased production of pro-oxidants (compounds that induce oxidative stress, either by generating reactive oxygen species or by inhibiting antioxidant systems), diminished antioxidant levels, depletion of essential dietary metal cofactors which potentiate the activity of antioxidant enzymes, and/or a failure in the repair of replacement systems [4]. It is thus generally accepted that mild elevations of ROS can be beneficial through hormesis {an adaptive response of cells and organisms to moderate (usually intermittent) stress} [16,32,37,38], where very high ROS levels are considered detrimental [20,21,39,40].

The OSTA generally shows age is positively correlated with oxidative damage [14,41,42], though some species, even when size is considered [41,43,44], contradict this correlation [20,25,26,45]. These exceptions include primates [46,47], birds [48–50], bats [51–54], as well as mole-rats [51,55–61]. These species are longer lived than others, likely through mechanisms that may mitigate the oxidative damage-induced changes usually associated with ageing [26,36,51,62] that affect the maximum lifespan potential (MLSP) [41,63,64]. The most well-known species to violate the OSTA is the naked mole-rat (*Heterocephalus glaber*) (NMR) [36,45,57,62,64–67]. Research on ageing in the NMR has demonstrated that they, even at a young age, exhibit extremely high levels of oxidative damage with no detriment to longevity, while the only finding of an oxidative damage limitation involved the insignificant increase across age [57,65]. Some factors have been identified which contribute to the longevity of the NMR, such as the poly-unsaturated fatty acid (PUFA) composition relative to saturated fatty acids to similarly aged mice [67,68], calorie-restriction-like symptoms such as a reduced metabolic rate, reduced body temperature, and reduced food consumption [69,70], and low circulating levels of methionine [69]. It is believed the main reason NMRs possess such unique characteristics is due to their exclusively subterranean lifestyle [71–73]. Despite the interest in NMRs as a conundrum in the oxidative stress theory of ageing [36,45,57,62,64–67], a comparison of their redox status with that of other Bathyergidae has not previously been carried out. This is surprising, as oxidative markers in tissues [45] and plasma [74] were previously found to differ, and Dammann [51] proposed that the importance of oxidative stress in bathyergids may be underestimated if only NMRs and mice are compared, or that the impact of oxidative stress may very well differ between the NMR and other African mole-rat species. Comparisons between the NMR and other African mole-rat species may allow for identifying the ecological and physiological factors involved in redox biology that are not apparent when considering only one species. Additionally, some other mole-rats, like the Damaraland mole-rat (*Fukomys damarensis*) (DMR), share multiple characteristics in common with the NMR but show a higher degree of similarity with humans in some aspects of their physiology, where this higher degree of similarity of the DMR to humans, in contrast to the NMR, may offer alternative insights to human biomedical research [75]. This may also be applicable to other African mole-rats and not just the NMR and the DMR.

Therefore, this study sought to investigate the comparison of the redox status between family members of Bathyergidae due to their unique physiology, life-history, and reproductive system as well as the current understanding of their oxidative ecology (Supplementary File S1) [22,51,55,71,74,76–113]. The current study included eight different subspecies and species of African mole-rats, namely the NMR, the DMR, the highveld mole-rat (*Cryptomys hottentotus pretoriae*) (CHP), the common mole-rat (*C. h. hottentotus*) (CHH), the Natal mole-rat (*C. h. natalensis*) (CHN), the Cape mole-rat (*Georychus capensis*) (GC), and the Cape Dune mole-rat (*Bathyergus suillus*) (BS) [77,80,114] (Table 1). These representatives were chosen due to their available data for the oxidative markers investigated, as well as representing some eusocial [96,97], social, and solitary species that vary in their resting metabolic rate (RMR) [76,115] (Table 1).

Table 1. Common names and characteristics of the seven mole-rat subspecies and species used in this study.

Subspecies and Species Name	Sociality	Aridity Index (Habitat Type)	Maximum Lifespan Potential (Years)	Body Mass (g)	Sample Size	Mass-Specific Resting Metabolic Rate (mL O_2 $h^{-1}g^{-1}$)	Oxidative Stress Data Source
Fukomys damarensis	Eusocial	0.10 (Arid) [76,115]	15.5	131 ± 26	15	0.87 [88]	[74]
Heterocephalus glaber	Eusocial	0.08 (Arid) [76]	31	46 ± 10	14	1 [116]	[74]
Cryptomys hottentotus pretoriae	Social	0.48 (Semi-arid) [76,115]	11	Wild-caught 108 ± 24	16	0.82 [88]	[79]
				Captive 133 ± 32	20	0.72 [92]	This study
Cryptomys hottentotus hottentotus	Social	0.19 (Arid) [76,115,117]	11	Wild-caught 53 ± 16	19	1.32 [88]	This study
				Captive 68 ± 20	18	0.92 [118]	This study
Cryptomys hottentotus natalensis	Social	1.33 (Hyper-mesic) [100]	11	103 ± 39	19	0.79 [88]	[94]
Georychus capensis	Solitary	0.51 (Mesic) [76,115]	11	209 ± 33	9	0.59 [119]	This study
Bathyergus suillus	Solitary	0.51 (Mesic) [76,115]	6	838 ± 209	7	0.47 [120]	This study

Data represent the mean ± SD. Maximum lifespan potential data obtained from AnAge [121] or laboratory husbandry information.

The main objective of this study was to establish how NMRs are different from, or similar to, other species from the family Bathyergidae with regards to their plasma oxidative markers, namely, total antioxidant capacity (TAC), total antioxidant status (TOS), and the ratio between these two (TOS:TAC) variables as an arbitrary measure of an oxidative stress index (OSI), and how these oxidative markers may vary with maximum lifespan potential (MLSP), RMR, the aridity index (AI), and sociality. These relationships were analysed through a phylogenetically controlled comparison of all captive mole-rat species (Table 1), wild-caught and captive comparisons (CHP and CHH), as well as wild comparisons of subspecies (CHN, CHH, and CHP) from habitats differing in their AI to determine the possible influences of the environment on oxidative markers. We also opted for a principal component analysis (PCA) to avoid analysing oxidative markers in a vacuum and determine their relationship simultaneously with variables of interest (MLSP, AI, RMR, and sociality). In order to make the data standardised, we opted to use captive individuals for the phylogenetic relationship and PCA analyses to avoid the complications of ecology (environmental factors) on plasma oxidative markers. We further opted to utilise the same markers analysed from the same laboratory, as markers measured from different laboratories are rarely comparable [57]. Also, due to the confounding effects known in social mole-rats species, namely a reproductive division of labour [74,79], we used only the non-breeding colony members of social mole-rats and solitary mole-rats individuals outside their breeding season for the current study. We realise that the markers investigated do not include DNA damage, an important aspect associated with ageing [24,28,122]; however, this study was undertaken as an initial overview of oxidative stress and ageing in this unique rodent family.

These analyses allow us to investigate: (1) Is there a phylogenetic relationship of plasma markers between African mole-rats? (2) To what extent does sociality explain the differences in oxidative markers? (3) Do oxidative markers correlate with the MLSP of

African mole-rats? (4) Do oxidative markers change between captive and wild-caught species? And (5) Does the aridity of the environment influence oxidative markers?

2. Materials and Methods

2.1. Ethics Statement

The University of Pretoria, Faculty of Veterinary Science Animal Ethics Committee, approved all experimental animal procedures under the NAS 068/2021 and NAS209/2021 project codes. In addition, DLRDD section 20 approval (SDAH-Epi-12/11/1/4/1 (1948 LH) and SDAH-Epi-21031811071) was also obtained along with relevant provincial animal capture permits (Western Cape: CN44-87-13780; Northern Cape: FAUNA 0419/2021, FAUNA 042/2021; Gauteng: CPF6-0124). All methods were performed following the relevant guidelines and regulations. In addition, all experimental procedures were carried out under the recommendations in the Guide for the Care and Use of Laboratory Animals of the National Institutes of Health [123].

2.2. Novel Data Analysis

2.2.1. Solitary Mole-Rats

Solitary mole-rats (GC and BS) were wild-caught during the non-breeding season (September–October) in 2021 close to the town of Darling (33°22′ S 15°25′ E) in the South-western Cape, South Africa. Animals were caught using Hickman live traps baited with sweet potatoes [124]. Once captured, the animals were transported to the mole-rat laboratory at the Department of Zoology and Entomology (25°45′13.3″ S, 28°13′50.9″ E), University of Pretoria, Hatfield, South Africa. These solitary species were kept in captivity for more than six months prior to sample collection (Table 1).

2.2.2. Social Mole-Rats

Wild-caught CHH used in the current study were captured at Klawer (31.7730° S, 18.6247° E) during the non-breeding season between November 2021 and March 2022. The social mole-rats were captured and transported as outlined for solitary mole-rats. A subset of animals (Table 1) was sampled within 72 h of being in captivity (similar to other studies [79,94]). An additional subset was maintained under laboratory conditions for more than six months before sampling.

All animals used in this study were considered adults and reproductively inactive (non-breeders) {see Bennett and Faulkes [125] on how reproductive status was determined} (Table 1). The use of non-breeders avoids complications between breeder and non-breeder comparisons associated with oxidative stress due to reproduction [74,79] and the effects of reproductive suppression differences [74,79,98–106,125,126] (Supplementary File S1).

2.2.3. Mole-Rat Blood Sample Collection

All plasma was collected between 08h00 and 13h00 to prevent the circadian rhythmicity of oxidative markers [127,128]. The body mass of each animal was recorded to the nearest 0.01 g (Scout Pro SPU123; Ohaus Corporation, Pine Brook, NJ, USA) (Table 1). All mole-rats were handheld, and venous blood samples were collected from the hindfoot, tail, or heart, if euthanised with an overdose of isoflurane. Approximately 300–500 µL of blood was collected from each animal. All blood was centrifuged at 13,300 rpm, and the resulting plasma was decanted and stored at −70–80 °C until further analysis.

2.2.4. Oxidative Stress Markers

The oxidative stress markers of TOS, TAC, and OSI were measured in captive and wild CHH, as well as captive CHP, captive GC, and BS (Table 1).

Reagents

Unless otherwise stated, all chemicals and reagents used in this study were obtained from Merck (Pty) Ltd. (Gauteng, South Africa).

Total Antioxidant Capacity (TAC) Assay

Plasma TAC levels were quantified using a commercially available kit (Antioxidant Assay Kit, Cayman Chemical Co., Ann Arbor, MI, USA) which measures the oxidation of ABTS (2,29-Azino-di-[3-ethybenzthiazoline sulphonate]) by metmyoglobin, which is inhibited by the non-enzymatic antioxidants contained in the sample. Oxidised ABTS is measured by spectrophotometry at a wavelength of 750 nm. The capacity of antioxidants in the sample to inhibit the oxidation of ABTS is compared with the capacity of known concentrations of Trolox, and the results are expressed as micromole Trolox equivalents per litre (μmol Trolox equivalents/L). Samples were run in duplicate and only once per plate with a repeatability of r = 0.96. The intra-assay variability (%CV) was 3.65%.

Total Oxidant Status (TOS) Assay

Plasma TOS levels were measured with Erel's method [129]. Briefly, this method is based on the oxidation of ferrous ions to ferric ions in the presence of various oxidative species. The oxidation reaction is enhanced by glycerol molecules, which are abundantly present in the reaction medium. In an acidic medium, the ferric ion makes a coloured complex with xylenol orange. The colour intensity, measured using a spectrophotometer, is related to the total amount of oxidant molecules that are present in the sample. The results are expressed in terms of micromole hydrogen peroxide equivalent per litre (μmol H_2O_2 equivalent/L). Samples were run in duplicate and not repeated once per plate with a repeatability of r = 0.99. The intra-assay variability (%CV) was 4.5%.

Oxidative Stress Index (OSI)

Oxidative stress was determined by the TOS:TAC ratio, which represents the oxidative stress index (OSI) arbitrary unit, which was calculated as follows: OSI = [(TOS, μmol H_2O_2 equivalent/L)/(TAC, mmol Trolox equivalent/L)].

2.3. Multi-Species Analyses

2.3.1. Animal Housing

All animals were housed in the mole-rat laboratory of the Department of Zoology and Entomology at the University of Pretoria, South Africa, and all animals were maintained under similar conditions. Animals were housed in climate-controlled rooms within their thermal neutral zone (TNZ) [76] and were maintained on a 12L:12D photoperiod with 50–60% relative humidity. The mole-rats were fed daily on sweet potatoes and apples ad libitum. All animals were fed the same variety of chopped vegetables and drank no free water.

2.3.2. Resting Metabolic Rate of Mole-Rats

The mass-specific resting metabolic rates (msRMR) for each mole-rat species for wild and captive mole-rats were obtained through several studies (Table 1). These msRMR values for a species were multiplied by the individual's body mass to estimate each individual RMR, hereafter referred to as ERMR.

2.4. Aridity Index Data

The AI was gathered using the methods outlined by Jacobs et al. [94]. Climate data for each species were retrieved from ERA5-Land of the European Centre for Medium-Range Weather Forecasts—the latest generation created by the Copernicus Climate Change Service [130]. The spatial resolution is $0.1 \times$ by 0.1. These data were used to calculate an annual *AI* (Equation (1)). Whereas the total precipitation (*tp*) was directly obtained from ERA5-Land, the potential evapotranspiration (*PET*) was calculated from the well-known Romanenko estimation (Equation (2)) [131]. For Equation (2), the relative humidity (*RH*) was calculated from ERA5-Land d2m (Equation (3)).

$$AI = \frac{tp}{PET} \tag{1}$$

$$PET\,0.0006 \times (100 - RH) \times (25 + Tair)^2 \tag{2}$$

$$RH = 100 \times 10^{7.591386\left(\frac{d2m}{d2m+240.7263} - \frac{Tair}{Tair+240.7263}\right)} \tag{3}$$

2.5. Phylogenetic Tree Determination

Cytochrome b gene sequence fasta files for each species were obtained from the existing GenBank data [132]. This involved the cytochrome b sequence for mole-rats, namely, *Heterocephalus glaber* Accession No. MT8453841 [133], *Fukomys damarensis* Accession No. AF012223.1 [134], *Cryptomys hottentotus pretoriae* Accession No. AF012236.1 [134], *Cryptomys hottentotus hottentotus* Accession No. MH186559.1 [135], *Bathyergus suillus* Accession No. KJ866687.1 [136], and *Georychus capensis* Accession No. U18837.1 [137]. We also obtained the cytochrome b sequences for the two outgroups, namely, *Rattus norvegicus* Accession No. KT024821.1 [138] and the Guinea pig *Cavia porcellus* Accession No. AY382793.1 [139]. Aligning was carried out using Mesquite version 2.75 [140]. Tree building was carried out using the Molecular Evolutionary Genetic Analysis Version 11 (MEGA11) program [141]. Mesquite files were converted to mega files and, using the "Phylogeny" menu, neighbour-joining trees were created using the Neighbour-Joining method [142]. The percentage of replicate trees in which the associated taxa clustered together in the bootstrap test (1000 replicates) are shown next to the branches [143]. The evolutionary history was inferred using the evolutionary distances computed using the Maximum Composite Likelihood method [144] and are in the units of the number of base substitutions per site. This analysis involved eight nucleotide sequences. The best DNA substitution model for each set of sequence alignments was found using the "Models" menu, and this was used to create the trees. All ambiguous positions were removed for each sequence pair (pairwise deletion option). There were a total of 1201 positions in the final dataset. Evolutionary analyses were conducted in MEGA11 [145]. The optimal tree is shown and drawn to scale, with the branch lengths the same units as those of evolutionary distances used to infer the phylogenetic tree (Figure 1).

2.6. Statistical Analysis

Unless specified, all calculations, statistical, and visual analyses were performed using the statistical software R version 4.2.2 [146] and GraphPad Prism 8.4.3. Data are presented as the mean ± standard error (s.e.m), and a p-value of ≤ 0.05 indicates significance. The normality of the response variables TOS, TAC, and OSI, was determined using Shapiro–Wilk tests, and the homogeneity of all dependent variables was checked with Levene's test.

2.6.1. Phylogenetic Analysis of Oxidative Variables

Individuals from five mole-rat species, of which two in the genus *Cryptomys* were two subspecies, (NMR, DMR, CHH, CHP, GC, and BS) that had been housed in captivity for longer than six months were used in the phylogenetic analysis. A generalised variance inflation factor (GVIF) was used to determine multicollinearity between the life history traits to account for the mix of continuous and categorical traits, undertaken in a stepwise fashion. A generalised linear mixed-effects model using a Markov chain Monte Carlo approach under a Bayesian statistical framework (MCMCglmm) was applied in the 'MCMCglmm' package [147]. This methodology was used to incorporate the multiple studies per subspecies or species that were present in the database. This approach fits the individual-level data whilst controlling for relationships in species traits due to common ancestry. A single consensus tree was used, and 130,000 iterations were applied with 100 thinning intervals and 30,000 burn-in. ERMR, AI, and sociality were used as predictor variables. The statistical significance of the genetic influence on TOS/TAC/OSI was assessed using 95% confidence intervals (CI) for the heritability estimates, which is the transmission of the phenotypic variability within a population from generation to generation, and a heritability value (H^2) was obtained. All calculations and statistical analyses were performed using the statistical software R version 4.2.2 [146].

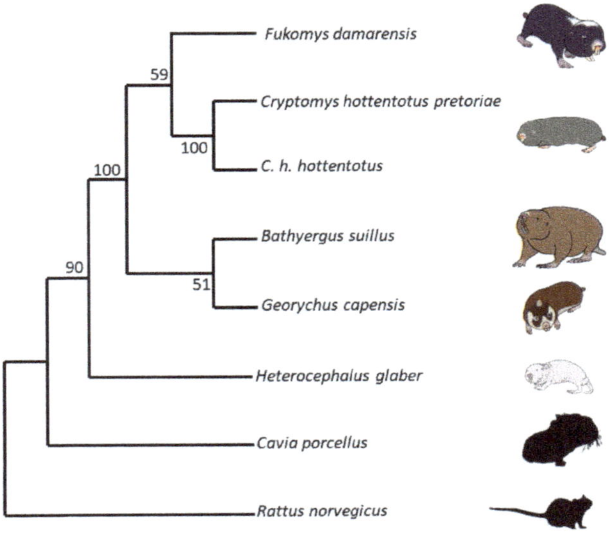

Figure 1. The optimal phylogenetic tree of five different captive mole-rat species and two subspecies, the naked mole-rat, *Heterocephalus glaber*, the Damaraland mole-rat, *Fukomys damarensis*, the highveld mole-rat, *Cryptomys hottentotus pretoriae*, the common mole-rat, *C. h. hottentotus*, the Cape mole-rat, *Georychus capensis*, and the Cape Dune mole-rat, *Bathyergus suillus*, as well as two outgroup species, the brown rat, *Rattus norvegicus* and the Guinea pig, *Cavia porcellus*. The tree is drawn to scale, with branch lengths being the same units as those of the evolutionary distances used to infer the phylogenetic tree. The evolutionary history was inferred using the Neighbour-Joining method [142]. The percentage of replicate trees in which the associated taxa clustered together in the bootstrap test (1000 replicates) are shown next to the branches [143]. Evolutionary distances were computed using the Maximum Composite Likelihood method [144] and are in the units of the number of base substitutions per site. This analysis involved eight nucleotide sequences. All ambiguous positions were removed for each sequence pair (pairwise deletion option). There were a total of 1201 positions in the final dataset.

2.6.2. Wild-Caught and Captive Mole-Rats Comparison

Wild-caught and captive mole-rat comparisons between the CHH and CHP used oxidative markers (TAC, TOS, and OSI) as the response variable and used the subspecies and status (wild-caught/captive) and their interaction as predictors, with ERMR as a covariate. All values violated the assumptions of normality and homoscedasticity; and as such, we used a generalised linear model (GZLM) with a gamma distribution with an inverse link function using the *lme4* package [148]. Models were model selected using the MuMIn package [149] and dredged with a delta function <2. If only one model was given, then that model was accepted as the best model. Significant variables in the linear modes were followed up with post-hoc comparisons, conducted using Tukey's HSD pairwise comparisons using the *emmeans* package [150]. Only relevant biological comparisons were presented in the results.

2.6.3. Wild-Caught Mole-Rat Comparison

For the analyses between the wild CHP, CHH, and CHN, we used subspecies as our predictor and ERMR as a covariate. This analysis was used to infer whether species differences exist due to these species occupying different AI habitats (Table 1). All values violated the assumptions of normality and homoscedasticity; as such, we used a GZLM with a gamma distribution with an inverse link function using the *lme4* package [148]. Models were model selected using the MuMIn package [149] and dredged with a delta function <2. If only one model was given, then that model was accepted as the best model.

Significant variables in the linear modes were followed up with post-hoc comparisons, conducted using Tukey's HSD pairwise comparisons using the *emmeans* package [150]. To prevent pseudo-replication when determining the aridity influence on wild-caught mole-rats' oxidative variables, the means of TAC, TOS, and OSI of each species were correlated using simple linear regressions with an AI from where the animals were caught.

2.6.4. Principal Component Analysis (PCA)

We opted to perform a PCA analysis due to the limitations of our variables being multicollinear. It has been proposed that PCAs can help investigate oxidative balance [151]. Briefly, this analysis allows for identifying the dimensionality of the oxidative variables within a context, with their loadings inferring how they interact within this dimensionality [151]. Furthermore, the axes, such as principal components (PCs), group several variables together, and that can be analysed as a new variable that can better represent the underlying process involved, compared to analysing the individual oxidative markers themselves [151]. Our study used the three oxidative markers (TAC, TOS, and OSI), AI, ERMR, and body mass. Furthermore, we also opted to use the MLSP of each species as determined through AnAge open-source dataset [121] and the mole-rat laboratory longevity data (Table 1) as our seventh component, which was applied to the six captive mole-rat species. The PCA was performed using the *prcomp* function using the ggplot2 package [152]. The number of PCs was determined through Eigenvalues, *skree* plots, and the cumulative proportion of variance explained by the PCs [153]. We used the Kaiser criterion to determine the number of PCs to analyse [153]. The functions *ggbiplot* from the ggplot2 package [152] and *cor* were used to determine the relationship of the variables within each component and the loading of the variables within each PC was visualised as a biplot [154]. The ellipse probability was set to 68%.

3. Results

3.1. Phylogenetic Analysis of TOS, TAC, and OSI

Phylogenetic heritability (H^2) or the lambda (λ) for TOS, TAC, and OSI suggest phylogeny does not strongly predict oxidative stress variables. Sociality and AI were not significant predictors for TOS, TAC, and OSI (Table 2). Similarly, ERMR was not a significant predictor for TOS and TAC but was a significant predictor of OSI (Table 2). A linear regression analysis between OSI and ERMR demonstrated a significant negative correlation ($R^2 = 0.2301$, $F_{(1,78)} = 23.23$, $p < 0.001$) (Figure 2). This linear regression remained significant when the BS was removed ($R^2 = 0.09$, $F_{(1,71)} = 6.844$, $p = 0.011$) (Figure 2).

Table 2. The best generalised linear mixed-effects models using a Markov chain Monte Carlo approach under a Bayesian statistical framework output for plasma oxidative markers, namely, the total oxidant status (TOS), total antioxidant activity (TAC), and oxidative stress index (OSI) for the five different captive mole-rat species and two subspecies, the naked mole-rat, *Heterocephalus glaber*, the Damaraland mole-rat, *Fukomys damarensis*, the highveld mole-rat, *Cryptomys hottentotus pretoriae*, the common mole-rat, *C. h. hottentotus*, the Cape mole-rat, *Georychus capensis*, and the Cape Dune mole-rat, *Bathyergus suillus* in response to sociality (eusocial:social:solitary), the aridity index (AI), and their estimated resting metabolic rate (ERMR).

	TOS		TAC		OSI	
H^2/λ	**0.72**		**0.46**		**0.33**	
	post.mean	p	post.mean	p	post.mean	p
Social~Solitary	0.123572	0.912	0.582482	0.350	−0.301008	0.864
Social~Eusocial	−3.118303	0.640	0.523573	0.272	−0.573683	0.776
Solitary~Eusocial	2.615357	0.516	−0.0039322	0.892	−0.029933	0.936
AI	−1.063284	0.830	−0.435481	0.684	2.632716	0.470
ERMR	−0.009047	0.210	−0.000211	0.838	−0.012625	0.038 *

An * signifies significance.

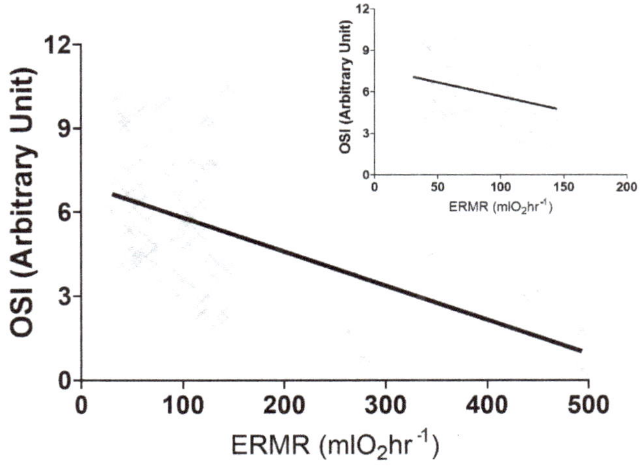

Figure 2. The significant negative linear regression output between the oxidative stress index (OSI) and the estimated resting metabolic rate (ERMR) for five different captive mole-rat species and two subspecies, the naked mole-rat, *Heterocephalus glaber*, the Damaraland mole-rat, *Fukomys damarensis*, the highveld mole-rat, *Cryptomys hottentotus pretoriae*, the common mole-rat, *C. h. hottentotus*, the Cape mole-rat, *Georychus capensis* and the Cape Dune mole-rat, *Bathyergus suillus*. The negative relationship remains significant when *B. suillus* is removed (figure insert).

3.2. Wild-Caught and Captive Mole-Rats Comparison

3.2.1. TOS

The best model did not retain ERMR for wild-caught and captive mole-rats (CHP and CHH) (Table 3). At the same time, species and captivity did not affect TOS (Table 3). The interaction between subspecies and status (wild-caught or captive) was significant (Table 3, Figure 3A). Post-hoc analyses revealed that wild-caught CHP had a significantly lower TOS compared to captive CHP ($t = -4.434$, $p = 0.0002$, Figure 3A). Additionally, wild-caught CHP had a significantly lower TOS as compared to wild-caught CHH ($t = -5.254$, $p < 0.0001$, Figure 3A). No significant differences were observed between the TOS of captive and wild-caught CHH ($t = -0.175$, $p = 0.9981$, Figure 3A).

Table 3. The best generalised linear model output for plasma oxidative markers, namely, the total oxidant status (TOS), total antioxidant activity (TAC), and oxidative stress index (OSI) for wild-caught and captive mole-rats, the highveld mole-rat, *Cryptomys hottentotus pretoriae*, and the common mole-rat, *C. h. hottentotus* in response to subspecies, status (wild-caught/captive), the interaction between subspecies and status, and the estimated resting metabolic rate (ERMR).

	TOS		TAC		OSI	
	t	*p*	t	*p*	t	*p*
Species	1.139′	0.26	nr	nr	−0.459	0.65
Status	0.175	0.86	6.169	<0.001 *	3.344	0.001 *
Species × Status	3.631	<0.001 *	nr	nr	4.244	<0.001 *
ERMR	nr	nr	nr	nr	nr	nr

nr: Variable not retained in the final model. An * signifies significance.

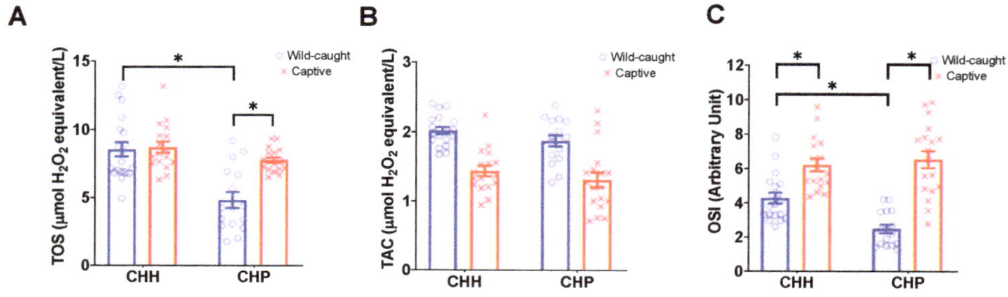

Figure 3. The plasma oxidative markers, (**A**) total oxidant status (TOS), (**B**) total antioxidant capacity (TAC), and (**C**) oxidative stress index (OSI) of wild-caught common mole-rats, *Cryptomys hottentotus hottentotus* and highveld mole-rats, *C. h. pretoriae* (blue bars and empty circles) compared to captive common mole-rats and highveld mole-rats (red bars and crosses). Bars represent the mean ± s.e.m. An asterisk (*) indicates significance ($p \leq 0.05$).

3.2.2. TAC

The best model did not retain ERMR for wild-caught and captive mole-rats (CHP and CHH) (Table 3). Only status (wild-caught vs. captive) was significant (Table 3), where wild-caught individuals had a higher TAC (1.96 ± 0.04 mmol Trolox equivalents/L) compared to captive individuals (1.37 ± 0.07 mmol Trolox equivalents/L) (Figure 3B). Subspecies or species and the interaction between these and status did not significantly influence TAC (Table 3; Figure 3B).

3.2.3. OSI

The ERMR for wild-caught and captive mole-rats (CHP and CHH) was not retained in the best model (Table 3). The best model demonstrated the significance of the main effect of status, where the wild-caught OSI was lower (3.49 ± 0.25) compared to captive mole-rats (6.42 ± 0.31), and the interaction between species and status was significant (Table 3, Figure 3C). Post-hoc analysis revealed that wild-caught CHP had a significantly lower OSI compared to captive CHP (t = −7.088, $p < 0.0001$, Figure 3C). Furthermore, wild-caught CHP had a significantly lower OSI than wild-caught CHH individuals (t = −4.472, $p = 0.0002$, Figure 3C). Interestingly, wild-caught and captive CHH significantly differed in their OSI, whereas wild-caught CHH had a significantly lower OSI compared to captive CHH (t = −3.344, $p = 0.0073$, Figure 3C).

3.3. Wild-Caught Mole-Rat Comparison

3.3.1. TOS

The ERMR for wild-caught mole-rats (CHP, CHH, and CHN) was not retained in the best model (Table 4). A significant effect of species was observed in the model (Table 4). Post-hoc comparisons between the three subspecies demonstrated that CHH had significantly higher TOS as compared to CHN (t = −5.328, df = 51, $p < 0.0001$) and CHH had a significantly higher TOS than CHP (t = −4.040, df = 51, $p = 0.0005$), but CHN and CHP did not significantly differ in TOS (t = 1.236, df = 51, $p = 0.44$) (Figure 4A). Simple linear regressions revealed that the TOS has a significant negative relationship with an increasing aridity index ($R^2 = 0.3221$, $F_{(1,52)} = 24.7$, $p < 0.0001$) (Figure 4B).

Table 4. The best generalised linear model output for plasma oxidative markers, namely, the total oxidant status (TOS), total antioxidant activity (TAC), and oxidative stress index (OSI) for wild-caught mole-rats, namely, the highveld mole-rat, *Cryptomys hottentotus pretoriae*, the common mole-rat, *C. h. hottentotus* and the Natal mole-rat, *C. h. natalensis*, in response to subspecies and the estimated resting metabolic rate (ERMR).

	TOS		TAC		OSI	
	z	*p*	z	*p*	t	*p*
Subspecies CHN	5.202	<0.0001	6.063	<0.0001	2.083	0.037
Subspecies CHP	3.906	<0.0001	0.885	0.38	3.773	<0.001
ERMR	nr	nr	nr	nr	nr	nr

nr: Variable not retained in the final model.

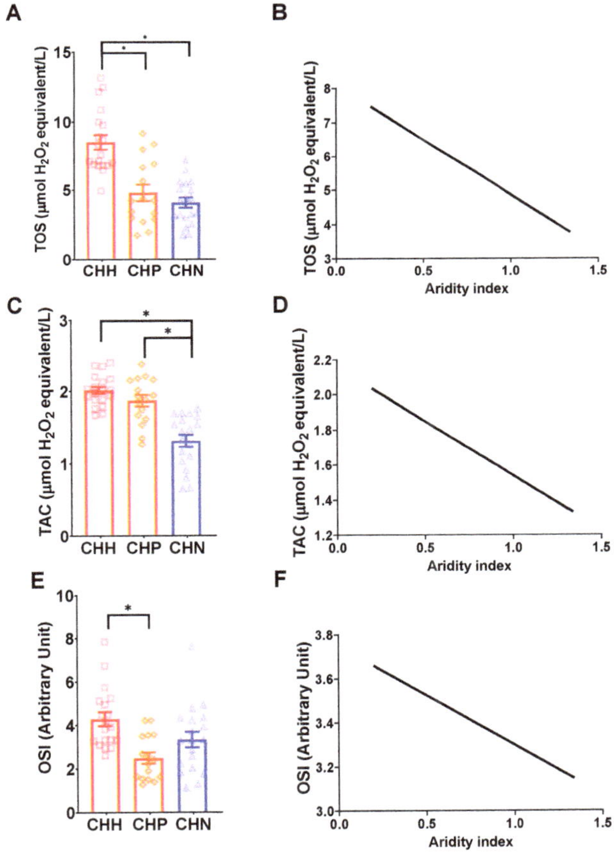

Figure 4. The relationship between oxidative markers and the aridity index between three different wild-caught mole-rats, namely common mole-rats, *Cryptomys hottentotus hottentotus* (CHH—red with squares), highveld mole-rats, *C. h. pretoriae* (CHP—orange with diamonds) and the Natal mole-rat, *C. h. natalensis* (CHN—blue with triangles) for (**A**,**B**) total oxidant status (TOS), (**C**,**D**) total antioxidant activity (TAC), and (**E**,**F**) oxidative stress index (OSI). Bars represent the mean ± s.e.m. An asterisk (*) indicates significance ($p \leq 0.05$).

3.3.2. TAC

The best model did not retain the ERMR for wild-caught mole-rats (CHP, CHH, and CHN) (Table 4). A significant effect of subspecies was observed in the model (Table 4), where post-hoc comparisons between the three subspecies demonstrated that CHN had a significantly lower TAC compared to CHH (t = −6.319, df = 51, $p < 0.0001$) and CHP (t = 5.122, df = 51, $p < 0.0001$) (Figure 4C). We also found that CHH had a higher TAC than CHP, but this was not significant (t = −1.083, df = 51, $p = 0.53$) (Figure 4C). Simple linear regressions revealed that TAC had a significant negative relationship with an increasing aridity index ($R^2 = 0.5145$, $F(_{1,52}) = 55.10$, $p < 0.0001$) (Figure 4D).

3.3.3. OSI

The ERMR for wild-caught mole-rats (CHP, CHH, and CHN) was not retained in the best model and was found to be significant (Table 4). A significant effect of subspecies was observed in the model (Table 4). Post-hoc comparisons between the three subspecies demonstrated that CHH had a significantly higher OSI as compared to CHP (t = −3.657, df = 51, $p = 0.002$) and higher OSI as compared to CHN, but this was not significant (t = −1.890, df = 51, $p = 0.15$) (Figure 4E). The OSI did not significantly differ between CHN and CHP (t = −2.061, df = 51, $p = 0.11$), whereas the CHN had a slightly higher OSI (Figure 4E). A simple linear regression found a non-significant relationship between the OSI and the aridity index ($R^2 = 0.022$, $F(_{1,52}) = 1.174$, $p = 0.28$) (Figure 4F).

3.4. Principal Component Analyses of Species

PCA plots revealed underlying factors in the plasma oxidative markers for species. Since sociality is not a continuous variable, species demarcation allows for the influence of eusociality, sociality, and solitary lifestyles for interpretation. The benefit of using a PCA over the previous analyses is the high likelihood of multicollinearity among variables, thus avoiding that complication and allowing for a better understanding of the relationship of these variables to each other.

Using the skreeplot, we identified two PCs with an Eigenvalue above one; as such, we utilised two PCs (PC1 and PC2). PC1 and PC2 cumulatively explained 75.3% of the variance (Figure 5). The loading of these variables within each PC was determined using linear correlations and visually represented in the direction of the arrows (Figure 5).

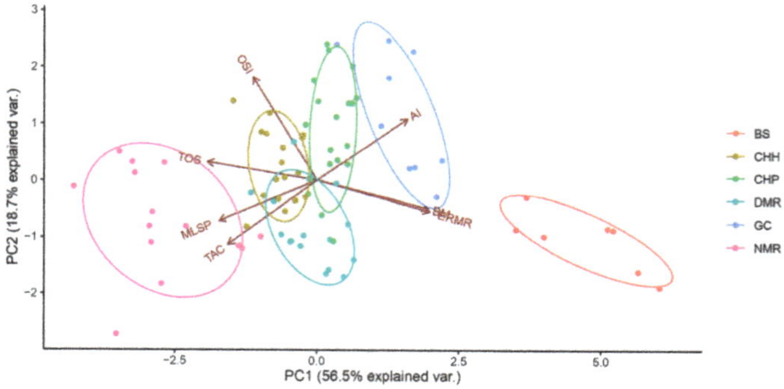

Figure 5. Principal component analysis illustrating the relationships between the total oxidant status (TOS-μmol H_2O_2 equivalents/L), total antioxidant capacity (TAC-mmol Trolox equivalents/L), oxidative stress index (OSI-arbitrary unit), maximum lifespan potential (MLSP), aridity index (AI), body mass (BM), and estimated resting metabolic rate (ERMR) for captive mole-rat species, namely, the naked mole-rat *Heterocephalus glaber* (NMR), the Damaraland mole-rat, *Fukomys damarensis* (DMR), the highveld mole-rat, *Cryptomys hottentotus pretoriae* (CHP), the common mole-rat, *C. h. hottentotus* (CHH), the Cape mole-rat, *Georychus capensis* (GC), and the Cape Dune mole-rat, *Bathyergus suillus* (BS). PC1 explained 55% of the variance and PC2 explained 20.3% of the variance for a cumulative variance of 75.4%. The direction of the arrows represents the loadings of the variables within a PC.

PC1 separated the species and subspecies mainly in two directions, where species with a higher body mass and ERMR were found in habitats with a higher aridity index (Figure 5). However, the direction of those three variables (BM, ERMR, and AI) negatively influenced TOS, TAC, OSI, and MLSP (Figure 5). This suggests that the more social species and subspecies (NMR, DMR, CHP, and CHH) orientated more to the left have a higher TOS, TAC, and OSI, but interestingly, would tend to have a higher MLSP (Figure 5). Contrastingly, the solitary species (GS and BS) are found on the right of the PCA, suggesting a lower TAC, TOS, and OSI, and tend to have a lower MLSP (Figure 5). PC2 separated species primarily on OSI, TAC, and the aridity index, implying that species and subspecies typically have a higher OSI and lower TAC as the aridity index increases (Figure 5). The other variables explained the minimal variance to these three factors (correlation < 0.25).

4. Discussion

This study investigated several factors which could have a possible influence on oxidative marker levels, such as phylogeny [155,156], sociality [72,157,158], wild-caught and captive differences [159], aridity [117], RMR [40,43,160], and MLSP [41,63,64]. Body mass and RMR have previously been shown to correlate with free radical production, where a higher metabolic rate would assume increased free radical production [12,160,161], with some exceptions [43,162,163]. Oxidative stress in species with increased MLSP is generally lower, whereby they either change the rate of production or have additional means of protection [13,41,164]. African mole-rats present a conundrum as they typically have lower metabolic rates and high oxidative stress, yet they are also long-lived. Our goal in this study was to identify, from the data available, how some oxidative markers correlate within and between subspecies and species of African mole-rats, where markers can be directly compared. Our current analyses provide insights into the critical processes that may influence the oxidative markers in African mole-rats. These observations include: (1) oxidative markers and sociality may explain some patterns associated with longevity but are likely outweighed by other factors, (2) a positive association between oxidative stress and MLSP, and (3) oxidative markers can vary due to changes in RMR and aridity, which may be more profound in some subspecies and species compared to others. There are some limitations to the current study, with the most notable including the limited number of species and subspecies of Bathyergidae investigated in this study. Some excluded species include the highly social giant mole-rat, *F. mechowii*, and some solitary mole-rats, namely the silvery mole-rat, *Heliophobius argenteocinereus*, and Namaqua Dune mole-rat, *B. janetta*. Namaqua Dune mole-rat data are hard to obtain due to species decline in the arid habitat (N.C. Bennett and D.W. Hart personal communication), resulting in their exclusion from this analysis. Additional shortcomings include the lack of additional oxidative markers, such as DNA-damage markers and enzymatic antioxidants.

Sociality and longevity are key aspects of the life history of a species [158,165,166], i.e., Williams and Shattuck [157] identified that eusociality and habitat play a prominent role in determining longevity. Zhu et al. [167] also provided further evidence for social subspecies and species having greater longevity in mammalian phylogenies as compared to solitary species. Previously, sociality has been correlated to longevity in insects [168]. It was proposed that African mole-rats' subterranean and social lifestyle contribute to their longevity [72,157], but Healy [169] proposed that being subterranean is not associated with MLSP. Our analysis emphasises habitat (determined through the AI) and sociality (social species tend to be more to the left of the PCA analysis) to be two components determining oxidative markers and orientation towards MLSP of the PCA. According to the aridity food distribution [170] and behavioural osmoregulation hypotheses [115], animals in habitats with increased aridity tend to be more social, and our data agree with this. However, this increased aridity congruently contributes to elevated oxidative stress, which seems to be promoting longevity in the social subspecies and species. One outlier in the study is that the solitary GC appears closer to the social species as opposed to the BS. Because the GC is a solitary species, it is expected to trend away from MLSP in our PCA. Since this was not the case, oxidative markers and the increased MLSP of a species are likely closely tied to other factors and not sociality, despite sociality contributing to the observation (social species tend to be more to the left in the direction of MLSP). Furthermore, Dammann, Šaffa and Šumbera [72] have also shown extended longevity in the solitary silvery mole-rat. They also agreed that sociality promotes longevity in this family [72]. Thus, our results agree with Williams and Shattuck [157] and partially disagree with Healy [169], emphasising that sociality plays a strong role in the correlations of oxidative markers to the MLSP in the family Bathyergidae but that habitat may have a greater influence.

Increased ROS species formation has been predicted as a significant determinant of the ageing process mediated through metabolic rate-producing free radicals [12,23,40,63]. Despite the simplicity of this hypothesis, recent evidence suggests that this link is not so straightforward and even contradicting. The general trend accepts that reducing oxidative stress promotes longevity, which is closely linked to body mass, metabolic rate, and the rate of free radical production [15,41]. Astoundingly, some cases report increased ROS formation to promote health and lifespan [40]. This phenomenon may likely manifest under a biphasic response of hormesis [32,37,38,171], where it has previously been shown that hormetic effects improve survival under chemical challenges such as hypoxia [32,172], where an elevated oxidative stress profile is expected in order to upregulate defences. This highlights a fundamental question: despite the high oxidative lipid peroxides in some social African mole-rats, what is the baseline to indicate a detrimental oxidative stress level to these mole-rats? Our data support the idea that increased oxidative stress benefits animals, resulting in increased MLSP.

Previously, it has been shown that the overall levels of antioxidants among mammalian species do not correlate with MLSP [45,173]. Since we compared captive individuals, food provisioning can affect antioxidant levels, as non-enzymatic antioxidants are primarily obtained from food [174,175]. Even with an extended time in captivity, these TAC differences are still present, suggesting species differences in the inter-organ transport of antioxidants [176,177]. Previously, NMRs have been suggested to have poor antioxidant activity [178,179], where the antioxidant activity was proposed to be independent of the high MLSP observed [45,180]. Munro et al. [60] found significantly higher antioxidant defences for NMRs, as NMRs consume hydrogen peroxide at a much higher rate in the matrix of mitochondria. In our study, NMRs had the highest TAC of all captive animals, which suggests NMR have the most effective non-enzymatic antioxidant activity (Supplementary File S2). Furthermore, the TAC in our PCA analyses contributes to the MLSP of the species. Importantly, NMRs were similar to the other social mole-rats and the solitary GC when considering OSI, while only the DMR and BS had a lower OSI. Lastly, since the GC demonstrated a similar OSI to the social species, but vastly different TOS and TAC, it suggests that the GC have a similar ecological mechanism as their OSI is similar to social mole-rats and maintains a higher MLSP than BS.

Emerging evidence suggests that an extended lifespan may be maintained by natural selection as a product of organismal adaption [181], where the majority of longevity attributes are under genetic regulation [182,183], and that these genetic regulators are constrained by environmental factors [184,185]. Hypoxia tolerance is an adaptation that may be one of the most relevant pathways promoting longevity in African mole-rats [73,88,89,186]. African mole-rats live in a subterranean environment, which is hypoxic and hypercapnic [72,125,157], promoting factors such as optimising oxygen uptake or reducing the oxygen requirement for metabolism [187]. The most well-known hypoxia-driven adaptation involves selection on the central metabolism, cellular respiration, haemoglobin-mediated oxygen transport, hypoxia-inducible factor pathways, and decreased thermogenesis [181,188–190]. Hypoxia adaptation induces hypothermia and hypo-metabolism, which are physiologically similar to calorie-restricted animals [70,191] and linked to extended lifespan [69,71]. The adaptation of living in a subterranean habitat gave way to these attributes [71,157], which suggests that for oxidative stress, hypoxia tolerance as an adaptation to a subterranean lifestyle promotes MLSP and disagrees with Healy [169]. This is supported by a recent study where mice underwent metabolic remodelling by exposure to NMR-like living conditions of hypercapnia and hypoxia (as experienced as a fossorial animal), ultimately leading to these mice living longer [70]. From our data, all species except for BS are hypoxia tolerant, with the NMR being the most extreme hypoxia-tolerant species [88,89]. The consequences of hypoxia generally result in increased antioxidants and increased ROS production, which can overall result in higher oxidative stress initiating hormesis, thus promoting longevity [32,37,171]. This also supports why some of the more hypoxia-tolerant species in this study have higher levels of all oxidative markers except for GC. Two factors separate the GC from the other African mole-rats, despite similarities in OSI. They live in a mesic habitat and are generally larger with a lower mass-specific RMR [76,192], both characteristics that have been shown to alter observed oxidative marker measurements. Additionally, despite being hypoxia tolerant like the social African mole-rats species, the GC does differ in some of their responses to hypoxia compared to some of the social species [88,193]. Lastly, Ivy et al. [88] suggested that these mechanisms were independent of body size and sociality, suggesting environmental pressures from their environment gave rise to these traits.

Oxidative markers, particularly non-enzymatic antioxidants, are influenced by the immediate environment an animal finds itself in due to changes in food availability, food types, and their antioxidant content [117,174,175,194,195]. Our study emphasises the comparison between captive and wild-caught species, as wild-caught species had a significantly higher TAC than captive animals' TAC. Our data likely support differences in diet under a wild context and possible arid-adapted mechanisms to obtain higher antioxidants, as both species had a similar TAC in captivity, supported by the significant negative relationship observed between AI and TAC. One additional mechanism that could affect differences observed in TAC is the differences in RMR, as elevated RMR may likely require additional antioxidants to combat the effects of elevated metabolism and other confounding effects of an increasingly arid environment such as water stress and thermal stress [92–94,196]. It may also be that antioxidants are not just readily available in foods in more arid environments to protect against droughts [194,197] but that animals in these more arid environments likely mobilise antioxidants more readily to deal with thermal challenges [176,177]. A similar observation was observed in the kidney tissues of CHH in differing arid environments [117]. Interestingly, this may suggest that oxidative kidney and plasma markers show similar changes in TAC, but this requires further investigation. Lastly, previous studies have observed differences in the RMR between wild-caught and captive species [88,92,120], but the current study did not observe significant differences in the RMR between wild-caught and captive mole-rats. This may likely be due to using an estimated measure of RMR which could not compensate for possible individual differences in RMR determination or that RMR differences between wild-caught and captive individuals are not large enough, as some species, such as the CHH, demonstrate a much larger increased in metabolic rate between captive and wild-caught individuals as opposed to the BS and CHP [88,92,120].

The study of the family Bathyergidae may have profound medical importance due to their cancer resistance, hypoxia-tolerance, and factors that promote longevity [59,66,75,198]. Other factors include the reliance on glycolysis instead of oxidative phosphorylation [186], which is enigmatic, as some African mole-rats demonstrate high oxidative stress despite the reduced ROS production from glycolysis instead of oxidative phosphorylation [199]. This highlights several vital factors for future research. Firstly, the determination of how other oxidative markers correlate, which can include other damage markers, in particular DNA, due to the relevance of DNA damage to ageing [200–203], and other antioxidant measures such as enzymatic activity [60,204] and other unconventional antioxidants such as uric acid [65,205,206] or melatonin [204,207]. Secondly, the rate of free radical production, instead of reliance on the measure of metabolism, as the metabolic rate may obscure the rate of free radical production [43]. Despite this, cellular ROS production does not correlate with longevity [208], although some studies have shown that where free radicals are produced is essential (near DNA) [35,164]. Thirdly, the susceptibility to oxidative stress and/or damage may include factors such as repair mechanisms [35,71], lipid membrane composition, which has previously also correlated to longevity [68,180], and protein structure and stability [64,209]. Lastly, excluding the domestic rat, there was a lack of plasma TOS and TAC oxidative markers available for phylogenetic-controlled redox comparisons outside the family Bathyergidae. The guinea pig had a single study [210] and there have been no studies on the African porcupine (*Hystrix africaeaustralis*), typical outgroup species used for mole-rat phylogenetic analyses [89,134]. Acquiring oxidative data from outgroup species is important to determine whether phylogeny is critical in redox physiology. As suggested, several avenues remain to be explored, not just how these factors may mediate oxidative markers but how they interact, to determine the mechanisms promoting longevity in species.

Wong, Freeman and Zhang [75] have previously stated that, despite the convergent evolution of the NMR and DMR, the DMR has demonstrated similar mechanisms to the NMR but has attracted less attention. Additionally, many physiological mechanisms for different adaptations in the NMR remain unexplored for the DMR [75]. From an oxidative stress perspective, our results show that the DMR is similar to the other social African mole-rat species, such as the *Cryptomys* sp. and GC, to some extent. In summary, we found a positive oxidative stress relationship with MLSP and a negative relationship between oxidative stress and RMR. This may be an evolutionary artefact present in African mole-rats, but we encourage further research to determine the similarities and differences of the NMR, not just to the DMR, but to the whole family of Bathyergidae.

Supplementary Materials: The following supporting information can be downloaded at: https://www.mdpi.com/article/10.3390/antiox12081486/s1. File S1: The family of Bathyergidae, their distribution, their life history, reproductive structure, reproductive suppression and oxidative ecology; File S2: Data.

Author Contributions: Conceptualization, P.J.J. and D.W.H.; Methodology, P.J.J.; Formal analysis, P.J.J., D.W.H. and H.N.M.; Investigation, P.J.J. and D.W.H.; Resources, N.C.B.; Data curation, H.N.M. and C.V.; Writing—original draft, P.J.J.; Writing—review & editing, P.J.J., D.W.H., H.N.M., C.V. and N.C.B.; Funding acquisition, N.C.B. All authors have read and agreed to the published version of the manuscript.

Funding: N.C.B. acknowledges funding from the SARChI chair of Mammalian Behavioural Ecology and Physiology from the DST-NRF South Africa, the National Research Foundation (grant no. 64756), and the Natural Environment Research Council under grant NE\L002485\1, and the University of Pretoria.

Institutional Review Board Statement: Permission to capture common mole rats was obtained from all landowners. A collecting permit was obtained from the relevant nature conservation authorities (permit No. FAUNA 0419/2021, FAUNA 042/2021, CPB6-1161, CPB6-1163, CNN44-87-17699). The Animal Use and Care Committee of the University of Pretoria evaluated and approved the experimental protocol (ethics clearance No. NAS016/2021) and DAFF section 20 approval (SDAH-Epi-21031811071).

Informed Consent Statement: Not applicable.

Data Availability Statement: Data are contained within the article or Supplementary Materials.

Acknowledgments: N.C.B. acknowledges funding from the SARChI chair of Mammalian Behavioural Ecology and Physiology from the DST-NRF South Africa, the National Research Foundation (grant no. 64756), and the University of Pretoria. We would like to thank Kerryn L. Grenfell for the drawing of the various African mole-rats.

Conflicts of Interest: The authors declare no conflict of interest.

References

1. Sessions, A.L.; Doughty, D.M.; Welander, P.V.; Summons, R.E.; Newman, D.K. The continuing puzzle of the great oxidation event. *Curr. Biol.* **2009**, *19*, R567–R574. [CrossRef]
2. Nisbet, E.; Sleep, N. The habitat and nature of early life. *Nature* **2001**, *409*, 1083–1091. [CrossRef]

3. Falkowski, P.G.; Katz, M.E.; Knoll, A.H.; Quigg, A.; Raven, J.A.; Schofield, O.; Taylor, F. The evolution of modern eukaryotic phytoplankton. *Science* **2004**, *305*, 354–360. [CrossRef]
4. Costantini, D. Oxidative stress in ecology and evolution: Lessons from avian studies. *Ecol. Lett.* **2008**, *11*, 1238–1251. [CrossRef] [PubMed]
5. Birnie-Gauvin, K.; Costantini, D.; Cooke, S.J.; Willmore, W.G. A comparative and evolutionary approach to oxidative stress in fish: A review. *Fish Fish.* **2017**, *18*, 928–942. [CrossRef]
6. Halliwell, B.; Gutteridge, J.M. *Free Radicals in Biology and Medicine*; Oxford University Press: New York, NY, USA, 2015.
7. Sanz, A. Mitochondrial reactive oxygen species: Do they extend or shorten animal lifespan? *Biochim. Biophys. Acta BBA—Bioenergetics* **2016**, *1857*, 1116–1126. [CrossRef]
8. Apel, K.; Hirt, H. Reactive oxygen species: Metabolism, oxidative stress, and signaling transduction. *Annu. Rev. Plant Biol.* **2004**, *55*, 373. [CrossRef] [PubMed]
9. Gerschman, R.; Gilbert, D.L.; Nye, S.W.; Dwyer, P.; Fenn, W.O. Oxygen poisoning and x-irradiation: A mechanism in common. *Science* **1954**, *119*, 623–626. [CrossRef]
10. Sies, H. *What Is Oxidative Stress?* Springer: Berlin/Heidelberg, Germany, 2000.
11. Sies, H. Oxidative stress: A concept in redox biology and medicine. *Redox. Biol.* **2015**, *4*, 180–183. [CrossRef]
12. Harman, D. Aging: A Theory Based on Free Radical and Radiation Chemistry. *J. Gerontol.* **1956**, *11*, 298–300. [CrossRef]
13. Barja, G. Updating the mitochondrial free radical theory of aging: An integrated view, key aspects, and confounding concepts. *Antioxid. Redox Signal.* **2013**, *19*, 1420–1445. [CrossRef] [PubMed]
14. Beckman, K.B.; Ames, B.N. The free radical theory of aging matures. *Physiol. Rev.* **1998**, *78*, 547–581. [CrossRef] [PubMed]
15. Finkel, T.; Holbrook, N.J. Oxidants, oxidative stress and the biology of ageing. *Nature* **2000**, *408*, 239. [CrossRef] [PubMed]
16. Costantini, D. Variation in oxidative stress threats and hormesis across environments. In *Oxidative Stress and Hormesis in Evolutionary Ecology and Physiology*; Springer: Berlin/Heidelberg, Germany, 2014; pp. 75–109. [CrossRef]
17. Costantini, D.; Rowe, M.; Butler, M.W.; McGraw, K.J. From molecules to living systems: Historical and contemporary issues in oxidative stress and antioxidant ecology. *Funct. Ecol.* **2010**, *24*, 950–959. [CrossRef]
18. Monaghan, P.; Metcalfe, N.B.; Torres, R. Oxidative stress as a mediator of life history trade-offs: Mechanisms, measurements and interpretation. *Ecol. Lett.* **2009**, *12*, 75–92. [CrossRef]
19. Liguori, I.; Russo, G.; Curcio, F.; Bulli, G.; Aran, L.; Della-Morte, D.; Gargiulo, G.; Testa, G.; Cacciatore, F.; Bonaduce, D. Oxidative stress, aging, and diseases. *Clin. Interv. Aging* **2018**, 757–772. [CrossRef]
20. Hekimi, S.; Lapointe, J.; Wen, Y. Taking a "good" look at free radicals in the aging process. *Trends Cell Biol.* **2011**, *21*, 569–576. [CrossRef]
21. Shields, H.J.; Traa, A.; van Raamsdonk, J.M. Beneficial and detrimental effects of reactive oxygen species on lifespan: A comprehensive review of comparative and experimental studies. *Front. Cell Dev. Biol.* **2021**, *9*, 181. [CrossRef]
22. Selman, C.; Blount, J.D.; Nussey, D.H.; Speakman, J.R. Oxidative damage, ageing, and life-history evolution: Where now? *Trends Ecol. Evol.* **2012**, *27*, 570–577. [CrossRef]
23. Shigenaga, M.K.; Hagen, T.M.; Ames, B.N. Oxidative damage and mitochondrial decay in aging. *Proc. Natl. Acad. Sci. USA* **1994**, *91*, 10771–10778. [CrossRef]
24. Wei, Y.-H.; Lee, H.-C. Oxidative stress, mitochondrial DNA mutation, and impairment of antioxidant enzymes in aging. *Exp. Biol. Med.* **2002**, *227*, 671–682. [CrossRef]
25. Stuart, J.A.; Maddalena, L.A.; Merilovich, M.; Robb, E.L. A midlife crisis for the mitochondrial free radical theory of aging. *Longev. Healthspan* **2014**, *3*, 1–15. [CrossRef] [PubMed]
26. Buffenstein, R.; Edrey, Y.H.; Yang, T.; Mele, J. The oxidative stress theory of aging: Embattled or invincible? Insights from non-traditional model organisms. *Age* **2008**, *30*, 99–109. [CrossRef] [PubMed]
27. Sanz, A.; Pamplona, R.; Barja, G. Is the mitochondrial free radical theory of aging intact? *Antioxid. Redox Signal.* **2006**, *8*, 582–599. [CrossRef]
28. Pacifici, R.E.; Davies, K.J. Protein, lipid and DNA repair systems in oxidative stress: The free-radical theory of aging revisited. *Gerontology* **1991**, *37*, 166–180. [CrossRef]
29. Droge, W. Free radicals in the physiological control of cell function. *Physiol. Rev.* **2002**, *82*, 47–95. [CrossRef]
30. Giordano, F.J. Oxygen, oxidative stress, hypoxia, and heart failure. *J. Clin. Investig.* **2005**, *115*, 500–508. [CrossRef]
31. Blair, A.S.; Hajduch, E.; Litherland, G.J.; Hundal, H.S. Regulation of Glucose Transport and Glycogen Synthesis in L6 Muscle Cells during Oxidative stress evidence for cross-talk between the insulin and sapk2/p38 mitogen-activated protein kinase signaling pathways. *J. Biol. Chem.* **1999**, *274*, 36293–36299. [CrossRef]
32. Oliveira, M.F.; Geihs, M.A.; França, T.F.; Moreira, D.C.; Hermes-Lima, M. Is "preparation for oxidative stress" a case of physiological conditioning hormesis? *Front. Physiol.* **2018**, *9*, 945. [CrossRef]
33. Burg, M.B.; Ferraris, J.D.; Dmitrieva, N.I. Cellular response to hyperosmotic stresses. *Physiol. Rev.* **2007**, *87*, 1441–1474. [CrossRef]
34. Chung, H.Y.; Sung, B.; Jung, K.J.; Zou, Y.; Yu, B.P. The molecular inflammatory process in aging. *Antioxid. Redox Signal.* **2006**, *8*, 572–581. [CrossRef] [PubMed]
35. Perez-Campo, R.; Lopez-Torres, M.; Cadenas, S.; Rojas, C.; Barja, G. The rate of free radical production as a determinant of the rate of aging: Evidence from the comparative approach. *J. Comp. Physiol. B* **1998**, *168*, 149–158. [CrossRef] [PubMed]

36. Rodriguez, K.A.; Wywial, E.; Perez, V.I.; Lambert, A.J.; Edrey, Y.H.; Lewis, K.N.; Grimes, K.; Lindsey, M.L.; Brand, M.D.; Buffenstein, R. Walking the oxidative stress tightrope: A perspective from the naked mole-rat, the longest-living rodent. *Curr. Pharm. Des.* **2011**, *17*, 2290–2307. [CrossRef]
37. Costantini, D. *Oxidative Stress and Hormesis in Evolutionary Ecology and Physiology: A Marriage between Mechanistic and Evolutionary Approaches*; Springer: Berlin/Heidelberg, Germany, 2014; pp. 19–22.
38. Hood, W.; Zhang, Y.; Mowry, A.; Hyatt, H.; Kavazis, A. Life history trade-offs within the context of mitochondrial hormesis. *Integr. Comp. Biol.* **2018**, *58*, 567–577. [CrossRef]
39. Sena, L.A.; Chandel, N.S. Physiological roles of mitochondrial reactive oxygen species. *Mol. Cell* **2012**, *48*, 158–167. [CrossRef]
40. Ristow, M.; Schmeisser, S. Extending life span by increasing oxidative stress. *Free Radic. Biol. Med.* **2011**, *51*, 327–336. [CrossRef]
41. Barja, G. Rate of generation of oxidative stress-related damage and animal longevity. *Free Radic. Biol. Med.* **2002**, *33*, 1167–1172. [CrossRef] [PubMed]
42. Golden, T.R.; Hinerfeld, D.A.; Melov, S. Oxidative stress and aging: Beyond correlation. *Aging Cell* **2002**, *1*, 117–123. [CrossRef]
43. Speakman, J.R. Body size, energy metabolism and lifespan. *J. Exp. Biol.* **2005**, *208*, 1717–1730. [CrossRef]
44. Buttemer, W.A.; Abele, D.; Costantini, D. From bivalves to birds: Oxidative stress and longevity. *Funct. Ecol.* **2010**, *24*, 971–983. [CrossRef]
45. Lewis, K.N.; Andziak, B.; Yang, T.; Buffenstein, R. The naked mole-rat response to oxidative stress: Just deal with it. *Antioxid. Redox. Sign.* **2013**, *19*, 1388–1399. [CrossRef] [PubMed]
46. Jones, J.H. Primates and the evolution of long, slow life histories. *Curr. Biol.* **2011**, *21*, R708–R717. [CrossRef]
47. Austad, S.N.; Fischer, K.E. Mammalian aging, metabolism, and ecology: Evidence from the bats and marsupials. *J. Gerontol.* **1991**, *46*, B47–B53. [CrossRef]
48. Holmes, D.; Flückiger, R.; Austad, S. Comparative biology of aging in birds: An update. *Exp. Gerontol.* **2001**, *36*, 869–883. [CrossRef] [PubMed]
49. Holmes, D.J.; Austad, S.N. Birds as animal models for the comparative biology of aging: A prospectus. *J. Gerontol. Ser. A Biol. Sci. Med. Sci.* **1995**, *50*, B59–B66. [CrossRef] [PubMed]
50. Holmes, D.; Ottinger, M. Birds as long-lived animal models for the study of aging. *Exp. Gerontol.* **2003**, *38*, 1365–1375. [CrossRef] [PubMed]
51. Dammann, P. Slow aging in mammals—Lessons from African mole-rats and bats. In *Seminars in Cell & Developmental Biology*; Academic Press: Cambridge, MA, USA, 2017; pp. 154–163.
52. Wilkinson, G.S.; Adams, D.M.; Haghani, A.; Lu, A.T.; Zoller, J.; Breeze, C.E.; Arnold, B.D.; Ball, H.C.; Carter, G.G.; Cooper, L.N. DNA methylation predicts age and provides insight into exceptional longevity of bats. *Nat. Commun.* **2021**, *12*, 1615. [CrossRef]
53. Wilkinson, G.S.; Adams, D.M. Recurrent evolution of extreme longevity in bats. *Biol. Lett.* **2019**, *15*, 20180860. [CrossRef]
54. Gorbunova, V.; Seluanov, A.; Kennedy, B.K. The world goes bats: Living longer and tolerating viruses. *Cell Metab.* **2020**, *32*, 31–43. [CrossRef]
55. Dammann, P.; Šumbera, R.; Maßmann, C.; Scherag, A.; Burda, H. Extended longevity of reproductives appears to be common in Fukomys mole-rats (Rodentia, Bathyergidae). *PLoS ONE* **2011**, *6*, e18757. [CrossRef]
56. Dammann, P.; Burda, H. Senescence patterns in African mole-rats (Bathyergidae, Rodentia). In *Subterranean Rodents*; Springer: Berlin/Heidelberg, Germany, 2007; pp. 251–263.
57. Buffenstein, R. Negligible senescence in the longest living rodent, the naked mole-rat: Insights from a successfully aging species. *J. Comp. Physiol. B* **2008**, *178*, 439–445. [CrossRef] [PubMed]
58. Andziak, B.; Buffenstein, R. Disparate patterns of age-related changes in lipid peroxidation in long-lived naked mole-rats and shorter-lived mice. *Aging Cell* **2006**, *5*, 525–532. [CrossRef] [PubMed]
59. Buffenstein, R. The naked mole-rat: A new long-living model for human aging research. *J. Gerontol. A Biol. Sci. Med. Sci.* **2005**, *60*, 1369–1377. [CrossRef] [PubMed]
60. Munro, D.; Baldy, C.; Pamenter, M.E.; Treberg, J.R. The exceptional longevity of the naked mole-rat may be explained by mitochondrial antioxidant defenses. *Aging Cell* **2019**, *18*, e12916. [PubMed]
61. Sahm, A.; Platzer, M.; Koch, P.; Henning, Y.; Bens, M.; Groth, M.; Burda, H.; Begall, S.; Ting, S.; Goetz, M. Increased longevity due to sexual activity in mole-rats is associated with transcriptional changes in the HPA stress axis. *eLife* **2021**, *10*, e57843. [CrossRef] [PubMed]
62. Zuo, W.; Tang, X.; Hou, C. Why Naked Mole-Rats Have High Oxidative Damage but Live a Long Life: A Simple Explanation Based on the Oxidative Stress Theory of Aging. *Adv. Geriatr. Med. Res.* **2020**, *2*, e200006.
63. Agarwal, S.; Sohal, R. Relationship between susceptibility to protein oxidation, aging, and maximum life span potential of different species. *Exp. Gerontol.* **1996**, *31*, 365–372. [CrossRef]
64. Pérez, V.I.; Buffenstein, R.; Masamsetti, V.; Leonard, S.; Salmon, A.B.; Mele, J.; Andziak, B.; Yang, T.; Edrey, Y.; Friguet, B. Protein stability and resistance to oxidative stress are determinants of longevity in the longest-living rodent, the naked mole-rat. *Proc. Natl. Acad. Sci. USA* **2009**, *106*, 3059–3064. [CrossRef]
65. Buffenstein, R.; Craft, W. The idiosyncratic physiological traits of the naked mole-rat; a resilient animal model of aging, longevity, and healthspan. In *The Extraordinary Biology of the Naked Mole-Rat*; Springer: Cham, Switzerland, 2021; pp. 221–254.
66. Lewis, K.N.; Buffenstein, R. The naked mole-rat: A resilient rodent model of aging, longevity, and healthspan. In *Handbook of the Biology of Aging*; Elsevier: Amsterdam, The Netherlands, 2016; pp. 179–204.

67. Mitchell, T.W.; Buffenstein, R.; Hulbert, A. Membrane phospholipid composition may contribute to exceptional longevity of the naked mole-rat (*Heterocephalus glaber*): A comparative study using shotgun lipidomics. *Exp. Gerontol.* **2007**, *42*, 1053–1062. [CrossRef]
68. Hulbert, A.J.; Faulks, S.C.; Buffenstein, R. Oxidation-resistant membrane phospholipids can explain longevity differences among the longest-living rodents and similarly-sized mice. *J. Gerontol. Ser. A Biol. Sci. Med. Sci.* **2006**, *61*, 1009–1018. [CrossRef]
69. Lewis, K.N.; Rubinstein, N.D.; Buffenstein, R. A window into extreme longevity; the circulating metabolomic signature of the naked mole-rat, a mammal that shows negligible senescence. *Geroscience* **2018**, *40*, 105–121. [CrossRef] [PubMed]
70. Tolstun, D.A.; Knyazer, A.; Tushynska, T.V.; Dubiley, T.A.; Bezrukov, V.V.; Fraifeld, V.E.; Muradian, K.K. Metabolic remodelling of mice by hypoxic-hypercapnic environment: Imitating the naked mole-rat. *Biogerontology* **2020**, *21*, 143–153. [CrossRef] [PubMed]
71. Fang, X.; Seim, I.; Huang, Z.; Gerashchenko, M.V.; Xiong, Z.; Turanov, A.A.; Zhu, Y.; Lobanov, A.V.; Fan, D.; Yim, S.H. Adaptations to a subterranean environment and longevity revealed by the analysis of mole rat genomes. *Cell Rep.* **2014**, *8*, 1354–1364. [CrossRef] [PubMed]
72. Dammann, P.; Šaffa, G.; Šumbera, R. Longevity of a solitary mole-rat species and its implications for the assumed link between sociality and longevity in African mole-rats (Bathyergidae). *Biol. Lett.* **2022**, *18*, 20220243. [CrossRef]
73. Pamenter, M.E.; Munro, D. Longevity or hypoxia: Who's driving? *Aging* **2019**, *11*, 5864. [CrossRef]
74. Jacobs, P.J.; Hart, D.W.; Bennett, N.C. Plasma oxidative stress in reproduction of two eusocial African mole-rat species, the naked mole-rat and the Damaraland mole-rat. *Front. Zool.* **2021**, *18*, 45. [CrossRef]
75. Wong, H.-S.; Freeman, D.A.; Zhang, Y. Not just a cousin of the naked mole-rat: Damaraland mole-rats offer unique insights into biomedicine. *Comp. Biochem. Physiol. Part B Biochem. Mol. Biol.* **2022**, 110772. [CrossRef]
76. Šumbera, R. Thermal biology of a strictly subterranean mammalian family, the African mole-rats (Bathyergidae, Rodentia)—A review. *J. Therm. Biol.* **2019**, *79*, 166–189. [CrossRef]
77. Faulkes, C.G.; Bennett, N.C. African mole-rats: Social and ecological diversity. In *Rodent Societies: An Ecological and Evolutionary Perspective*; University of Chicago Press: Chicago, IL, USA, 2007; pp. 427–437.
78. Holtze, S.; Braude, S.; Lemma, A.; Koch, R.; Morhart, M.; Szafranski, K.; Platzer, M.; Alemayehu, F.; Goeritz, F.; Hildebrandt, T.B. The microenvironment of naked mole-rat burrows in East Africa. *Afr. J. Ecol.* **2018**, *56*, 279–289. [CrossRef]
79. Jacobs, P.J.; Hart, D.W.; Suess, T.; Janse van Vuuren, A.K.; Bennett, N.C. The Cost of Reproduction in a Cooperatively Breeding Mammal: Consequences of Seasonal Variation in Rainfall, Reproduction, and Reproductive Suppression. *Front. Physiol.* **2021**, 2116. [CrossRef]
80. Visser, J.H.; Bennett, N.C.; van Vuuren, B.J. Phylogeny and biogeography of the African Bathyergidae: A review of patterns and processes. *PeerJ* **2019**, *7*, e7730. [CrossRef]
81. Schmidt, C.M.; Jarvis, J.U.; Bennett, N.C. The long-lived queen: Reproduction and longevity in female eusocial Damaraland mole-rats (Fukomys damarensis). *Afr. Zool.* **2013**, *48*, 193–196. [CrossRef]
82. Schmidt, C.M.; Blount, J.D.; Bennett, N.C. Reproduction is associated with a tissue-dependent reduction of oxidative stress in eusocial female Damaraland mole-rats (*Fukomys damarensis*). *PloS ONE* **2014**, *9*, e103286. [CrossRef] [PubMed]
83. Burda, H.; Šumbera, R.; Begall, S. Microclimate in burrows of subterranean rodents—Revisited. In *Subterranean Rodents: News from Underground*; Springer: Berlin/Heidelberg, Germany, 2007; pp. 21–33.
84. Begall, S.; Burda, H.; Schleich, C.E. Subterranean rodents: News from underground. In *Subterranean Rodents: News from Underground*; Springer, 2007; pp. 3–9.
85. Lacey, E.A. *Life Underground: The Biology of Subterranean Rodents*; University of Chicago Press: Chicago, IL, USA, 2000.
86. Nevo, E. Adaptive convergence and divergence of subterranean mammals. *Annu. Rev. Ecol. Syst.* **1979**, *10*, 269–308. [CrossRef]
87. Roper, T.; Bennett, N.C.; Conradt, L.; Molteno, A. Environmental conditions in burrows of two species of African mole-rat, Georhychus capensis and Cryptomys damarensis. *J. Zool.* **2001**, *254*, 101–107. [CrossRef]
88. Ivy, C.M.; Sprenger, R.J.; Bennett, N.C.; van Jaarsveld, B.; Hart, D.W.; Kirby, A.M.; Yaghoubi, D.; Storey, K.B.; Milsom, W.K.; Pamenter, M.E. The hypoxia tolerance of eight related African mole-rat species rivals that of naked mole-rats, despite divergent ventilatory and metabolic strategies in severe hypoxia. *Acta Physiol.* **2020**, *228*, e13436. [CrossRef]
89. Logan, S.M.; Szereszewski, K.E.; Bennett, N.C.; Hart, D.W.; van Jaarsveld, B.; Pamenter, M.E.; Storey, K.B. The brains of six African mole-rat species show divergent responses to hypoxia. *J. Exp. Biol.* **2020**, *223*, jeb215905. [CrossRef]
90. Lovegrove, B. The cost of burrowing by the social mole rats (Bathyergidae) Cryptomys damarensis and Heterocephalus glaber: The role of soil moisture. *Physiol. Zool.* **1989**, *62*, 449–469. [CrossRef]
91. Vleck, D. The energy cost of burrowing by the pocket gopher Thomomys bottae. *Physiol. Zool.* **1979**, *52*, 122–136. [CrossRef]
92. Wallace, K.M.; van Jaarsveld, B.; Bennett, N.C.; Hart, D.W. The joint effect of micro-and macro-climate on the thermoregulation and heat dissipation of two African mole-rat (Bathyergidae) sub-species, Cryptomys hottentotus mahali and C. h. pretoriae. *J. Therm. Biol.* **2021**, 103025. [CrossRef]
93. Hart, D.W.; van Jaarsveld, B.; Lasch, K.G.; Grenfell, K.L.; Oosthuizen, M.K.; Bennett, N.C. Ambient Temperature as a Strong Zeitgeber of Circadian Rhythms in Response to Temperature Sensitivity and Poor Heat Dissipation Abilities in Subterranean African Mole-Rats. *J. Biol. Rhythm.* **2021**, 07487304211034287. [CrossRef]
94. Jacobs, P.J.; Finn, K.T.; van Vuuren, A.K.J.; Suess, T.; Hart, D.W.; Bennett, N.C. Defining the link between oxidative stress, behavioural reproductive suppression and heterothermy in the Natal mole-rat (Cryptomys hottentotus natalensis). *Comp. Biochem. Physiol. Part B Biochem. Mol. Biol.* **2022**, 110753. [CrossRef]

95. Nevo, E.; Reig, O. *Evolution of Subterranean Mammals at the Organismal and Molecular Levels: Proceedings of the Fifth International Theriological Congress Held in Rome, Italy, August 22–29, 1989*; Wiley-Liss: Hoboken, NJ, USA, 1990.
96. Sherman, P.W.; Lacey, E.A.; Reeve, H.K.; Keller, L. The eusociality continuum. *Behav. Ecol.* **1995**, *6*, 102–108. [CrossRef]
97. Michener, C.D. Comparative social behavior of bees. *Annu. Rev. Entomol.* **1969**, *14*, 299–342. [CrossRef]
98. Bennett, N.C.; Faulkes, C.G.; Molteno, A.J. Reproductive suppression in subordinate, non-breeding female Damaraland mole-rats: Two components to a lifetime of socially induced infertility. *Proc. Biol. Sci.* **1996**, *263*, 1599–1603. [CrossRef]
99. Lutermann, H.; Young, A.J.; Bennett, N.C. Reproductive status and testosterone among females in cooperative mole-rat societies. *Gen. Comp. Endocr.* **2013**, *187*, 60–65. [CrossRef] [PubMed]
100. Oosthuizen, M.K.; Bennett, N.C.; Lutermann, H.; Coen, C. Reproductive suppression and the seasonality of reproduction in the social Natal mole-rat (Cryptomys hottentotus natalensis). *Gen. Comp. Endocrinol.* **2008**, *159*, 236–240. [CrossRef]
101. Bennett, N.C.; Ganswindt, A.; Ganswindt, S.B.; Jarvis, J.; Zöttl, M.; Faulkes, C. Evidence for contrasting roles for prolactin in eusocial naked mole-rats, *Heterocephalus glaber* and Damaraland mole-rats, *Fukomys damarensis*. *Biol. Lett.* **2018**, *14*, 20180150. [CrossRef] [PubMed]
102. Medger, K.; Bennett, N.C.; Ganswindt, S.B.; Ganswindt, A.; Hart, D.W. Changes in prolactin, cortisol and testosterone concentrations during queen succession in a colony of naked mole-rats (*Heterocephalus glaber*): A case study. *Sci. Nat.* **2019**, *106*, 1–7. [CrossRef]
103. Blecher, A.S.; Bennett, N.C.; Medger, K.; Hagenah, N.; Ganswindt, A.; Oosthuizen, M.K. Effect of colony disruption and social isolation on naked mole-rat endocrine correlates. *Gen. Comp. Endocrinol.* **2020**, *295*, 113520. [CrossRef]
104. Hart, D.W.; van Vuuren, A.J.; Erasmus, A.; Süess, T.; Hagenah, N.; Ganswindt, A.; Bennett, N.C. The endocrine control of reproductive suppression in an aseasonally breeding social subterranean rodent, the Mahali mole-rat (Cryptomys hottentotus mahali). *Horm. Behav.* **2022**, *142*, 105155. [CrossRef] [PubMed]
105. Young, A.J.; Oosthuizen, M.K.; Lutermann, H.; Bennett, N.C. Physiological suppression eases in Damaraland mole-rat societies when ecological constraints on dispersal are relaxed. *Horm. Behav.* **2010**, *57*, 177–183. [CrossRef]
106. Van der Walt, L.; Bennett, N.C.; Schoeman, S. Reproductive suppression and pituitary sensitivity to exogenous GnRH in the highveld mole-rat (Cryptomys hottentotus pretoriae). *J. Zool.* **2001**, *254*, 177–184. [CrossRef]
107. Jarvis, J.U.; O'Riain, M.J.; Bennett, N.C.; Sherman, P.W. Mammalian eusociality: A family affair. *Trends Ecol. Evol.* **1994**, *9*, 47–51. [CrossRef] [PubMed]
108. Molteno, A.; Bennett, N.C. Rainfall, dispersal and reproductive inhibition in eusocial Damaraland mole-rats (*Cryptomys damarensis*). *J. Zool.* **2002**, *256*, 445–448. [CrossRef]
109. Scantlebury, M.; Speakman, J.R.; Oosthuizen, M.K.; Roper, T.J.; Bennett, N.C. Energetics reveals physiologically distinct castes in a eusocial mammal. *Nature* **2006**, *440*, 795–797. [CrossRef]
110. Spinks, A.; van der Horst, G.; Bennett, N.C. Influence of breeding season and reproductive status on male reproductive characteristics in the common mole-rat, Cryptomys hottentotus hottentotus. *Reproduction* **1997**, *109*, 79–86. [CrossRef]
111. Spinks, A.; Bennett, N.C.; Jarvis, J. Regulation of reproduction in female common mole-rats (Cryptomys hottentotus hottentotus): The effects of breeding season and reproductive status. *J. Zool.* **1999**, *248*, 161–168. [CrossRef]
112. Janse van Rensburg, L. The Reproductive Biology of *Cryptomys hottentotus pretoriae* (Rodentia: Bathyergidae). Ph.D. Thesis, University of Pretoria, Pretoria, South Africa, 2006.
113. Bennett, N.C.; Jarvis, J.; Faulkes, C.; Millar, R. LH responses to single doses of exogenous GnRH by freshly captured Damaraland mole-rats, Cryptomys damarensis. *Reproduction* **1993**, *99*, 81–86. [CrossRef]
114. Bennett, N.C.; Faulkes, C.G.; Hart, L.; Jarvis, J.U. Bathyergus suillus (Rodentia: Bathyergidae). *Mamm. Species.* **2009**, 1–7. [CrossRef]
115. Hart, D.W.; Bennett, N.C.; Oosthuizen, M.K.; Waterman, J.; Hambly, C.; Scantlebury, D. Energetics and water flux in the subterranean rodent family Bathyergidae. *Front. Ecol. Evol.* **2022**, *10*, 867350. [CrossRef]
116. Buffenstein, R.; Yahav, S. Is the naked mole-rat Hererocephalus glaber an endothermic yet poikilothermic mammal? *J. Therm. Biol.* **1991**, *16*, 227–232. [CrossRef]
117. Jacobs, P.J.; Hart, D.W.; Merchant, H.N.; Janse van Vuuren, A.K.; Faulkes, C.G.; Portugal, S.J.; van Jaarsveld, B.; Bennett, N.C. Tissue Oxidative Ecology along an Aridity Gradient in a Mammalian Subterranean Species. *Antioxidants* **2022**, *11*, 2290. [CrossRef]
118. Bennett, N.C.; Clarke, B.; Jarvis, J. A comparison of metabolic acclimation in two species of social mole-rats (Rodentia, Bathyergidae) in southern Africa. *J. Arid. Environ.* **1992**, *23*, 189–198. [CrossRef]
119. Lovegrove, B. Thermoregulation in the subterranean rodent Georychus capensis (Rodentia: Bathyergidae). *Physiol. Zool.* **1987**, *60*, 174–180. [CrossRef]
120. Lovegrove, B. Thermoregulation of the subterranean rodent genus Bathyergus (Bathyergidae). *S. Afr. J. Zool.* **1986**, *21*, 283–288. [CrossRef]
121. Tacutu, R.; Craig, T.; Budovsky, A.; Wuttke, D.; Lehmann, G.; Taranukha, D.; Costa, J.; Fraifeld, V.E.; de Magalhaes, J.P. Human Ageing Genomic Resources: Integrated databases and tools for the biology and genetics of ageing. *Nucleic Acids Res.* **2012**, *41*, D1027–D1033. [CrossRef]
122. MacRae, S.L.; Croken, M.M.; Calder, R.; Aliper, A.; Milholland, B.; White, R.R.; Zhavoronkov, A.; Gladyshev, V.N.; Seluanov, A.; Gorbunova, V. DNA repair in species with extreme lifespan differences. *Aging.* **2015**, *7*, 1171. [CrossRef]
123. Council, N.R. *Guide for the Care and Use of Laboratory Animals*; National Academies Press: Washington, DC, USA, 2010.

124. Hickman, G. A live-trap and trapping technique for fossorial mammals. *S. Afr. J. Zool.* **1979**, *14*, 9–12. [CrossRef]
125. Bennett, N.C.; Faulkes, C.G. *African Mole-Rats: Ecology and Eusociality*; Cambridge University Press: Cambridge, UK, 2000.
126. Bennett, N.; Faulkes, C.; Spinks, A. LH responses to single doses of exogenous GnRH by social Mashona mole–rats: A continuum of socially induced infertility in the family Bathyergidae. *Proc. R. Soc. Lond. Ser. B Biol. Sci.* **1997**, *264*, 1001–1006. [CrossRef]
127. Hardeland, R.; Coto-Montes, A.; Poeggeler, B. Circadian rhythms, oxidative stress, and antioxidative defense mechanisms. *Chronobiol. Int.* **2003**, *20*, 921–962. [CrossRef] [PubMed]
128. Wilking, M.; Ndiaye, M.; Mukhtar, H.; Ahmad, N. Circadian rhythm connections to oxidative stress: Implications for human health. *Antioxid. Redox Signal.* **2013**, *19*, 192–208. [CrossRef] [PubMed]
129. Erel, O. A new automated colorimetric method for measuring total oxidant status. *Clin. Biochem.* **2005**, *38*, 1103–1111. [CrossRef] [PubMed]
130. Muñoz-Sabater, J.; Dutra, E.; Agustí-Panareda, A.; Albergel, C.; Arduini, G.; Balsamo, G.; Boussetta, S.; Choulga, M.; Harrigan, S.; Hersbach, H. ERA5-Land: A state-of-the-art global reanalysis dataset for land applications. *Earth Syst. Sci. Data Discuss.* **2021**, 1–50. [CrossRef]
131. Romanenko, V. Computation of the autumn soil moisture using a universal relationship for a large area. *Proc. Ukr. Hydrometeorol. Res. Inst.* **1961**, *3*, 12–25.
132. Clark, K.; Karsch-Mizrachi, I.; Lipman, D.J.; Ostell, J.; Sayers, E.W. GenBank. *Nucleic Acids Res.* **2016**, *44*, D67–D72. [CrossRef]
133. Zemlemerova, E.D.; Kostin, D.S.; Lebedev, V.S.; Martynov, A.A.; Gromov, A.R.; Alexandrov, D.Y.; Lavrenchenko, L.A. Genetic diversity of the naked mole-rat (Heterocephalus glaber). *J. Zool. Syst. Evol. Res.* **2021**, *59*, 323–340. [CrossRef]
134. Faulkes, C.; Bennett, N.C.; Bruford, M.W.; O'brien, H.; Aguilar, G.; Jarvis, J. Ecological constraints drive social evolution in the African mole–rats. *Proc. R. Soc. Lond. Ser. B Biol. Sci.* **1997**, *264*, 1619–1627. [CrossRef]
135. Visser, J.H.; Bennett, N.C.; van Vuuren, B.J. Evolutionary and ecological patterns within the South African Bathyergidae: Implications for taxonomy. *Mol. Phylogenet. Evol.* **2019**, *130*, 181–197. [CrossRef]
136. Visser, J.H.; Bennett, N.C.; Jansen van Vuuren, B. Local and regional scale genetic variation in the Cape dune mole-rat, Bathyergus suillus. *PloS ONE* **2014**, *9*, e107226. [CrossRef]
137. Honeycutt, R.L.; Nedbal, M.A.; Adkins, R.M.; Janecek, L.L. Mammalian mitochondrial DNA evolution: A comparison of the cytochrome b and cytochrome c oxidase II genes. *J. Mol. Evol.* **1995**, *40*, 260–272. [CrossRef] [PubMed]
138. Guo, G.; Sheng, J.; Wu, X.; Wang, Y.; Guo, L.; Zhang, X.; Yao, H. Seoul virus in the brown rat (Rattus norvegicus) from Ürümqi, Xinjiang, northwest of China. *J. Wildl. Dis.* **2016**, *52*, 705–708. [CrossRef] [PubMed]
139. Spotorno, Á.E.; Valladares, J.P.; Marín, J.C.; Zeballos, H. Molecular diversity among domestic guinea-pigs (*Cavia porcellus*) and their close phylogenetic relationship with the Andean wild species Cavia tschudii. *Rev. Chil. Hist. Nat.* **2004**, *77*, 243–250. [CrossRef]
140. Maddison, W.P.; Maddison, D.R. Mesquite: A Modular System for Evolutionary Analysis. Version 3.81. 2023. Available online: http://www.mesquiteproject.org (accessed on 3 April 2023).
141. Kumar, S.; Stecher, G.; Li, M.; Knyaz, C.; Tamura, K. MEGA X: Molecular evolutionary genetics analysis across computing platforms. *Mol. Biol. Evol.* **2018**, *35*, 1547–1549. [CrossRef]
142. Saitou, N.; Nei, M. The neighbor-joining method: A new method for reconstructing phylogenetic trees. *Mol. Biol. Evol.* **1987**, *4*, 406–425.
143. Felsenstein, J. Confidence limits on phylogenies: An approach using the bootstrap. *Evolution* **1985**, *39*, 783–791. [CrossRef]
144. Tamura, K.; Nei, M.; Kumar, S. Prospects for inferring very large phylogenies by using the neighbor-joining method. *Proc. Natl. Acad. Sci. USA* **2004**, *101*, 11030–11035. [CrossRef]
145. Tamura, K.; Stecher, G.; Kumar, S. MEGA11: Molecular evolutionary genetics analysis version 11. *Mol. Biol. Evol.* **2021**, *38*, 3022–3027. [CrossRef]
146. Team, R.D.C. R Foundation for Statistical Computing, Vienna. 2018. Available online: www.r-project.org (accessed on 3 April 2023).
147. Hadfield, J.D. MCMC methods for multi-response generalized linear mixed models: The MCMCglmm R package. *J. Stat. Softw.* **2010**, *33*, 1–22. [CrossRef]
148. Bates, D.; Mächler, M.; Bolker, B.; Walker, S. Fitting linear mixed-effects models using lme4. *arXiv* **2014**, arXiv:1406.5823.
149. Barton, K. MuMIn: Multi-Model Inference, Version 1.43. 6. Available online: http://CRAN.R-poject.org/package=MuMln (accessed on 3 April 2023).
150. Lenth, R.; Singmann, H.; Love, J.; Buerkner, P.; Herve, M. Emmeans: Estimated Marginal Means, aka Least-Squares Means. *R Package Version* **2018**, *1*, 3. Available online: https://CRAN.R-project.org/package=emmeans (accessed on 3 April 2023).
151. Hõrak, P.; Cohen, A. How to measure oxidative stress in an ecological context: Methodological and statistical issues. *Funct. Ecol.* **2010**, *24*, 960–970. [CrossRef]
152. Wickham, H.; Chang, W.; Wickham, M.H. Package 'ggplot2'. Create elegant data visualisations using the grammar of graphics. *Version* **2016**, *2*, 1–189.
153. Abdi, H.; Williams, L.J. Principal component analysis. *Wiley Interdiscip. Rev. Comput. Stat.* **2010**, *2*, 433–459. [CrossRef]
154. Gabriel, K.R. The biplot graphic display of matrices with application to principal component analysis. *Biometrika* **1971**, *58*, 453–467. [CrossRef]

155. Jimenez, A.G.; O'Connor, E.S.; Tobin, K.J.; Anderson, K.N.; Winward, J.D.; Fleming, A.; Winner, C.; Chinchilli, E.; Maya, A.; Carlson, K.; et al. Does Cellular Metabolism from Primary Fibroblasts and Oxidative Stress in Blood Differ between Mammals and Birds? The (Lack-thereof) Scaling of Oxidative Stress. *Integr. Comp. Biol.* **2019**, *59*, 953–969. [CrossRef] [PubMed]
156. Cohen, A.A.; McGraw, K.J.; Wiersma, P.; Williams, J.B.; Robinson, W.D.; Robinson, T.R.; Brawn, J.D.; Ricklefs, R.E. Interspecific associations between circulating antioxidant levels and life-history variation in birds. *Am. Nat.* **2008**, *172*, 178–193. [CrossRef] [PubMed]
157. Williams, S.A.; Shattuck, M.R. Ecology, longevity and naked mole-rats: Confounding effects of sociality? *Proc. R. Soc. B Biol. Sci.* **2015**, *282*, 20141664. [CrossRef]
158. Lucas, E.R.; Keller, L. The co-evolution of longevity and social life. *Funct. Ecol.* **2020**, *34*, 76–87. [CrossRef]
159. Costantini, D.; Møller, A.P. Does immune response cause oxidative stress in birds? A meta-analysis. *Comp. Biochem. Physiol. Part A Mol. Integr. Physiol.* **2009**, *153*, 339–344. [CrossRef]
160. Frisard, M.; Ravussin, E. Energy metabolism and oxidative stress: Impact on the metabolic syndrome and the aging process. *Endocrine* **2006**, *29*, 27–32. [CrossRef]
161. Alonso-Alvarez, C.; Bertrand, S.; Devevey, G.; Prost, J.; Faivre, B.; Sorci, G. Increased susceptibility to oxidative stress as a proximate cost of reproduction. *Ecol. Lett.* **2004**, *7*, 363–368. [CrossRef]
162. Brzęk, P.; Książek, A.; Ołdakowski, Ł.; Konarzewski, M. High basal metabolic rate does not elevate oxidative stress during reproduction in laboratory mice. *J. Exp. Biol.* **2014**, *217*, 1504–1509. [CrossRef]
163. Salin, K.; Auer, S.K.; Rudolf, A.M.; Anderson, G.J.; Cairns, A.G.; Mullen, W.; Hartley, R.C.; Selman, C.; Metcalfe, N.B. Individuals with higher metabolic rates have lower levels of reactive oxygen species in vivo. *Biol. Lett.* **2015**, *11*, 20150538. [CrossRef]
164. Barja, G.; Cadenas, S.; Rojas, C.; Perez-Campo, R.; Lopez-Torres, M. Low mitochondrial free radical production per unit O_2 consumption can explain the simultaneous presence of high longevity and high aerobic metabolic rate in birds. *Free. Radic. Res.* **1994**, *21*, 317–327. [CrossRef] [PubMed]
165. Bourke, A.F. Kin selection and the evolutionary theory of aging. *Annu. Rev. Ecol. Evol. Syst.* **2007**, *38*, 103–128. [CrossRef]
166. Kirkwood, T.B.; Holliday, R. The evolution of ageing and longevity. *Proc. R. Soc. Lond. B* **1979**, *205*, 531–546. [PubMed]
167. Zhu, P.; Liu, W.; Zhang, X.; Li, M.; Liu, G.; Yu, Y.; Li, Z.; Li, X.; Du, J.; Wang, X. Correlated evolution of social organization and lifespan in mammals. *Nat. Commun.* **2023**, *14*, 372. [CrossRef] [PubMed]
168. Keller, L.; Genoud, M. Extraordinary lifespans in ants: A test of evolutionary theories of ageing. *Nature* **1997**, *389*, 958–960. [CrossRef]
169. Healy, K. Eusociality but not fossoriality drives longevity in small mammals. *Proc. R. Soc. B Biol. Sci.* **2015**, *282*, 20142917. [CrossRef]
170. Spinks, A.C.; Jarvis, J.U.; Bennett, N.C. Comparative patterns of philopatry and dispersal in two common mole-rat populations: Implications for the evolution of mole-rat sociality. *J. Anim. Ecol.* **2000**, *69*, 224–234. [CrossRef]
171. Ristow, M.; Schmeisser, K. Mitohormesis: Promoting health and lifespan by increased levels of reactive oxygen species (ROS). *Dose-Response Publ. Int. Hormesis Soc.* **2014**, *12*, 288–341. [CrossRef] [PubMed]
172. Calabrese, V.; Cornelius, C.; Dinkova-Kostova, A.T.; Iavicoli, I.; di Paola, R.; Koverech, A.; Cuzzocrea, S.; Rizzarelli, E.; Calabrese, E.J. Cellular stress responses, hormetic phytochemicals and vitagenes in aging and longevity. *Biochim. Biophys. Acta BBA—Mol. Basis Dis.* **2012**, *1822*, 753–783. [CrossRef] [PubMed]
173. Sohal, R.; Sohal, B.H.; Brunk, U.T. Relationship between antioxidant defenses and longevity in different mammalian species. *Mech. Ageing Dev.* **1990**, *53*, 217–227. [CrossRef]
174. Kumar, S. Free radicals and antioxidants: Human and food system. *Adv. Apllied Sci. Res.* **2011**, *2*, 129–135.
175. Song, W.; Derito, C.M.; Liu, M.K.; He, X.; Dong, M.; Liu, R.H. Cellular antioxidant activity of common vegetables. *J. Agric. Food Chem.* **2010**, *58*, 6621–6629. [CrossRef]
176. Banerjee, A.K.; Mandal, A.; Chanda, D.; Chakraborti, S. Oxidant, antioxidant and physical exercise. *Mol. Cell. Biochem.* **2003**, *253*, 307–312. [CrossRef] [PubMed]
177. Ji, L.L. Exercise and oxidative stress: Role of the cellular antioxidant systems. *Exerc. Sport Sci. Rev.* **1995**, *23*, 135–166. [CrossRef]
178. Andziak, B.; O'Connor, T.P.; Qi, W.; DeWaal, E.M.; Pierce, A.; Chaudhuri, A.R.; van Remmen, H.; Buffenstein, R. High oxidative damage levels in the longest-living rodent, the naked mole-rat. *Aging Cell* **2006**, *5*, 463–471. [CrossRef]
179. Saldmann, F.; Viltard, M.; Leroy, C.; Friedlander, G. The naked mole rat: A unique example of positive oxidative stress. *Oxidative Med. Cell. Longev.* **2019**, *2019*. [CrossRef]
180. Hulbert, A.; Pamplona, R.; Buffenstein, R.; Buttemer, W. Life and death: Metabolic rate, membrane composition, and life span of animals. *Physiol. Rev.* **2007**, *87*, 1175–1213. [CrossRef]
181. Omotoso, O.; Gladyshev, V.N.; Zhou, X. Lifespan extension in long-lived vertebrates rooted in ecological adaptation. *Front. Cell Dev. Biol.* **2021**, 2663. [CrossRef] [PubMed]
182. Guarente, L.; Kenyon, C. Genetic pathways that regulate ageing in model organisms. *Nature* **2000**, *408*, 255–262. [CrossRef] [PubMed]
183. Magalhães, J.P.d.; Costa, J.; Church, G.M. An analysis of the relationship between metabolism, developmental schedules, and longevity using phylogenetic independent contrasts. *J. Gerontol. Ser. Biol. Sci. Med. Sci.* **2007**, *62*, 149–160. [CrossRef]
184. Clare, M.J.; Luckinbill, L.S. The effects of gene-environment interaction on the expression of longevity. *Heredity* **1985**, *55*, 19–26. [CrossRef]

185. Jobson, R.W.; Nabholz, B.; Galtier, N. An evolutionary genome scan for longevity-related natural selection in mammals. *Mol. Biol. Evol.* **2010**, *27*, 840–847. [CrossRef] [PubMed]
186. Nathaniel, T.I.; Otukonyong, E.; Abdellatif, A.; Soyinka, J.O. Effect of hypoxia on metabolic rate, core body temperature, and c-fos expression in the naked mole rat. *Int. J. Dev. Neurosci.* **2012**, *30*, 539–544. [CrossRef] [PubMed]
187. Pamenter, M.E.; Hall, J.E.; Tanabe, Y.; Simonson, T.S. Cross-species insights into genomic adaptations to hypoxia. *Front. Genet.* **2020**, *11*, 743. [CrossRef] [PubMed]
188. Pamenter, M.E. Adaptations to a hypoxic lifestyle in naked mole-rats. *J. Exp. Biol.* **2022**, *225*, jeb196725. [CrossRef]
189. Ding, D.; Liu, G.; Hou, L.; Gui, W.; Chen, B.; Kang, L. Genetic variation in PTPN1 contributes to metabolic adaptation to high-altitude hypoxia in Tibetan migratory locusts. *Nat. Commun.* **2018**, *9*, 4991. [CrossRef]
190. Simonson, T.S. Altitude adaptation: A glimpse through various lenses. *High Alt. Med. Biol.* **2015**, *16*, 125–137. [CrossRef] [PubMed]
191. Walsh, M.E.; Shi, Y.; van Remmen, H. The effects of dietary restriction on oxidative stress in rodents. *Free Radic. Biol. Med.* **2014**, *66*, 88–99. [CrossRef] [PubMed]
192. Bennett, N.C.; Maree, S.; Faulkes, C.G. Georychus capensis. *Mammalian Species* **2006**, 1–4. [CrossRef]
193. Hart, D.W.; Bennett, N.C.; Best, C.; van Jaarsveld, B.; Cheng, H.; Ivy, C.M.; Kirby, A.M.; Munro, D.; Sprenger, R.J.; Storey, K.B. The relationship between hypoxia exposure and circulating cortisol levels in social and solitary African mole-rats: An initial report. *Gen. Comp. Endocrinol.* **2023**, *339*, 114294. [CrossRef]
194. Kabtni, S.; Sdouga, D.; Bettaib Rebey, I.; Save, M.; Trifi-Farah, N.; Fauconnier, M.-L.; Marghali, S. Influence of climate variation on phenolic composition and antioxidant capacity of Medicago minima populations. *Sci. Rep.* **2020**, *10*, 8293.
195. Wolfe, K.L.; Kang, X.; He, X.; Dong, M.; Zhang, Q.; Liu, R.H. Cellular antioxidant activity of common fruits. *J. Agric. Food Chem.* **2008**, *56*, 8418–8426. [CrossRef] [PubMed]
196. McGowan, N.E.; Scantlebury, D.M.; Bennett, N.C.; Maule, A.G.; Marks, N.J. Thermoregulatory differences in African mole-rat species from disparate habitats: Responses and limitations. *J. Therm. Biol.* **2020**, *88*, 102495. [CrossRef]
197. Daniels, C.; Rautenbach, F.; Marnewick, J.L.; Valentine, A.; Babajide, O.; Mabusela, W. Environmental stress effect on the phytochemistry and antioxidant activity of a South African bulbous geophyte, Gethyllis multifolia L. Bolus. *S. Afr. J. Bot.* **2015**, *96*, 29–36. [CrossRef]
198. Pamenter, M.E.; Cheng, H. Supermole-rat to the rescue: Does the naked mole-rat offer a panacea for all that ails us? *Comp. Biochem. Physiol. Part A Mol. Integr. Physiol.* **2022**, *266*, 111139. [CrossRef]
199. Alhayaza, R.; Haque, E.; Karbasiafshar, C.; Sellke, F.W.; Abid, M.R. The Relationship Between Reactive Oxygen Species and Endothelial Cell Metabolism. *Front. Chem.* **2020**, *8*. [CrossRef]
200. Fraga, C.G.; Shigenaga, M.K.; Park, J.-W.; Degan, P.; Ames, B.N. Oxidative damage to DNA during aging: 8-hydroxy-2′-deoxyguanosine in rat organ DNA and urine. *Proc. Natl. Acad. Sci. USA* **1990**, *87*, 4533–4537. [CrossRef] [PubMed]
201. Ames, B.N.; Shigenaga, M.K. Oxidants are a major contributor to aging. *Ann. N. Y. Acad. Sci.* **1992**, *663*, 85–96. [CrossRef] [PubMed]
202. Richter, C.; Gogvadze, V.; Laffranchi, R.; Schlapbach, R.; Schweizer, M.; Suter, M.; Walter, P.; Yaffee, M. Oxidants in mitochondria: From physiology to diseases. *Biochim. Biophys. Acta BBA—Mol. Basis Dis.* **1995**, *1271*, 67–74. [CrossRef] [PubMed]
203. Richter, C.; Park, J.-W.; Ames, B.N. Normal oxidative damage to mitochondrial and nuclear DNA is extensive. *Proc. Natl. Acad. Sci. USA* **1988**, *85*, 6465–6467. [CrossRef]
204. Bharti, V.K.; Srivastava, R.; Kumar, H.; Bag, S.; Majumdar, A.; Singh, G.; Pandi-Perumal, S.; Brown, G.M. Effects of melatonin and epiphyseal proteins on fluoride-induced adverse changes in antioxidant status of heart, liver, and kidney of rats. *Adv. Pharmacol. Sci.* **2014**, *2014*. [CrossRef]
205. Tasaki, E.; Sakurai, H.; Nitao, M.; Matsuura, K.; Iuchi, Y. Uric acid, an important antioxidant contributing to survival in termites. *PLoS ONE* **2017**, *12*, e0179426. [CrossRef]
206. Waring, W. Uric acid: An important antioxidant in acute ischaemic stroke. *Qjm* **2002**, *95*, 691–693. [CrossRef]
207. Rodriguez, C.; Mayo, J.C.; Sainz, R.M.; Antolín, I.; Herrera, F.; Martín, V.; Reiter, R.J. Regulation of antioxidant enzymes: A significant role for melatonin. *J. Pineal Res.* **2004**, *36*, 1–9. [CrossRef]
208. Labinskyy, N.; Csiszar, A.; Orosz, Z.; Smith, K.; Rivera, A.; Buffenstein, R.; Ungvari, Z. Comparison of endothelial function, $O_2^{-}\cdot$ and H_2O_2 production, and vascular oxidative stress resistance between the longest-living rodent, the naked mole rat, and mice. *Am. J. Physiol.-Heart Circ. Physiol.* **2006**, *291*, H2698–H2704. [CrossRef]
209. De Waal, E.M.; Liang, H.; Pierce, A.; Hamilton, R.T.; Buffenstein, R.; Chaudhuri, A.R. Elevated protein carbonylation and oxidative stress do not affect protein structure and function in the long-living naked-mole rat: A proteomic approach. *Biochem. Biophys. Res. Commun.* **2013**, *434*, 815–819. [CrossRef]
210. Cenesiz, M.; Atakişi, O.; Akar, A.; Önbilgin, G.; Ormancı, N. Effects of 900 and 1800 MHz electromagnetic field application on electrocardiogram, nitric oxide, total antioxidant capacity, total oxidant capacity, total protein, albumin and globulin levels in guinea pigs. *Kafkas Univ. Vet. Fak. Derg.* **2011**, *17*, 357–362.

MDPI
St. Alban-Anlage 66
4052 Basel
Switzerland
www.mdpi.com

Antioxidants Editorial Office
E-mail: antioxidants@mdpi.com
www.mdpi.com/journal/antioxidants

www.ingramcontent.com/pod-product-compliance
Lightning Source LLC
Chambersburg PA
CBHW061505210926
51850CB00008B/1823
9783036590127